A Categorical Approach to Imprimitivity Theorems for C^*-Dynamical Systems

MEMOIR

A Calculated Approach to Estimating the
Theorem for Compaction Systems

of the
American Mathematical Society

Number 850

A Categorical Approach to Imprimitivity Theorems for C^*-Dynamical Systems

Siegfried Echterhoff
S. Kaliszewski
John Quigg
Iain Raeburn

March 2006 • Volume 180 • Number 850 (fourth of 5 numbers) • ISSN 0065-9266

American Mathematical Society
Providence, Rhode Island

2000 *Mathematics Subject Classification.* Primary 46L55.

Library of Congress Cataloging-in-Publication Data
A categorical approach to imprimitivity theorems for C^*-dynamical systems/Siegfried Echterhoff... [et al.].
    p. cm. — (Memoirs of the American Mathematical Society, ISSN 0065-9266 ; no. 850)
Includes bibliographical references and index.
ISBN 0-8218-3857-1 (alk. paper)
    1. C^*-algebras. 2. Categories (Mathematics) I. Echterhoff, Siegfried, 1960– II. Series.

QA3.A57 no. 850
[QA326]
510 s—dc22
[512′.556]
    2005057160

Memoirs of the American Mathematical Society

This journal is devoted entirely to research in pure and applied mathematics.

Subscription information. The 2006 subscription begins with volume 179 and consists of six mailings, each containing one or more numbers. Subscription prices for 2006 are US$624 list, US$499 institutional member. A late charge of 10% of the subscription price will be imposed on orders received from nonmembers after January 1 of the subscription year. Subscribers outside the United States and India must pay a postage surcharge of US$31; subscribers in India must pay a postage surcharge of US$43. Expedited delivery to destinations in North America US$35; elsewhere US$130. Each number may be ordered separately; *please specify number* when ordering an individual number. For prices and titles of recently released numbers, see the New Publications sections of the *Notices of the American Mathematical Society*.

Back number information. For back issues see the *AMS Catalog of Publications*.

Subscriptions and orders should be addressed to the American Mathematical Society, P. O. Box 845904, Boston, MA 02284-5904, USA. *All orders must be accompanied by payment*. Other correspondence should be addressed to 201 Charles Street, Providence, RI 02904-2294, USA.

Copying and reprinting. Individual readers of this publication, and nonprofit libraries acting for them, are permitted to make fair use of the material, such as to copy a chapter for use in teaching or research. Permission is granted to quote brief passages from this publication in reviews, provided the customary acknowledgment of the source is given.

Republication, systematic copying, or multiple reproduction of any material in this publication is permitted only under license from the American Mathematical Society. Requests for such permission should be addressed to the Acquisitions Department, American Mathematical Society, 201 Charles Street, Providence, Rhode Island 02904-2294, USA. Requests can also be made by e-mail to `reprint-permission@ams.org`.

Memoirs of the American Mathematical Society is published bimonthly (each volume consisting usually of more than one number) by the American Mathematical Society at 201 Charles Street, Providence, RI 02904-2294, USA. Periodicals postage paid at Providence, RI. Postmaster: Send address changes to Memoirs, American Mathematical Society, 201 Charles Street, Providence, RI 02904-2294, USA.

© 2006 by the American Mathematical Society. All rights reserved.
Copyright of this publication reverts to the public domain 28 years
after publication. Contact the AMS for copyright status.
This publication is indexed in *Science Citation Index*®, *SciSearch*®, *Research Alert*®,
CompuMath Citation Index®, *Current Contents*®/*Physical, Chemical & Earth Sciences*.
Printed in the United States of America.

∞ The paper used in this book is acid-free and falls within the guidelines
established to ensure permanence and durability.
Visit the AMS home page at `http://www.ams.org/`

10 9 8 7 6 5 4 3 2 1   11 10 09 08 07 06

Contents

Introduction	1
Outline	3
Epilogue	6
Chapter 1. Right-Hilbert Bimodules	9
1.1. Right-Hilbert bimodules and partial imprimitivity bimodules	9
1.2. Multiplier bimodules and homomorphisms	13
1.3. Tensor products	21
1.4. The C-multiplier bimodule $M_C(X \otimes C)$	24
1.5. Linking algebras	27
Chapter 2. The Categories	33
2.1. C^*-Algebras	33
2.2. Group actions	35
2.3. Group coactions	37
2.4. Actions and coactions	43
2.5. Actions and coactions on linking algebras	44
2.6. Standard factorization of morphisms	45
2.7. Morphisms and induced representations	51
Chapter 3. The Functors	57
3.1. Crossed products	57
3.2. Restriction and inflation	68
3.3. Decomposition	71
3.4. Induced actions	75
3.5. Combined functors	77
Chapter 4. The Natural Equivalences	79
4.1. Statement of the main results	79
4.2. Some further linking algebra techniques	81
4.3. Green's Theorem for induced algebras	86
4.4. Green's Theorem for induced representations	90
4.5. Mansfield's Theorem	94
Chapter 5. Applications	101
5.1. Equivariant triangles	101
5.2. Restriction and induction	111
5.3. Symmetric imprimitivity	114
Appendix A. Crossed Products by Actions and Coactions	117
A.1. Tensor products	117

A.2.	Actions and their crossed products	121
A.3.	Coactions	126
A.4.	Slice maps and nondegeneracy	130
A.5.	Covariant representations and crossed products	132
A.6.	Dual actions and decomposition coactions	138
A.7.	Normal coactions and normalizations	139
A.8.	The duality theorems of Imai-Takai and Katayama	143
A.9.	Other definitions of coactions	148

Appendix B. The Imprimitivity Theorems of Green and Mansfield 151
   B.1. Imprimitivity theorems for actions 151
   B.2. Mansfield's imprimitivity bimodule 155

Appendix C. Function Spaces 159
   C.1. The spaces $C_c(T, \mathcal{X})$ for locally convex spaces $\mathcal{X}$ 159
   C.2. Functions in multiplier algebras and multiplier bimodules 161

Appendix. Bibliography 167

Abstract

Imprimitivity theorems provide a fundamental tool for studying the representation theory and structure of crossed-product C^*-algebras. In this work, we show that the Imprimitivity Theorem for induced algebras, Green's Imprimitivity Theorem for actions of groups, and Mansfield's Imprimitivity Theorem for coactions of groups can all be viewed as natural equivalences between various crossed-product functors among certain equivariant categories.

The categories involved have C^*-algebras with actions or coactions (or both) of a fixed locally compact group G as their objects, and equivariant equivalence classes of right-Hilbert bimodules as their morphisms. Composition is given by the balanced tensor product of bimodules.

The functors involved arise from taking crossed products; restricting, inflating, and decomposing actions and coactions; inducing actions; and various combinations of these.

Several applications of this categorical approach are also presented, including some intriguing relationships between the Green and Mansfield bimodules, and between restriction and induction of representations.

Received by the editor June 24, 2002, and in revised form February 2, 2005.

2000 *Mathematics Subject Classification.* 46L55.

Key words and phrases. Morita equivalence, C^*-dynamical systems, coactions, crossed products.

This research was partially supported by National Science Foundation Grant #DMS9401253, the Australian Research Council, the Deutsche Forschungsgesellschaft (SFB 478), the European Union (RTN QSNG, Contract No. HPRN-CT-2002-00280), Arizona State University, and the University of Newcastle, Australia.

Introduction

Given a dynamical system (A, G, α) in which a locally compact group G acts by automorphisms of a C^*-algebra A, Mackey and Takesaki's *induction* process allows us to construct representations of (A, G, α) from representations of the system $(A, H, \alpha|_H)$ associated to any closed subgroup H of G. Much is known about induction: there are imprimitivity theorems which allow us to recognize induced representations, and the process is functorial with respect to intertwining operators.

In the modern framework of Rieffel, one introduces the crossed product $A \times_\alpha G$, which is a C^*-algebra encapsulating the representation theory of (A, G, α), and induces instead from $A \times_\alpha H$ to $A \times_\alpha G$; induction of representations from one C^*-algebra D to another C is achieved by tensoring the underlying Hilbert space with a Hilbert bimodule $_C X_D$, which has a D-valued inner product and in which the left action of C is by adjointable operators. An *imprimitivity theorem* tells us how to expand the left action of C to one of a larger algebra E in such a way that $_E X_D$ is an imprimitivity bimodule — that is, reversible. The theorem then says that a representation of C is equivalent to one induced from D if and only if there is a compatible representation of E.

Duality tells us how to recover a dynamical system (A, G, α) from its crossed product $A \times_\alpha G$. When G is abelian, the crossed product carries a canonical dual action $\widehat\alpha$ of the dual group $\widehat G$, and the Takesaki-Takai Duality Theorem says that the double dual system $((A \times_\alpha G) \times_{\widehat\alpha} \widehat G, G, \widehat{\widehat\alpha})$ is Morita equivalent to the original one. For nonabelian G, one has to use instead the dual coaction of G, and recover the system from the crossed product by this dual coaction. For duality to be a useful tool, one has to understand these coactions and their crossed products, and a good deal of progress has been made in the past 15 years. (An overview of this area has been provided in an Appendix; see also [52] for a recent survey.) Crucial for us is Mansfield's theory of induction for crossed products by coactions: he provides a Hilbert bimodule which allows us to induce representations from crossed products by quotient groups, and an imprimitivity theorem which characterizes these induced representations.

Induction and duality interact in deep and mysterious ways. One general principle appears to be that duality swaps induction of representations with restriction of representations. This is enormously appealing: restriction of representations (for example, passing from a representation U of G to the representation $U|_H$ of a subgroup H) is ostensibly a trivial process. Theorems making this induction-restriction duality precise have been proved, first for abelian groups in [14], and later for arbitrary groups in [29, 18]. We have gradually learned that it is best to prove such theorems by manipulating the Hilbert bimodules which implement the various induction and restriction processes; however, the bimodules involved are hard to work with — especially Mansfield's — and the results can safely be described as "technically challenging". To make things worse, applications frequently require that various isomorphisms and equivalences are equivariant, and one is continually having to construct compatible coactions on bimodules and check that they carry through complicated arguments. So it is definitely of interest to find a more systematic approach.

Our goal here is to provide such a systematic approach and to use it to complete our program of induction-restriction duality. We shall show that many of the key technical problems in this area amount to asking for functoriality of some construction or naturality of some equivalence between functors. Asking for equivalences to be equivariant amounts to asking for an equivalence in a different category, one which includes coactions or actions in its objects and morphisms. We have found that functoriality of the various crossed-product constructions encompasses many results of the kind "Morita equivalent systems have Morita equivalent crossed products", and naturality of the equivalences many results of the kind "induction is compatible with Morita equivalence".

To help see how our approach works, we consider one of our main theorems. It concerns the generalization of Green's Imprimitivity Theorem to crossed products of induced algebras, which is, loosely speaking, the analogue of the imprimitivity theorem for actions α of a subgroup H which do not extend to actions of G. The induced algebra $\operatorname{Ind}_H^G(A, \alpha)$ is a subalgebra of $C_b(G, A)$ which carries a left action τ of G by translation, and the generalization says that the crossed product $\operatorname{Ind}_H^G(A, \alpha) \times_\tau G$ is Morita equivalent to $A \times_\alpha H$. We shall prove that this equivalence is natural, and that it is equivariant for the dual coaction $\widehat{\tau}$ of G on $\operatorname{Ind}_H^G(A, \alpha) \times_\tau G$ and the inflation $\operatorname{Inf} \widehat{\alpha}$ to G of the dual coaction on $A \times_\alpha H$. To make this precise, we have to set up categories $\mathcal{C}$ of C^*-algebras, $\mathcal{A}(G)$ of dynamical systems (A, G, α), and $\mathcal{C}(G)$ of cosystems (A, G, δ) in which δ is a coaction of G on A. We then prove that $(A, G, \alpha) \mapsto (\operatorname{Ind}_H^G(A, \alpha) \times_\tau G, \widehat{\tau})$ and $(A, G, \alpha) \mapsto (A \times_\alpha H, \operatorname{Inf} \widehat{\alpha})$ are the object maps for functors from $\mathcal{A}(G)$ to $\mathcal{C}(G)$, so that it makes sense to say that they are naturally equivalent.

When we assert that, for example, $(A, G, \alpha) \mapsto (A \times_\alpha G, \widehat{\alpha})$ is a functor, we are completely ignoring the morphisms, and we cannot appreciate what naturality means until we deal with them too: a natural equivalence T between two functors $F, G \colon \mathcal{A} \to \mathcal{B}$ assigns to each object A of $\mathcal{A}$ an equivalence $T(A) \colon F(A) \to G(A)$ (that is, an invertible morphism $T(A)$ in the category $\mathcal{B}$) such that, for each morphism $\varphi \colon A \to B$ in $\mathcal{A}$, the diagram

$$\begin{array}{ccc} F(A) & \xrightarrow{T(A)} & G(A) \\ F(\varphi) \downarrow & & \downarrow G(\varphi) \\ F(B) & \xrightarrow{T(B)} & G(B) \end{array}$$

commutes in $\mathcal{B}$. In our categories, the morphisms will be based on Hilbert bimodules; in $\mathcal{A}(G)$, for example, a morphism from (A, G, α) to (B, G, β) will be given by a Hilbert bimodule ${}_A X_B$ with a compatible action γ of G. The composition of morphisms will be based on the balanced tensor product of bimodules, so that a diagram

$$\begin{array}{ccc} A & \xrightarrow{X} & B \\ {\scriptstyle Y}\downarrow & & \downarrow{\scriptstyle W} \\ C & \xrightarrow{Z} & D \end{array}$$

of Hilbert bimodules commutes if $Y \otimes_C Z \cong X \otimes_B W$ as Hilbert $A - D$ bimodules; in $\mathcal{A}(G)$ or $\mathcal{C}(G)$ this isomorphism has to be appropriately equivariant. The equivalences in these categories are the morphisms which are given by imprimitivity bimodules, so to prove that two of our functors F, G are naturally equivalent amounts to finding imprimitivity bimodules $_{F(A)}X(A)_{G(A)}$ such that

$$X(A) \otimes_{G(A)} G(Y) \cong F(Y) \otimes_{F(B)} X(B)$$

as Hilbert $F(A) - G(B)$ bimodules for each Hilbert bimodule $_AY_B$. The modules are the usual ones, but many of the details needed to establish these isomorphisms and their properties are new.

This paper, like any other in which coactions appear, involves some gritty technical arguments. We will therefore begin by outlining the main new issues which we face in this program, and how we have dealt with them. Those who are interested in seeing how the categorical ideas impact when there are no coactions around are encouraged to read our previous paper [17] first. Indeed, this might help even those who are already coaction-compliant!

Outline

We begin in Chapter 1 with a detailed discussion of the Hilbert bimodules on which our morphisms are based. The axioms are intrinsically asymmetric; to see why, note that a homomorphism $\varphi \colon A \to B$ gives B the structure of an A-module, but not the other way round. Our modules $_AX_B$ will be right Hilbert B-modules with a left action of A given by a nondegenerate homomorphism κ of A into the C^*-algebra $\mathcal{L}(X_B)$ of adjointable operators on X. As in [29], we shall call these *right-Hilbert bimodules* to emphasize that the Hilbert-module structure is on the right; we have stuck with this name because the alternatives (C^*-correspondences or Hilbert bimodules) do not carry the same sense of direction. The theory of right-Hilbert bimodules is similar to that of imprimitivity bimodules, but there seem to be enough subtle differences to warrant a detailed discussion.

The first section contains the basic facts about multiplier bimodules and homomorphisms between bimodules. These are used repeatedly: a coaction on a bimodule X, for example, is by definition a homomorphism of X into the multiplier bimodule $M(X \otimes C^*(G))$. Our treatment is similar to that of imprimitivity bimodules in [20]. Section 1.1.3 is about the balanced tensor products which are used to define the composition of morphisms; we need to know in particular how this process extends to multipliers. We also discuss external tensor products, which are crucial for the definition of coactions on Hilbert bimodules. The last section of Chapter 1 is about linking algebras. These are used primarily as a technical tool in the proofs of naturality (an idea lifted from [21], and expounded in an easier setting in [17]).

In Chapter 2 we describe the categories in which we work. The basic category $\mathcal{C}$ of C^*-algebras appears in [17]; we review the main facts in Section 2.1. The objects are C^*-algebras and the morphisms from A to B are the isomorphism classes of right-Hilbert $A-B$ bimodules: we have to pass to isomorphism classes to ensure that the composition law $[_CY_B] \circ [_AX_B] = [_A(X \otimes_C Y)_B]$ has the required properties. The other categories $\mathcal{A}(G)$, $\mathcal{C}(G)$ and $\mathcal{AC}(G)$ are associated to a fixed locally compact group G, and are obtained by adding, respectively, actions of G, coactions of G, and both actions and coactions to the objects and morphisms of $\mathcal{C}$. Adding actions is relatively routine, but (as will be no surprise to those familiar with them) adding coactions is a little harder. (Coactions on Hilbert bimodules first appeared in [**2**].) We show that in each of these categories, the equivalences (that is, the invertible morphisms) are the morphisms in which the underlying bimodules are imprimitivity bimodules, and then that every morphism is a composition of a morphism coming from a nondegenerate homomorphism $\varphi \colon A \to M(C)$ and a morphism based on an imprimitivity bimodule $_CX_B$.

In Chapter 3 we show that the various crossed products appearing in our theorems define functors between appropriate categories. There are two main problems. The first is to define suitable crossed products. We are interested here in coactions and nonabelian duality, which is basically a theory about reduced crossed products, so we have decided to give in gracefully and use reduced crossed products throughout. (This is definitely a choice: we have already proved the naturality of Green's Imprimitivity Theorem for full crossed products in [**17**], and providing we were willing to omit all statements about the coactions, we could presumably do the same here.) But because the objects in our categories are C^*-algebras rather than isomorphism classes of C^*-algebras, it is important that we don't just *choose* a regular representation willy-nilly. So we shall discuss a specific realization of the reduced product. The second main problem is to define crossed products of the Hilbert bimodules which define the morphisms. We do this differently for actions and coactions; for actions we make heavy use of the convenience of C_c-functions, and for coactions we realize the crossed product inside a certain multiplier bimodule. For imprimitivity bimodules, it is handy to recognize that if $L(X)$ is the linking algebra of X, then the bimodule crossed product $X \times G$ embeds as the top right corner of $L(X) \times G$, and we have the important relation $L(X) \times G = L(X \times G)$ almost by definition. We should mention that defining these crossed products and establishing their properties has been done before; see [**2**], [**7**], [**6**], [**20**], and [**30**].

We gather all the necessary functors in Chapter 3; even though some are easy, it is convenient to deal with them all at once. The key difficulty is the same in each case: it is not obvious that crossed products preserve composition. This amounts to proving things like

$$(X \otimes_B Y) \times G \cong (X \times G) \otimes_{B \times G} (Y \times G),$$

and again our techniques are different for actions and coactions.

Our main theorems are in Chapter 4. We have already discussed the first, which is about crossed products of induced algebras, and which we prove in Section 4.1. The proofs of this and our other main theorems follow the same general pattern. We factor each morphism $_AX_B$ as a composition of a nondegenerate homomorphism

$\varphi\colon A \to M(C)$ and an imprimitivity bimodule ${}_CY_B$. To prove that

$$\begin{array}{ccc} F(A) & \xrightarrow{T(A)} & G(A) \\ {\scriptstyle F(\varphi)}\downarrow & & \downarrow{\scriptstyle G(\varphi)} \\ F(C) & \xrightarrow{T(C)} & G(C) \end{array}$$

commutes, we extend the homomorphisms $(F(\varphi), G(\varphi))$ to a homomorphism of imprimitivity bimodules $T(A) \to M(T(C))$, and use a general lemma which says this suffices. To prove that

$$\begin{array}{ccc} F(C) & \xrightarrow{T(C)} & G(C) \\ {\scriptstyle F(Y)}\downarrow & & \downarrow{\scriptstyle G(Y)} \\ F(B) & \xrightarrow{T(B)} & G(B) \end{array}$$

commutes, we realize $T(C)$ and $T(D)$ as the diagonal corners in an imprimitivity bimodule Z over the linking algebras $L(F(Y))$ and $L(G(Y))$, and use a general lemma from [21] which identifies both $F(Y) \otimes_{F(B)} T(B)$ and $T(C) \otimes_{G(C)} G(Y)$ with the top off-diagonal corner in Z. The hard part in both halves is to build the compatible coaction.

It may be known that this theorem about crossed products of induced algebras is a generalization of Green's Imprimitivity Theorem, but it does not appear to be well-documented. We therefore give a careful derivation, which could be of some independent interest (see the discussion preceding Theorem B.3 in Appendix B). We then use this to deduce our second main theorem, which is a natural and equivariant version of the Imprimitivity Theorem itself. There are many possible variations on this theme, depending on choices of full and reduced crossed products and on whether or not the subgroup is normal. Here we have already decided to use reduced crossed products, and we have further chosen to discuss what happens for normal subgroups. We have made this choice because in this case there are several more actions and coactions in play, and the theorem has something to say about all of them. To see what is happening here, recall that if N is normal, we can view the imprimitivity algebra $(A \otimes C_0(G/N)) \times_{\alpha \otimes \tau} G$ in Green's theorem as the crossed product $(A \times_\alpha G) \times_{\widehat{\alpha}|} G/N$ by the restriction of the dual coaction; thus this imprimitivity algebra carries a dual action $(\widehat{\alpha}|)\widehat{}$ of G/N as well as a dual coaction $(\alpha \otimes \tau)\widehat{}$ of G. Our theorem says that Green's imprimitivity bimodule matches $(\widehat{\alpha}|)\widehat{}$ with the so-called decomposition action of G on $A \times_\alpha N$ and $(\alpha \otimes \tau)\widehat{}$ with the inflation to G of the dual coaction of N on $A \times_\alpha N$. This observation seems to be new. Indeed, we believe that the equivariance and the naturality are both potentially important new pieces of information about Green's theorem.

Our third main theorem is a version of Mansfield's Imprimitivity Theorem. This has all the same features as the version of Green's theorem which we have just discussed: Mansfield's Morita equivalence of $(A \times_\delta G) \times_{\widehat{\delta}} N$ with $A \times_{\delta|} G/N$ is natural and equivariant for canonical actions and coactions on the crossed products. For this theorem, the difficult part of the proof is establishing the naturality with respect to ordinary homomorphisms $\varphi\colon A \to M(C)$; we have to work hard to build compatible homomorphisms on Mansfield's bimodule.

In Chapter 5 we give some applications to our motivating problem of understanding the relationships between induction and duality. In Section 5.1, we uncover some new and very intriguing relationships between Green and Mansfield induction. Important special cases of these results say that the Green bimodules $X^G_{\{e\}}(A)$ and Mansfield bimodules $Y^G_{G/G}(A)$ are in duality:

$$X^G_{\{e\}}(A) \times G \cong Y^G_{G/G}(A \times G)$$

and
$$X^G_{\{e\}}(A \times G) \cong Y^G_{G/G}(A) \times G.$$

Results of this type require several applications of our main theorems, and it is vital that we know everything is appropriately equivariant. Our main new application to induction-restriction duality is Theorem 5.16, which completes the program of [**14, 29, 18**] by handling the restriction of representations from $A \times_\alpha G$ to $A \times_{\alpha|} N$. We close with a new application of linking-algebra techniques to the Symmetric Imprimitivity Theorem of [**51**].

Since this project is intrinsically involved with nonabelian duality, we have necessarily made heavy use of coactions and their crossed products. There are several different sets of definitions available: the subject is stabilizing, but some key questions of a fundamental nature remain unresolved, and hence this is taking longer than one might have wished. So we have included as an appendix a survey of the area, which outlines what we believe to be the most satisfactory approach and describes how this approach relates to the others in the literature.

A second appendix collects the precise versions of the imprimitivity theorems we need; various formulations appear in the literature, so we felt it would be handy to record exactly what we want.

Finally, the third appendix contains some technical results on function spaces with values in locally convex spaces which are used throughout the text to construct multipliers of bimodules. In applications, the locally convex spaces will be multiplier algebras or bimodules with the strict topology: we need to know, for example, that strictly continuous functions of compact support from G to $M(X)$ define multipliers of $X \times G$, and that they do so in an orderly fashion.

Epilogue

Although this paper has turned out much longer than we intended, we have made all sorts of simplifying assumptions to keep the length down, and these are probably logically unnecessary. First of all, we have deliberately excised twisted crossed products, though some residual traces remain in the presence of the decomposition actions and coactions. Any serious application of these ideas to the Mackey machine — which was, after all, our original motivation [**14**] — will require that we can handle twisted crossed products. Second, we have used reduced crossed products throughout. For our present applications involving nonabelian duality and crossed products by coactions, this makes sense: the current duality theorems all factor through the reduced crossed product. But for applications to ordinary crossed products this is not necessarily desirable, and there are surely versions of Theorem 4.1 and Theorem 4.2 for full crossed products. We have already described a version of Green's theorem in [**17**], but we neglected questions of equivariance there. Third, we have considered only some of the important Morita equivalences. The others, such as the Symmetric Imprimitivity Theorem and the Stabilization

Trick for twisted crossed products, should be natural too. (Working in the context of general locally compact quantum groups was not even an issue, since there are currently no imprimitivity theorems available in that generality!)

On the other hand, we have taken the liberty of treating actions separately from coactions — rather than viewing actions of G as coactions of $C_0(G)$ — although this would have led to a much shorter exposition. Our main reasons for this are that we think that actions are much easier to understand than coactions, and that we feel there may be more general interest in the action case than in the coaction case.

We hope that we have given convincing evidence that issues involving functoriality, naturality and equivariance are likely to occur frequently in our subject, and that it will pay for us get in the habit of dealing with them as we go. We also hope that we have made a few other points along the way: our view of induced C^*-algebras as an obstruction to imprimitivity, our heavy use of linking-algebra techniques to identify imprimitivity bimodules, and the seemingly deep and strange relations between induction and duality, should all have applications elsewhere. For instance, the strong connection between these ideas and equivariant KK-theory is well-documented in [2] and [30]. Several of the ideas are also present in the approach of [9] and [8] towards a Mackey machine for the Baum-Connes conjecture, and are applied to the Connes-Kasparov conjecture in [10].

CHAPTER 1

Right-Hilbert Bimodules

In this chapter we gather together the basic theory of right-Hilbert bimodules. We start with the basic definitions and some important notation which shall be used throughout this work.

1.1. Right-Hilbert bimodules and partial imprimitivity bimodules

Let B be a C^*-algebra. Recall that a *Hilbert B-module* is a vector space X which is a right B-module equipped with a positive definite B-valued sesquilinear form $\langle \cdot, \cdot \rangle_B$ satisfying

(1.1) $\quad \langle x, y \cdot b \rangle_B = \langle x, y \rangle_B b \quad \text{and} \quad \langle x, y \rangle_B^* = \langle y, x \rangle_B \quad$ for all $x, y \in X, b \in B,$

and which is complete in the norm $\|x\| = \|\langle x, x \rangle_B\|^{1/2}$. Our primary reference for Hilbert modules is [**54**], and a secondary reference is [**33**]. Some notational conventions: we often omit the dot ($\cdot$) when writing module actions; and in general, if $(u,v) \mapsto uv \colon U \times V \to W$ is a pairing among vector spaces, then for $P \subseteq U$ and $Q \subseteq V$ we write PQ to mean the *linear span* of the set $\{uv \mid u \in P, v \in Q\}$.

DEFINITION 1.1. Let A and B be C^*-algebras. A *right-Hilbert A–B bimodule* is a Hilbert B-module X which is also a nondegenerate left A-module (*i.e.*, $AX = X$) satisfying

(1.2a) $\qquad\qquad a \cdot (x \cdot b) = (a \cdot x) \cdot b \qquad$ and

(1.2b) $\qquad\qquad \langle a \cdot x, y \rangle_B = \langle x, a^* \cdot y \rangle_B$

for all $a \in A$, $x, y \in X$, and $b \in B$. We write ${}_A X_B$ to indicate all the data, and we call X *full* if it is full as a Hilbert B-module, *i.e.*, $\overline{\langle X, X \rangle}_B = B$. In general, if X is not full, we shall write B_X for the closed ideal $\overline{\langle X, X \rangle}_B \subseteq B$, and we call B_X the *range* of the inner product on X.

REMARK 1.2. (1) In recent years, objects very similar to right-Hilbert bimodules have been introduced into the literature: for example, the $A - B$ correspondences of [**39**]. In many cases (as in [**39**]), the left module action is permitted to be degenerate; we require it to be nondegenerate so that we can extend it to the multiplier algebra $M(A)$ (see below).

(2) Note that if X is an $A - B$ correspondence, then $\overline{AX}$ is a closed $A - B$ sub-bimodule of X, and therefore becomes a right-Hilbert $A - B$ bimodule. In fact we have $\overline{AX} = AX = \{ax \mid a \in A, x \in X\}$, since it follows from Cohen's factorization theorem that $\overline{AX} = A\overline{AX} \subseteq AX$ (we refer to [**54**, Proposition 2.33] for a statement and an easy proof of Cohen's factorization theorem in the case where A is a C^*-algebra). More generally, a similar application of Cohen's theorem implies that for any C^*-subalgebras C and D of A and B, respectively, we have $\overline{CX} = \{cx \mid c \in C, x \in X\}$ and $\overline{XD} = \{xd \mid x \in X, d \in D\}$.

EXAMPLE 1.3. If B is a C^*-algebra, then B becomes a full right-Hilbert $B - B$ bimodule in a natural way by putting
$$a \cdot b \cdot c = abc \quad \text{and} \quad \langle a, b \rangle_B = a^*b \quad \text{for } a, b, c \in B.$$
If $\varphi \colon A \to M(B)$ is a nondegenerate C^*-algebra homomorphism, then B becomes a full right-Hilbert $A - B$ bimodule with left action given by
$$a \cdot b = \varphi(a)b.$$
More generally, if $\varphi \colon A \to M(B)$ is an arbitrary (possibly degenerate) $*$-homomorphism, then B becomes an $A - B$ correspondence and, therefore, $X = \varphi(A)B$ is a right-Hilbert $A - B$ bimodule. We call a right-Hilbert bimodule $_AX_B$ arising in this way *standard*. If $\varphi \colon A \to M(B)$ is nondegenerate, i.e., if $X = \varphi(A)B = B$, then we say that $_AB_B$ is a *nondegenerate* standard right-Hilbert bimodule.

REMARK 1.4. It is clear that a nondegenerate standard right-Hilbert bimodule $_AB_B$ is full. The converse is not true in general. To see an example let $B = M_2(\mathbb{C})$ and let $\varphi \colon \mathbb{C} \to M_2(\mathbb{C}); \varphi(\lambda) = \begin{pmatrix} \lambda & 0 \\ 0 & 0 \end{pmatrix}$. Then $M_2(\mathbb{C})\varphi(\mathbb{C})M_2(\mathbb{C}) = M_2(\mathbb{C})$ and $X = \varphi(\mathbb{C})M_2(\mathbb{C}) \cong \mathbb{C}^2$ is a full right-Hilbert $\mathbb{C} - M_2(\mathbb{C})$ bimodule, but φ is degenerate.

If X and Y are Hilbert B-modules, $\mathcal{L}_B(X, Y)$ denotes the set of maps $T \colon X \to Y$ which are adjointable in the sense that there exists $T^* \colon Y \to X$ such that
$$\langle Tx, y \rangle_B = \langle x, T^*y \rangle_B \quad \text{for all } x \in X, y \in Y.$$
Such T are automatically bounded and B-linear [**54**, Lemma 2.18]. The notation is shortened to $\mathcal{L}(X, Y)$ if B is understood, and $\mathcal{L}_B(X)$ (or just $\mathcal{L}(X)$) if $X = Y$. In the latter case $\mathcal{L}(X)$ is a C^*-algebra with the operator norm $\|T\| = \sup\{\|Tx\| \mid \|x\| \leq 1\}$ [**54**, Proposition 2.1].

Now, if $_AX_B$ is a right-Hilbert $A - B$ bimodule then for each $a \in A$ the map $x \mapsto a \cdot x$ is adjointable (consequently the associativity condition (1.2a) is redundant), so we get a homomorphism $\kappa \colon A \to \mathcal{L}_B(X)$ such that
$$\kappa(a)x = a \cdot x,$$
and which is nondegenerate in the sense that $\kappa(A)X = X$. Conversely, every right-Hilbert bimodule arises in this way: If X is a Hilbert B-module and $\kappa \colon A \to \mathcal{L}(X)$ is a nondegenerate homomorphism, then X becomes a right-Hilbert $A - B$ bimodule via
$$a \cdot x = \kappa(a)x.$$
Thus a right-Hilbert $A - B$ bimodule is nothing more nor less than a Hilbert B-module X together with a nondegenerate homomorphism $A \to \mathcal{L}(X)$.

If X and Y are Hilbert B-modules, $\mathcal{K}(X, Y)$ denotes the *compact operators* from X to Y: by definition, it is the closed span in $\mathcal{L}(X, Y)$ of the maps $z \mapsto y\langle x, z \rangle_B$ for $x \in X$ and $y \in Y$. $\mathcal{K}(X) = \mathcal{K}(X, X)$ is a closed ideal in $\mathcal{L}(X)$, and in fact $\mathcal{L}(X) \cong M(\mathcal{K}(X))$ [**54**, Corollary 2.54]. In particular, if X is a Hilbert B-module, then the formula
$$(1.3) \qquad _{\mathcal{K}(X)}\langle x, y \rangle z = x \langle y, z \rangle_B$$
defines a full $\mathcal{K}(X)$-valued inner product on X, which gives X the structure of a *left* Hilbert $\mathcal{K}(X)$-module. Then B acts via adjointable operators on the right of $_{\mathcal{K}(X)}X$, and $B_X = \overline{\langle X, X \rangle}_B$ identifies with the compact operators of the left Hilbert $\mathcal{K}(X)$-module $_{\mathcal{K}(X)}X$.

We pause to clear up an apparent ambiguity: Suppose $_A X_B$ is a right-Hilbert bimodule, and let $\kappa\colon A \to \mathcal{L}_B(X)$ be the associated homomorphism. By definition κ is nondegenerate in the sense that $\kappa(A)X = X$. Is κ still nondegenerate when viewed as a homomorphism of A into the multiplier algebra $M(\mathcal{K}(X))$? Yes, because $\kappa(A)\mathcal{K}(X)$ contains all the "rank-one" maps $_{\mathcal{K}(X)}\langle x, y\rangle$: just factor $x = \kappa(a)z$ for some $a \in A$ and $z \in X$, and then

$$_{\mathcal{K}(X)}\langle x, y\rangle = {}_{\mathcal{K}(X)}\langle \kappa(a)z, y\rangle = \kappa(a)_{\mathcal{K}(X)}\langle z, y\rangle.$$

Thus κ extends uniquely to a strictly continuous homomorphism $\bar{\kappa}\colon M(A) \to M(\mathcal{K}(X))$. Since the strict topology on $M(\mathcal{K}(X)) = \mathcal{L}(X)$ is stronger than the strong operator topology [**54**, Proposition C.7], $\bar{\kappa}$ is also strict-strong operator continuous into $\mathcal{L}(X)$. In particular, X is a left $M(A)$-module, in fact a right-Hilbert $M(A) - M(B)$ bimodule.

In this work, imprimitivity bimodules will play a very important rôle, since they will represent the equivalences in our categories (see Chapter 2 below). But we shall also need the more general notion of partial imprimitivity bimodules:

DEFINITION 1.5. Suppose that A and B are C^*-algebras. A *partial $A - B$ imprimitivity bimodule* is a complex vector space X which is a right Hilbert B-module and a left Hilbert A-module such that

$$a \cdot (x \cdot b) = (a \cdot x) \cdot b \quad \text{and} \quad {}_A\langle x, y\rangle \cdot z = x \cdot \langle y, z\rangle_B$$

for all $a \in A$, $b \in B$, and $x, y, z \in X$. If $_A\overline{\langle X, X\rangle} = A$, we say that X is a *right-partial imprimitivity bimodule*. If $\overline{\langle X, X\rangle}_B = B$, X is a *left-partial imprimitivity bimodule*. If both $_A\langle\cdot,\cdot\rangle$ and $\langle\cdot,\cdot\rangle_B$ are full, then X is called an *$A - B$ imprimitivity bimodule*. A and B are called *Morita equivalent* if there exists at least one $A - B$ imprimitivity bimodule.

Imprimitivity bimodules for C^*-algebras were introduced by Rieffel in [**56**]. As a general reference we recommend [**54**]. Our partial imprimitivity bimodules are often called Hilbert bimodules in the literature. However, this terminology would be much too close to "right-Hilbert bimodules", so we decided to introduce new terminology to prevent confusion.

The reason for choosing the notion of partial imprimitivity bimodules comes from the trivial fact that if X is a partial $A - B$ imprimitivity bimodule, then X becomes an $A_X - B_X$ imprimitivity bimodule for the ideals $A_X = {}_A\overline{\langle X, X\rangle}$ and $B_X = \overline{\langle X, X\rangle}_B$ of A and B, respectively. In the special case where $A = C_0(V)$ and $B = C_0(W)$ are commutative, this just means that we have a homeomorphism between the open sets V_X and W_X corresponding to the ideals A_X and B_X, i.e., a partial homeomorphism between V and W (see [**54**, Corollary 3.33]).

EXAMPLE 1.6. (1) Suppose that X is a right-Hilbert $A - B$ bimodule. Let $_{\mathcal{K}(X)}\langle\cdot,\cdot\rangle$ be the left $\mathcal{K}(X)$-valued inner product on X as defined in (1.3). Then X becomes a right-partial $\mathcal{K}(X) - B$ imprimitivity bimodule. In particular, if X is a full Hilbert B-module, then X is a $\mathcal{K}(X) - B$ imprimitivity bimodule.

(2) More generally, suppose that X is a right-Hilbert $A - B$ bimodule such that the corresponding homomorphism $\kappa\colon A \to \mathcal{L}_B(X)$ is injective and $\kappa(A) \supseteq \mathcal{K}(X)$. Define a left A-valued inner product on X by

$$_A\langle x, y\rangle = \kappa^{-1}({}_{\mathcal{K}(X)}\langle x, y\rangle)$$

for all $x, y \in X$. Then X becomes a partial $A - B$ imprimitivity bimodule. In particular, if X is a full right-Hilbert $A-B$ bimodule, then X is an $A-B$ imprimitivity bimodule if and only if $\kappa \colon A \to \mathcal{L}(X)$ is injective and $\kappa(A) = \mathcal{K}(X)$.

(3) Suppose that X is a partial $A - B$ imprimitivity bimodule. Let $\widetilde{X} = \{\tilde{x} \mid x \in X\}$ with left B-action and right A-action defined by

$$b \cdot \tilde{x} = \widetilde{x \cdot b^*} \quad \text{and} \quad \tilde{x} \cdot a = \widetilde{a^* \cdot x}$$

and B- and A-valued inner products given by

$$_B\langle \tilde{x}, \tilde{y}\rangle = \langle x, y\rangle_B \quad \text{and} \quad \langle \tilde{x}, \tilde{y}\rangle_A = {}_A\langle x, y\rangle,$$

for $x, y \in X$. Then $\widetilde{X}$ is a partial $B-A$ imprimitivity bimodule, called the *conjugate* (also called the *reverse*, or *dual* in the literature) of X. If X is a right-partial imprimitivity bimodule, then $\widetilde{X}$ is a left-partial imprimitivity bimodule and *vice versa*. If X is an $A - B$ imprimitivity bimodule, then $\widetilde{X}$ is a $B - A$ imprimitivity bimodule.

(4) If X is a partial $A-B$ imprimitivity bimodule, then via the extension of the actions to the multiplier algebras, X becomes a partial $M(A)-M(B)$ imprimitivity bimodule.

Another important example is given by the multiplier bimodule $M(X)$ of a right-partial imprimitivity bimodule X, which we shall define in Section 1.2 below. We now recall the notion of induced ideals. If not explicitly stated otherwise, by an ideal of an algebra we always mean a two-sided ideal.

DEFINITION 1.7. Suppose that X is a right-Hilbert $A - B$ bimodule (or just an $A-B$ correspondence). Let I be a closed ideal in B. Then call XI the *closed $A-B$ submodule of X corresponding to I*. The closed ideal $X\text{-Ind}\, I = \{a \in A \mid aX \subseteq XI\}$ is called the *ideal of A induced from I*.

If X is actually an $A - B$ imprimitivity bimodule then we have the following basic result. For a proof we refer to [**54**, Theorem 3.22, Lemma 3.23 and Proposition 3.24].

PROPOSITION 1.8 (Rieffel Correspondence). *Suppose that X is an $A - B$ imprimitivity bimodule. Then there exist inclusion-preserving one-to-one correspondences between*

  (i) *the closed ideals of B,*
  (ii) *the closed $A - B$ submodules of X, and*
  (iii) *the closed ideals of A.*

If I is a closed ideal of B, then XI is the corresponding closed $A - B$ submodule of X and $X\text{-Ind}\, I$ is the corresponding closed ideal of A. Moreover, we have the identities $X\text{-Ind}\, I = {}_A\overline{\langle XI, XI\rangle}$, $(X\text{-Ind}\, I)X = XI$, and $\overline{\langle XI, XI\rangle}_B = I$.

REMARK 1.9. If X is an $A - B$ imprimitivity bimodule, then it follows from the Rieffel correspondence that if I is a closed ideal of B, then XI becomes a $X\text{-Ind}\, I - I$ imprimitivity bimodule via restricting all actions and inner products to the subspaces. With a little bit more work one can also show that X/XI can be made into an $A/(X\text{-Ind}\, I) - B/I$ imprimitivity bimodule with respect to the obvious actions and inner products (see [**54**, Proposition 3.25] for more details).

1.2. Multiplier bimodules and homomorphisms

A theory of multipliers of imprimitivity bimodules was developed in [20]. We cannot adopt that theory directly, as our right-Hilbert bimodules are decidedly left-challenged; we need something more like the one-sided theory of [2] and [40]. Our gadgets and objectives are not exactly the same as in these references; for this reason and for completeness we give a self-contained treatment.

If X is a Hilbert B-module, then $\mathcal{L}(X)$ is a C^*-algebra, hence is a right-Hilbert $\mathcal{L}(X) - \mathcal{L}(X)$ bimodule. This extends to adjointable maps between different modules:

PROPOSITION 1.10. *Let X and Y be Hilbert B-modules. Then $\mathcal{L}(X,Y)$ is a right-Hilbert $\mathcal{L}(Y) - \mathcal{L}(X)$ bimodule with operations*

$$P \cdot T = P \circ T, \quad T \cdot Q = T \circ Q, \quad \text{and} \quad \langle T, S \rangle_{\mathcal{L}(X)} = T^* \circ S$$

for $P \in \mathcal{L}(Y)$, $T, S \in \mathcal{L}(X,Y)$, and $Q \in \mathcal{L}(X)$. Moreover, $\mathcal{L}(X,Y)$ becomes a partial $\mathcal{L}(Y) - \mathcal{L}(X)$ imprimitivity bimodule if we define the $\mathcal{L}(Y)$-valued inner product on $\mathcal{L}(X,Y)$ by $_{\mathcal{L}(Y)}\langle T, S \rangle = T \circ S^*$.

PROOF. Of course $PT \in \mathcal{L}(X,Y)$, with $(PT)^* = T^*P^*$, and similarly for the other compositions. Composition is associative, so $\mathcal{L}(X,Y)$ is an $\mathcal{L}(Y) - \mathcal{L}(X)$ bimodule, and easy calculations verify the sesquilinearity of $\langle \cdot, \cdot \rangle_{\mathcal{L}(X)}$ and the consistency relations (1.1) and (1.2b) (recall that (1.2a) is redundant).

For positivity, note that for each $x \in X$,

$$\langle x, T^*Tx \rangle_B = \langle Tx, Tx \rangle_B \geq 0,$$

so that $T^*T \geq 0$ in $\mathcal{L}(X)$ by [54, Lemma 2.28]. The above equality also implies that $\|Tx\|^2 = 0$ for all $x \in X$ whenever $T^*T = 0$, which gives definiteness of the inner product.

Since $\mathcal{L}(Y)$ is unital, it certainly acts nondegenerately on $\mathcal{L}(X,Y)$, so it only remains to check that $\mathcal{L}(X,Y)$ is complete in the norm

$$\|T\| = \|\langle T, T \rangle_{\mathcal{L}(X)}\|^{1/2} = \|T^*T\|_{\mathcal{L}(X)}^{1/2}.$$

First we claim that this norm coincides with the operator norm on the Banach space $\mathcal{B}(X,Y)$ of all bounded linear maps of X into Y. Since $T^*T \geq 0$ in $\mathcal{L}(X)$, we can find $Q \in \mathcal{L}(X)$ with $Q^*Q = T^*T$. Then

$$\|T\|_{\mathcal{B}(X,Y)}^2 = \sup_{\|x\| \leq 1} \{\|Tx\|_Y^2\} = \sup_{\|x\| \leq 1} \{\|\langle Tx, Tx \rangle_B\|\} = \sup_{\|x\| \leq 1} \{\|\langle x, T^*Tx \rangle_B\|\}$$

$$= \sup_{\|x\| \leq 1} \{\|\langle x, Q^*Qx \rangle_B\|\} = \sup_{\|x\| \leq 1} \{\|Qx\|_X^2\} = \|Q\|_{\mathcal{L}(X)}^2$$

$$= \|Q^*Q\|_{\mathcal{L}(X)} = \|T^*T\|_{\mathcal{L}(X)} = \|T\|^2.$$

Next, the adjoint map $T \mapsto T^*$ is isometric from $\mathcal{L}(X,Y)$ into $\mathcal{L}(Y,X)$:

$$\|T^*\| = \|T^*\|_{\mathcal{B}(Y,X)} = \sup_{\|y\| \leq 1} \{\|T^*y\|_X\} = \sup_{\|x\|, \|y\| \leq 1} \{\|\langle x, T^*y \rangle_B\|\}$$

$$= \sup_{\|x\|, \|y\| \leq 1} \{\|\langle Tx, y \rangle_B\|\} = \sup_{\|x\| \leq 1} \{\|Tx\|_Y\} = \|T\|_{\mathcal{B}(X,Y)} = \|T\|.$$

Now a standard argument shows that $\mathcal{L}(X,Y)$ is closed in $\mathcal{B}(X,Y)$, and therefore complete: suppose that $\{T_i\}$ is Cauchy in $\mathcal{L}(X,Y)$, so $T_i \to T$ in $\mathcal{B}(X,Y)$ and

$T_i^* \to S$ in $\mathcal{B}(Y,X)$. Then S is an adjoint for T:
$$\langle Tx, y\rangle_B = \lim_i \langle T_i x, y\rangle_B = \lim_i \langle x, T_i^* y\rangle_B = \langle x, Sy\rangle_B.$$
The last assertion now follows from the complete symmetry of the situation. □

REMARK 1.11. The $\mathcal{L}(Y)$- and $\mathcal{L}(X)$-valued inner products on $\mathcal{L}(X,Y)$ are in general not full, so $\mathcal{L}(X,Y)$ is in general not an $\mathcal{L}(Y) - \mathcal{L}(X)$ imprimitivity bimodule. To see an example, let $X = \mathcal{H}$ be an infinite-dimensional Hilbert space, viewed as a Hilbert $\mathbb{C}$-module, and let $Y = \mathbb{C}$. Then one can check that $\mathcal{L}(\mathcal{H}, \mathbb{C}) \cong \mathcal{H}$ and $_{\mathcal{L}(\mathcal{H})}\langle \mathcal{H}, \mathcal{H}\rangle = \mathcal{K}(\mathcal{H})$. But $\mathcal{L}(\mathcal{H})$ coincides with the much larger algebra $\mathcal{B}(\mathcal{H})$.

REMARK 1.12. In what follows, we are primarily interested in $\mathcal{L}_B(B, X)$, where X is a Hilbert B-module, and B is viewed as a Hilbert B-module in the usual way. In this case $B \cong \mathcal{K}_B(B)$ and (hence) $M(B) \cong \mathcal{L}_B(B)$.

COROLLARY 1.13. Let X be a right-Hilbert $A - B$ bimodule. Then $\mathcal{L}_B(B, X)$ is a right-Hilbert $M(A) - M(B)$ bimodule with operations
$$(1.4) \quad (n \cdot T)b = n \cdot (Tb), \quad (T \cdot m)b = T(mb), \quad \text{and} \quad \langle T, S\rangle_{M(B)} = T^*S$$
for $n \in M(A)$, $T, S \in \mathcal{L}_B(B, X)$, $b \in B$, and $m \in M(B)$.

PROOF. By Proposition 1.10 and Remark 1.12, $\mathcal{L}(B, X)$ is a right-Hilbert $\mathcal{L}(X) - M(B)$ bimodule, so we only need to produce the compatible $M(A)$-action. But this is easy: letting $\kappa \colon A \to \mathcal{L}(X)$ be the associated nondegenerate homomorphism, we have the canonical extension $\bar{\kappa} \colon M(A) \to \mathcal{L}(X)$, so $M(A)$ acts on the left of $\mathcal{L}(B, X)$ by $m \cdot T = \bar{\kappa}(m)T$. Unraveling the identifications, one sees that this left action is given by the first formula in (1.4). □

DEFINITION 1.14. We call $\mathcal{L}_B(B, X)$ the *multiplier bimodule* of the right-Hilbert $A - B$ bimodule X, and we denote it by $M(X)$.

REMARK 1.15. (1) If X is a right-partial $A - B$ imprimitivity bimodule, then the left action of $M(A)$ on X identifies $M(A)$ with $M(\mathcal{K}(X)) \cong \mathcal{L}_B(X)$. Thus it follows from Proposition 1.10 that $M(X)$ carries a left $M(A)$-valued inner product which makes $M(X)$ into a partial $M(A) - M(B)$ imprimitivity bimodule.

(2) It follows from [**20**, Proposition 1.3] that if $_A X_B$ is an $A - B$ imprimitivity bimodule, then the above-defined multiplier bimodule $M(X)$ is canonically isomorphic to the multiplier bimodule defined in [**20**, Definition 1.1].

DEFINITION 1.16. Let $_A X_B$ and $_C Y_D$ be right-Hilbert bimodules, let $\varphi \colon A \to M(C)$ and $\psi \colon B \to M(D)$ be homomorphisms, and let $\Phi \colon X \to M(Y)$ be a linear map. We call Φ a $\varphi - \psi$ *compatible right-Hilbert bimodule homomorphism* if
  (i) $\Phi(a \cdot x) = \varphi(a) \cdot \Phi(x)$,
  (ii) $\Phi(x \cdot b) = \Phi(x) \cdot \psi(b)$, and
  (iii) $\psi(\langle x, z\rangle_B) = \langle \Phi(x), \Phi(z)\rangle_{M(D)}$ for all $a \in A$, $x, z \in X$ and $b \in B$.

We call φ and ψ the *coefficient maps* of Φ; we write $_\varphi\Phi_\psi \colon {_A X_B} \to M(_C Y_D)$ to indicate all the data.

We say that Φ is *nondegenerate* if $\overline{\Phi(X)D} = Y$ (i.e., the closed linear span of $\{\Phi(x)d \mid x \in X, d \in D\}$ equals Y) and both φ and ψ are nondegenerate.

If φ and ψ are isomorphisms of A and B onto C and D, respectively, and Φ is a bijection onto Y, we call Φ a *right-Hilbert bimodule isomorphism* of X onto Y.

If $A = C$, $B = D$, and both φ and ψ are identity maps, we call Φ a *right-Hilbert $A - B$ bimodule homomorphism*. If in addition the map Φ is bijective we call it a *right-Hilbert $A - B$ bimodule isomorphism*.

REMARK 1.17. (1) Note that a right-Hilbert bimodule homomorphism $\Phi \colon X \to M(Y)$ is always norm-decreasing, since (with the above notation) for each $x \in X$ we have

$$\|\Phi(x)\|^2 = \|\langle \Phi(x), \Phi(x) \rangle_{M(D)}\| = \|\psi(\langle x, x \rangle_B)\| \leq \|\langle x, x \rangle_B\| = \|x\|^2.$$

Of course, this also shows that a right-Hilbert bimodule isomorphism is necessarily isometric.

(2) Condition (ii) is actually implied by condition (iii) and the linearity of ψ:

$$\begin{aligned}
&\|\Phi(xb) - \Phi(x)\psi(b)\|^2 \\
&= \big\| \langle \Phi(xb), \Phi(xb) \rangle_{M(D)} - \langle \Phi(xb), \Phi(x)\psi(b) \rangle_{M(D)} \\
&\quad + \langle \Phi(x)\psi(b), \Phi(x)\psi(b) \rangle_{M(D)} - \langle \Phi(x)\psi(b), \Phi(xb) \rangle_{M(D)} \big\| \\
&= \big\| \langle \Phi(xb), \Phi(xb) \rangle_{M(D)} - \langle \Phi(xb), \Phi(x) \rangle_{M(D)} \psi(b) \\
&\quad + \psi(b)^* \langle \Phi(x), \Phi(x) \rangle_{M(D)} \psi(b) - \psi(b)^* \langle \Phi(x), \Phi(xb) \rangle_{M(D)} \big\| \\
&= \big\| \psi\big(\langle xb, xb \rangle_{M(B)} - \langle xb, x \rangle_{M(B)} b + b^* \langle x, x \rangle_{M(B)} b - b^* \langle x, xb \rangle_{M(B)} \big) \big\| \\
&= 0.
\end{aligned}$$

In fact, a similar computation shows that if $\Phi \colon X \to M(Y)$ is only a *map* satisfying (i) and (iii) then Φ is automatically linear, hence is a right-Hilbert bimodule homomorphism (see, for example, [**62**, Lemma 2.5]).

If $_AX_B$ and $_CY_D$ are partial imprimitivity bimodules, a nondegenerate right-Hilbert bimodule homomorphism automatically preserves this extra structure (as well as possible). Recall that in this situation the action of C on Y determines a homomorphism $\kappa_C \colon M(C) \to \mathcal{L}_D(Y)$. If Y is a right-partial $C - D$ imprimitivity bimodule we have $M(C) \cong \mathcal{L}_D(Y)$ via κ_C, and we used this identification to define an $M(C)$-valued inner product on $M(Y)$.

LEMMA 1.18. *Suppose that $_AX_B$ is a partial imprimitivity bimodule, $_CY_D$ is a right-Hilbert bimodule and $_\varphi\Phi_\psi \colon X \to M(Y)$ is a right-Hilbert bimodule homomorphism such that $\overline{\Phi(X)D} = Y$. Then*

$$_{\mathcal{L}(Y)}\langle \Phi(x), \Phi(z) \rangle = \kappa_C \circ \varphi(_A\langle x, z \rangle),$$

for all $x, z \in X$. If, in addition, $_CY_D$ is a right-partial imprimitivity bimodule, then $_\varphi\Phi_\psi \colon X \to M(Y)$ is a partial imprimitivity bimodule homomorphism, that is, Φ is a right-Hilbert bimodule homomorphism with the additional property that

$$_{M(C)}\langle \Phi(x), \Phi(z) \rangle = \varphi(_A\langle x, z \rangle)$$

for all $x, z \in X$.

PROOF. The second assertion follows directly from the first assertion and the definition of $_{M(C)}\langle \cdot, \cdot \rangle$. For the first assertion we have to check that

$$\varphi(_A\langle x, z \rangle)y = {}_{\mathcal{L}(Y)}\langle \Phi(x), \Phi(z) \rangle y$$

for all $y \in Y$. Since $Y = \overline{\Phi(X)D}$, we may assume without loss of generality that $y = \Phi(w)d$ for some $w \in X$, $d \in D$, and then

$$\varphi(_A\langle x, z\rangle) \cdot \Phi(w)d = \Phi(_A\langle x, z\rangle \cdot w)d = \Phi(x \cdot \langle z, w\rangle_B)d$$
$$= \Phi(x) \cdot \langle \Phi(z), \Phi(w)\rangle_{M(D)}d = \Phi(x) \circ \Phi(z)^* \circ \Phi(w) \cdot d$$
$$= {}_{\mathcal{L}(Y)}\langle \Phi(x), \Phi(z)\rangle \cdot \Phi(w) \cdot d.$$

$\square$

REMARK 1.19. (1) If $_AX_B$ and $_CY_D$ are imprimitivity bimodules, it follows from the above lemma that a nondegenerate right-Hilbert bimodule homomorphism $_\varphi\Phi_\psi \colon X \to M(Y)$ is automatically a nondegenerate imprimitivity bimodule homomorphism in the sense of [**20**, Definition 1.8].

Curiously, the above statement has a converse: if Φ is a nondegenerate imprimitivity bimodule homomorphism, then Φ is automatically a nondegenerate right-Hilbert bimodule homomorphism, since then $\overline{\Phi(X)D} = Y$ by [**29**, Lemma 5.1]. However, we must be careful in applying this remark: if X and Y are imprimitivity bimodules, but we only know that $\Phi \colon X \to M(Y)$ is a right-Hilbert bimodule homomorphism, then nondegeneracy of the coefficient homomorphisms does *not* imply nondegeneracy of Φ — for example, if $\mathcal{H}$ is a Hilbert space of dimension greater than one, and if $\xi \in \mathcal{H}$, we get a right-Hilbert bimodule homomorphism $_\varphi\Phi_\psi \colon {}_\mathbb{C}\mathbb{C}_\mathbb{C} \to {}_{\mathcal{K}(\mathcal{H})}\mathcal{H}_\mathbb{C}$ by defining

$$\Phi(c) = c\xi, \quad \varphi(c) = c1_\mathcal{H}, \quad \text{and} \quad \psi(c) = c.$$

Both φ and ψ are nondegenerate, but Φ is degenerate. Of course, Φ is *not* an imprimitivity bimodule homomorphism. For nondegeneracy of Φ to follow from that of the coefficient maps, we must know Φ preserves *both* inner products. As a reward for checking both inner products, we can avoid checking the left module action, as a computation similar to that in item (2) of Remark 1.17 shows.

(2) Several times we will need the following adaptation to our context of [**20**, Example 1.10]: if $_\varphi\Phi_\psi \colon {}_AX_B \to M(_CY_D)$ is a nondegenerate right-Hilbert bimodule homomorphism, then there exists a unique map $\mu \colon \mathcal{K}_B(X) \to M(\mathcal{K}_D(Y))$ such that $_\mu\Phi_\psi$ is a nondegenerate right-Hilbert bimodule homomorphism from $_{\mathcal{K}(X)}X_B$ to $M(_{\mathcal{K}(Y)}Y_D)$. Note that it follows from Lemma 1.18 above that $_\mu\Phi_\psi$ is then automatically a right-Hilbert bimodule homomorphism, so that μ is determined by the equation

$$\mu(_{\mathcal{K}(X)}\langle x, y\rangle) = {}_{M(\mathcal{K}_D(Y))}\langle \Phi(x), \Phi(y)\rangle \qquad \text{for all } x, y \in X.$$

It is useful to note the following relation between right-partial imprimitivity bimodule homomorphisms and induced ideals.

LEMMA 1.20. *Let* $_\varphi\Phi_\psi \colon {}_AX_B \to M(_CY_D)$ *be a right-Hilbert bimodule homomorphism. Then* $\ker \Phi = X \cdot \ker \psi$. *In particular,* Φ *is injective if* ψ *is. If, in addition,* $_AX_B$ *and* $_CY_D$ *are right-partial imprimitivity bimodules, we also have* $\ker \varphi = X\text{-}\operatorname{Ind}(\ker \psi)$. *Thus, if* $_AX_B$ *is an imprimitivity bimodule, then* $\ker \varphi$, $\ker \Phi$, *and* $\ker \psi$ *correspond to each other via the Rieffel correspondence.*

PROOF. Since Φ is a bimodule map, it follows that $X \cdot \ker \psi \subseteq \ker \Phi$. Conversely, if $Y = \ker \Phi$, then Y is a closed $A - B$ submodule of X such that $\langle Y, Y\rangle_B \subseteq \ker \psi$. It follows that $Y = Y \cdot \overline{\langle Y, Y\rangle_B} \subseteq X \cdot \ker \psi$.

Suppose now that $A = \mathcal{K}(X)$ and $C = \mathcal{K}(Y)$. As above we first see that $(\ker\varphi)X \subseteq \ker\Phi$, so we only have to check that $aX \subseteq \ker\Phi$ implies $a \in \ker\varphi$ for all $a \in \mathcal{K}(X)$. But, using Lemma 1.18, $aX \subseteq \ker\Phi$ implies that

$$\varphi(a\mathcal{K}(X)) = \varphi\big(a(_{\mathcal{K}(X)}\overline{\langle X, X \rangle})\big) = \varphi(_{\mathcal{K}(X)}\overline{\langle aX, X \rangle}) = {}_{M(\mathcal{K}(Y))}\overline{\langle \Phi(aX), \Phi(X) \rangle} = \{0\},$$

and hence $\varphi(a) = 0$. The final assertions now follow from Proposition 1.8. □

REMARK 1.21. One useful application of the above lemma and the Rieffel correspondence is the following result: If ${}_\varphi\Phi_\psi \colon {}_AX_B \to {}_CY_D$ is a homomorphism between imprimitivity bimodules, and if φ and ψ are isomorphisms, then so is Φ. The injectivity of Φ follows from the injectivity of ψ, and the surjectivity follows from the Rieffel correspondence and the fact that $\Phi(X)$ is a closed $C-D$ submodule of Y such that $\overline{\langle \Phi(X), \Phi(X) \rangle}_D = \psi(B) = D$.

In Proposition 1.34 below, and elsewhere, we need to construct a right-Hilbert bimodule homomorphism by extending from a pre-right-Hilbert bimodule. We pause to make this precise, and streamline the process with an elementary lemma.

DEFINITION 1.22. Let A_0 and B_0 be dense *-subalgebras of C^*-algebras A and B, respectively, and let X_0 be an $A_0 - B_0$ bimodule. We say X_0 is a *pre-right-Hilbert $A_0 - B_0$ bimodule* if it has a B_0-valued pre-inner product (with positivity interpreted in B), so that (1.1) holds, and also

$$\langle a \cdot x, a \cdot x \rangle_{B_0} \leq \|a\|^2 \langle x, x \rangle_{B_0} \qquad \text{for all } a \in A_0, x \in X_0.$$

As with pre-Hilbert modules (*cf.* [**54**, Lemma 2.16]), the completion X of X_0 becomes a right-Hilbert $A - B$ bimodule by taking limits of the operations.

LEMMA 1.23. *Let X_0 be a pre-right-Hilbert $A_0 - B_0$ bimodule, with completion ${}_AX_B$. Suppose we have homomorphisms $\varphi \colon A \to M(C)$ and $\psi \colon B \to M(D)$, and a map $\Phi_0 \colon X_0 \to M(Y)$ such that:*
  (i) $\Phi_0(a \cdot x) = \varphi(a) \cdot \Phi_0(x)$ *and*
  (ii) $\langle \Phi_0(x), \Phi_0(y) \rangle_{M(D)} = \psi(\langle x, y \rangle_{B_0})$

for all $a \in A_0$ and $x, y \in X_0$. Then Φ extends uniquely to a right-Hilbert bimodule homomorphism $\Phi \colon X \to M(Y)$.

PROOF. Similarly to Remark 1.17, the inner product property (ii) implies first of all that Φ_0 is linear, then that it is bounded, hence extends uniquely to a bounded linear map $\Phi \colon X \to M(Y)$. The compatibility conditions (i) and (iii) of Definition 1.16 follow from continuity and density. (Recall from Remark 1.17 that Definition 1.16 (ii) is redundant.) □

Occasionally it is convenient to take advantage of certain labor-saving devices when we are dealing with partial imprimitivity bimodules. In the language we have already introduced, we observe that a pre-imprimitivity bimodule can be defined as a pre-right-Hilbert bimodule ${}_{A_0}(X_0)_{B_0}$ which also has a left A_0-valued pre-inner product (with positivity interpreted in the completion A) such that for all $x, y, z \in X_0$ and $b \in B_0$ we have

$${}_{A_0}\langle xb, xb \rangle \leq \|b\|^2 {}_{A_0}\langle x, x \rangle \quad \text{and} \quad {}_{A_0}\langle x, y \rangle z = x \langle y, z \rangle_{B_0}.$$

LEMMA 1.24. *Let X_0 be a pre-imprimitivity $A_0 - B_0$ bimodule, with completion ${}_AX_B$. Suppose we have homomorphisms $\varphi \colon A \to M(C)$ and $\psi \colon B \to M(D)$, and a map $\Phi_0 \colon X_0 \to M(Y)$ such that:*

(i) $_{M(C)}\langle \Phi_0(x), \Phi_0(y)\rangle = \varphi(_{A_0}\langle x, y\rangle)$ and
(ii) $\langle \Phi_0(x), \Phi_0(y)\rangle_{M(D)} = \psi(\langle x, y\rangle_{B_0})$

for all $x, y \in X_0$. Then Φ extends uniquely to an imprimitivity bimodule homomorphism $\Phi: X \to M(Y)$.

Moreover, if φ and ψ are nondegenerate, then so is Φ.

PROOF. Similarly to the proof of the preceding lemma, the inner product properties (i)–(ii) imply first that Φ_0 is linear, then that it is bounded, hence extends uniquely to a bounded linear map $\Phi: X \to M(Y)$ which preserves both inner products by continuity and density; Remark 1.17 assures us that this suffices to give the unique extension Φ. We have already mentioned in Remark 1.19 that nondegeneracy of φ and ψ guarantee nondegeneracy of Φ. □

DEFINITION 1.25. Let $_A X_B$ be a right-Hilbert bimodule. The *strict topology* on $M(X)$ is that generated by the seminorms $m \mapsto \|Tm\|, \|mb\|$ for $T \in \mathcal{K}_B(X)$ and $b \in B$.

REMARK 1.26. (1) Thus the strict topology on $M(_A X_B)$ has nothing to do with the left A-module action and only depends on the right-partial imprimitivity bimodule structure of $_{\mathcal{K}(X)} X_B$.

(2) What [27] and [40] call the strict topology on $M(X)$ is really the $*$-strong topology. This is easily seen to be weaker than the strict topology.[1] Indeed, if $m_i \to m$ strictly in $M(X)$, then $m_i b \to mb$ for all $b \in B$ by definition. Thus we only have to check that $m_i^* x \to m^* x$ for all $x \in X$, which follows from factoring $x = cy$ for some $c \in \mathcal{K}(X)$, $y \in X$ and computing

$$m_i^*(cy) = \langle m_i, cy\rangle_{M(B)} = \langle c^* m_i, y\rangle_{M(B)} \to \langle c^* m, y\rangle_{M(B)} = m(cy).$$

The above definition is modeled after [20, Definition 1.5], where it is discussed for imprimitivity bimodules. If X is full, the strict topology on $M(X)$ only depends upon the $\mathcal{K}(X) - B$ imprimitivity bimodule structure on X and all results from [20] are available. However, we need to generalize several results of [20] to more general right-Hilbert bimodules.

Note first that we have a canonical right-Hilbert $\mathcal{L}(X) - M(B)$ bimodule embedding of X into $M(X)$ given by $x \mapsto m_x$, where $m_x b = x \cdot b$ and $m_x^*(y) = \langle x, y\rangle_B$. It follows directly from the definitions that this embedding preserves the left and right actions and the $M(B)$-valued inner product. Moreover, since the left action of $M(A)$ on X and $M(X)$ is given via the same homomorphism $\kappa: M(A) \to \mathcal{L}(X)$, we see that $x \to m_x$ is also a right-Hilbert $M(A) - M(B)$ bimodule homomorphism.

PROPOSITION 1.27 (cf. [20, Proposition 1.6]). *Let X be a right-Hilbert $A - B$ bimodule and let us view X as an $\mathcal{L}(X) - M(B)$ sub-bimodule of $M(X)$ via the above embedding. Then:*

(i) $\mathcal{K}(X) \cdot M(X) \subseteq X$ *and* $M(X) \cdot B \subseteq X$.
(ii) $M(X)$ *is the strict completion of X.*
(iii) *The pairings*

$$\mathcal{L}(X) \times M(X) \to M(X), \quad M(A) \times M(X) \to M(X) \text{ and}$$
$$M(X) \times M(B) \to M(X), \quad \langle \cdot, \cdot \rangle_{M(B)}: M(X) \times M(X) \to M(B)$$

are separately strictly continuous, where we identify $\mathcal{L}(X)$ with $M(\mathcal{K}(X))$.

---

[1] Is it always the same? Presumably not, but we do not know a counterexample.

PROOF. Let $m \in M(X)$. Factoring $b \in B$ as $b = cd$ for $c, d \in B$, it follows that $mb = (mc)(d) \in X$. On the other side, we have $_{\mathcal{K}(X)}\langle x, y\rangle m = x \langle y, m\rangle_{M(B)} \in X$ for all $x, y \in X$, $m \in M(X)$. This proves (i).

For the proof of (ii), let $(u_i)_{i \in I}$ be an approximate unit in B. Then it follows from (i) that mu_i converges strictly to m. Thus X is strictly dense in $M(X)$. Conversely, assume that $(m_i)_{i \in I}$ is a strict Cauchy net in $M(X)$. Then it follows from Remark 1.26 that $(m_i)_{i \in I}$ is also a $*$-strong Cauchy net. This implies that we can define $m \in M(X)$ by $m(b) = \lim_{i \in I} m_i(b)$, $m^*(x) = \lim_{i \in I} m_i^*(x)$. Now let $c \in \mathcal{K}(X)$. Since $(cm_i)_{i \in I}$ is a Cauchy net in X by assumption, it follows that cm_i converges to some $y \in X$, and since $yb = \lim_{i \in I}(cm_i)b = \lim_{i \in I} c(m_i b) = c(mb) = (cm)b$ for all $b \in B$, it follows that $y = cm$. Thus $M(X)$ is complete in the strict topology.

The proof of (iii) now follows from an application of Remark 1.26 and the norm-continuity of the pairings. For example, if $m_i \to m$ strictly in $M(X)$ and $n \in M(X)$, then it follows for all $b \in B$ that

$$\langle m_i, n\rangle_{M(B)} b = m_i^*(n(b)) \to m^*(n(b)) = \langle m, n\rangle_{M(B)} b,$$

by $*$-strong convergence of $(m_i)_{i \in I}$. Similarly,

$$b\langle m_i, n\rangle_{M(B)} = \langle m_i(b^*), n\rangle_{M(B)} \to \langle m(b^*), n\rangle_{M(B)} = b\langle m, n\rangle_{M(B)}.$$

It follows that $\langle m_i, n\rangle_{M(B)} \to \langle m, n\rangle_{M(B)}$ strictly. The other assertions follow from similar arguments. $\square$

In fact, $M(X)$ is maximal with respect to the above properties:

PROPOSITION 1.28. *Let X be a right-Hilbert $A - B$ bimodule. Suppose M is a right-Hilbert $M(A) - M(B)$ bimodule containing X as a right-Hilbert $M(A) - M(B)$ sub-bimodule (extending the operations on X in the usual way) such that*

$$M \cdot B \subseteq X.$$

Then M embeds as an $M(A) - M(B)$ sub-bimodule of $M(X)$.

PROOF. For $m \in M$ define $\Phi(m) \colon B \to X$ by

(1.5) $$\Phi(m)b = mb.$$

Then $\Phi(m)$ is adjointable, with

(1.6) $$\Phi(m)^* x = \langle m, x\rangle_{M(B)}.$$

To see this, note that $\langle m, x\rangle_{M(B)} \in B$, since we may factor $x = yb$ for some $y \in X$ and $b \in B$, and then

$$\langle m, x\rangle_{M(B)} = \langle m, yb\rangle_{M(B)} = \langle m, y\rangle_{M(B)} b \in B.$$

Checking the adjoint property, for $b \in B$ and $x \in X$ we have

$$\langle b, \langle m, x\rangle_{M(B)}\rangle_B = b^* \langle m, x\rangle_{M(B)} = \langle mb, x\rangle_{M(B)} = \langle mb, x\rangle_B = \langle \Phi(m)b, x\rangle_B.$$

Thus we get a map $\Phi \colon M \to M(X)$.

For each $b \in B$, $k \in M(A)$, and $m, n \in M$ we have

$$\Phi(km)b = kmb = k\Phi(m)b$$

and

$$\langle \Phi(m), \Phi(n)\rangle_{M(B)} b = \Phi(m)^* \Phi(n) b = \Phi(m)^* nb = \langle m, nb\rangle_{M(B)} = \langle m, n\rangle_{M(B)} b.$$

Thus $\Phi\colon M \to M(X)$ is a right-Hilbert $M(A) - M(B)$ bimodule homomorphism which is necessarily isometric; of course Φ restricts to the usual embedding of X in $M(X)$.

For uniqueness, suppose that $\Psi\colon M \to M(X)$ is another right-Hilbert $M(A) - M(B)$-bimodule homomorphism extending the embedding of X. Then, for all $b \in B$, $(\Phi(m) - \Psi(m))(b) = mb - \Psi(mb) = 0$ since $mb \in X$. Thus $\Psi = \Phi$. □

REMARK 1.29. When X is an $A - B$ imprimitivity bimodule, we have two ostensibly different $M(A) - M(B)$ bimodules: $M(_AX) = \mathcal{L}_A(A, X)$ and $M(X_B) = \mathcal{L}_B(B, X)$. In fact these turn out to be naturally isomorphic. The link is the multiplier bimodule $M(_AX_B)$ studied in [**20**], which consists of pairs (m_A, m_B) in which $m_A\colon A \to X$ is A-linear, $m_B\colon B \to X$ is B-linear, and
$$m_A(a) \cdot b = a \cdot m_B(b) \qquad \text{for all } a \in A, b \in B.$$
The maps $(m_A, m_B) \mapsto m_A$ and $(m_A, m_B) \mapsto m_B$ give isomorphisms of $M(_AX_B)$ onto $M(_AX)$ and $M(X_B)$, respectively [**20**, Proposition 1.3]. This is not obvious: there is *a priori* no assertion of adjointability in the definition of $M(_AX_B)$, and as a result the abstract characterization of $M(_AX_B)$ in [**20**, Proposition 1.2] looks rather different from its one-sided analogue in Proposition 1.28. Unfortunately, a similar result does not hold for arbitrary partial imprimitivity bimodules, so that we definitely have a one-sided theory of multiplier bimodules for partial imprimitivity bimodules.

We will now show that nondegeneracy allows us to extend right-Hilbert bimodule homomorphisms to the multiplier bimodules.

THEOREM 1.30. *Let $_\varphi\Phi_\psi\colon {_AX_B} \to M(_CY_D)$ be a nondegenerate right-Hilbert bimodule homomorphism. Then Φ extends uniquely to a right-Hilbert bimodule homomorphism*
$$_{\bar\varphi}\bar\Phi_{\bar\psi}\colon {_{M(A)}M(X)_{M(B)}} \to {_{M(C)}M(Y)_{M(D)}},$$
where $\bar\varphi\colon M(A) \to M(C)$ and $\bar\psi\colon M(B) \to M(D)$ are the unique extensions of φ and ψ. Moreover, $\bar\Phi$ is strictly continuous.

PROOF. We first aim to show that Φ is continuous from the relative strict topology of X to the strict topology of $M(Y)$. By Remark 1.19 there exists a map $\mu\colon \mathcal{K}(X) \to M(\mathcal{K}(Y))$ such that $_\mu\Phi_\psi\colon {_{\mathcal{K}(X)}X_B} \to M(_{\mathcal{K}(Y)}Y_D)$ is a nondegenerate right-Hilbert bimodule homomorphism. Let $x_i \to 0$ strictly in X. Then for all $d \in D$ we can factor $d = \psi(b)d'$ for some $b \in B$ and $d' \in D$, and then
$$\|\Phi(x_i)d\| = \|\Phi(x_i)\psi(b)d'\| = \|\Phi(x_i \cdot b)d'\| \le \|\Phi(x_i \cdot b)\|\|d'\| \le \|x_i \cdot b\|\|d'\| \to 0.$$
Similarly, for all $T \in \mathcal{K}(Y)$ we can factor $T = T'\mu(S)$ for some $T' \in \mathcal{K}(Y)$ and $S \in \mathcal{K}(X)$, and then
$$\|T\Phi(x_i)\| = \|T'\mu(S)\Phi(x_i)\| = \|T'\Phi(Sx_i)\| \le \|T'\|\|\Phi(Sx_i)\| \le \|T'\|\|Sx_i\| \to 0.$$
Thus $\Phi(x_i) \to 0$ strictly in $M(Y)$, as desired.

Therefore Φ certainly has a unique strictly continuous extension $\bar\Phi\colon M(X) \to M(Y)$. Since $\bar\varphi$ and $\bar\psi$ are also strictly continuous, it follows from the strict continuity of all bimodule operations that $\bar\Phi$ is a $\bar\mu - \bar\psi$ compatible right-Hilbert $\mathcal{K}(X) - B$ bimodule homomorphism.

Finally, for the uniqueness, suppose $\Psi\colon M(X) \to M(Y)$ is any right-Hilbert bimodule homomorphism extending Φ. Since the extensions of φ and ψ to the multiplier algebras are unique, Ψ must be $\bar\varphi - \bar\psi$ compatible. For all $m \in M(X)$ and $d \in D$ we can factor $d = \psi(b)d'$ for some $b \in B$ and $d' \in D$, and then

$$\Psi(m)d = \Psi(m)\psi(b)d' = \Psi(mb)d' = \Phi(mb)d' = \bar\Phi(m)d.$$

Therefore $\Psi(m) = \bar\Phi(m)$. $\square$

The above result allows us to compose nondegenerate right-Hilbert bimodule homomorphisms: if $\Phi\colon {}_A X_B \to M({}_C Y_D)$ and $\Psi\colon {}_C Y_D \to M({}_E Z_F)$ are nondegenerate homomorphisms, then we have a nondegenerate composition $\Psi\circ\Phi\colon X \to M(Z)$.

1.3. Tensor products

Let ${}_A X_B$ and ${}_B Y_C$ be right-Hilbert bimodules. Then the algebraic tensor product $X \odot Y$ becomes a pre-right-Hilbert $A - C$ bimodule with operations

$$a \cdot (x \otimes y) = a \cdot x \otimes y, \quad (x \otimes y) \cdot b = x \otimes y \cdot b, \quad \text{and}$$

$$\langle x \otimes y, z \otimes w \rangle_C = \langle y, \langle x, z \rangle_B \cdot w \rangle_C$$

[54, Proposition 3.16]. The completion is a right-Hilbert $A - C$ bimodule $X \otimes_B Y$. Note that if X and Y are full, then so is $X \otimes_B Y$, since

$$\overline{\langle X \otimes_B Y, X \otimes_B Y \rangle}_C = \overline{\langle Y, \langle X, X \rangle_B \cdot Y \rangle}_C = \overline{\langle Y, B \cdot Y \rangle}_C = \overline{\langle Y, Y \rangle}_C = C.$$

DEFINITION 1.31. We call $X \otimes_B Y$ the *balanced tensor product* of ${}_A X_B$ and ${}_B Y_C$.

Given two right-Hilbert bimodule homomorphisms $\Phi\colon {}_A X_B \to M({}_D Z_E)$ and $\Psi\colon {}_B Y_C \to M({}_E W_F)$, we would like to form a tensor product homomorphism

$$\Phi \otimes_B \Psi \colon X \otimes_B Y \to M(Z \otimes_E W).$$

There are two obstructions to this utopia: first, understandably we have to require that the coefficient maps at B coincide. More significantly, the obvious map $x \otimes_B y \mapsto \Phi(x) \otimes_{M(E)} \Psi(y)$ will be into $M(Z) \otimes_{M(E)} M(W)$. Fortunately, using the abstract characterization of Proposition 1.28, we can show that $M(Z) \otimes_{M(E)} M(W)$ always embeds in $M(Z \otimes_E W)$:

LEMMA 1.32. *Let ${}_D Z_E$ and ${}_E W_F$ be right-Hilbert bimodules. There exists an isometric right-Hilbert $M(D) - M(F)$ bimodule homomorphism $\Upsilon\colon M(Z) \otimes_{M(E)} M(W) \to M(Z \otimes_E W)$ such that*

$$(1.7) \qquad \Upsilon(m)f = m \cdot f \quad \text{and} \quad \Upsilon(m)^* x = \langle m, x \rangle_{M(F)}$$

for each $m \in M(Z) \otimes_{M(E)} M(W)$, $f \in F$, and $x \in Z \otimes_E W$.

PROOF. Note that $M(Z) \otimes_{M(E)} M(W)$ is a right-Hilbert $M(D) - M(F)$ bimodule which contains a copy of $Z \otimes_E W = Z \otimes_{M(E)} W$. Moreover,

$$M(Z) \otimes_{M(E)} M(W) \cdot F = M(Z) \otimes_{M(E)} W = M(Z) \otimes_{M(E)} E \cdot W$$
$$= M(Z) \cdot E \otimes_{M(E)} W = Z \otimes_{M(E)} W = Z \otimes_E W.$$

Hence Proposition 1.28 provides an isometric right-Hilbert $M(D) - M(F)$ bimodule homomorphism $\Upsilon\colon M(Z) \otimes_{M(E)} M(W) \to M(Z \otimes_E W)$ which is the identity on $Z \otimes W$, and which by (1.5) and (1.6) in the proof of Proposition 1.28 satisfies (1.7). $\square$

REMARK 1.33. In the future we will suppress the map Υ, using the above lemma to identify $M(Z) \otimes_{M(E)} M(W)$ with its image in $M(Z \otimes_E W)$. Note that for $m \in M(Z)$ and $n \in M(W)$ the adjointable map $m \otimes n \colon F \to Z \otimes_E W$ is given by $(m \otimes n)f = m \otimes n \cdot f$.

PROPOSITION 1.34. *Let* $_\varphi \Phi_\psi \colon {_A}X_B \to M(_D Z_E)$ *and* $_\psi \Psi_\theta \colon {_B}Y_C \to M(_E W_F)$ *be right-Hilbert bimodule homomorphisms. Then there exists a $\varphi - \theta$ compatible right-Hilbert bimodule homomorphism* $\Phi \otimes_B \Psi \colon {_A}(X \otimes_B Y)_C \to M(_D(Z \otimes_E W)_F)$ *such that*

(1.8) $$(\Phi \otimes_B \Psi)(x \otimes y) = \Phi(x) \otimes \Psi(y).$$

If Φ and Ψ are nondegenerate, then $\Phi \otimes_B \Psi$ is too.

PROOF. The pairing $(x, y) \mapsto \Phi(x) \otimes \Psi(y)$ is bilinear, so determines a unique linear map $\Phi \odot \Psi \colon X \odot Y \to M(Z \otimes_E W)$. We verify conditions (i) and (ii) of Lemma 1.23, and it suffices to do this for elementary tensors: for all $a \in A$, $x, z \in X$, and $y, w \in Y$ we have

$$(\Phi \odot \Psi)(a \cdot (x \otimes y)) = (\Phi \odot \Psi)(a \cdot x \otimes y) = \Phi(a \cdot x) \otimes \Psi(y)$$
$$= \varphi(a) \cdot \Psi(x) \otimes \Psi(y) = \varphi(a) \cdot (\Phi(x) \otimes \Psi(y))$$

and

$$\langle (\Phi \odot \Psi)(x \otimes y), (\Phi \odot \Psi)(z \otimes w) \rangle_{M(F)} = \langle \Phi(x) \otimes \Psi(y), \Phi(z) \otimes \Psi(w) \rangle_{M(F)}$$
$$= \langle \Psi(y), \langle \Phi(x), \Phi(x) \rangle_{M(E)} \cdot \Psi(w) \rangle_{M(F)} = \langle \Psi(y), \psi(\langle x, z \rangle_B) \cdot \Psi(w) \rangle_{M(F)}$$
$$= \langle \Psi(y), \Psi(\langle x, z \rangle_B \cdot w) \rangle_{M(F)} = \theta(\langle y, \langle x, z \rangle_B \cdot w \rangle_C)$$
$$= \theta(\langle x \otimes y, z \otimes w \rangle_C).$$

Now suppose Φ and Ψ are nondegenerate. Then we have

$$\overline{(\Phi \otimes_B \Psi)(X \otimes_B Y) \cdot F} = \overline{(\Phi(X) \odot \Psi(Y))F} = \overline{\Phi(X) \odot \Psi(Y) \cdot F}$$
$$= \overline{\Phi(X) \odot W} = \overline{\Phi(X) \odot E \cdot W} = \overline{\Phi(X) \cdot E \odot W} = Z \otimes_E W.$$

Since φ and θ are nondegenerate by assumption, this shows that $\Phi \otimes_B \Psi$ is nondegenerate. □

We will need the exterior tensor product as well. Let $_A X_B$ and $_C Y_D$ be right-Hilbert bimodules. Then the algebraic tensor product $X \odot Y$ becomes a pre-right-Hilbert $(A \odot C) - (B \odot D)$ bimodule with operations

$$(a \otimes c) \cdot (x \otimes y) = a \cdot x \otimes c \cdot y, \quad (x \otimes y) \cdot (b \otimes d) = x \cdot b \otimes y \cdot d, \quad \text{and}$$
$$\langle x \otimes y, z \otimes w \rangle_{B \otimes D} = \langle x, z \rangle_B \otimes \langle y, w \rangle_D$$

[**54**, Proposition 3.36]. The completion is a right-Hilbert $(A \otimes C) - (B \otimes D)$ bimodule $X \otimes Y$, where "$\otimes$" always denotes the *minimal* tensor product for C^*-algebras. If X and Y are full, then so is $X \otimes Y$, since

$$\overline{\langle X \otimes Y, X \otimes Y \rangle}_{B \otimes D} = \overline{\langle X, X \rangle_B \otimes \langle Y, Y \rangle_D} = B \otimes D.$$

DEFINITION 1.35. We call $X \otimes Y$ the *exterior tensor product* of $_A X_B$ and $_C Y_D$.

As for balanced tensor products, we want to "exterior tensor" bimodule homomorphisms, and again we first need to know that things go into the right place:

1.3. TENSOR PRODUCTS

LEMMA 1.36. *Let $_EZ_F$ and $_GW_H$ be right-Hilbert bimodules. There exists an isometric right-Hilbert bimodule homomorphism $\Xi\colon M(Z)\otimes M(W) \to M(Z\otimes W)$ with coefficient maps the canonical homomorphisms $M(E)\otimes M(G) \to M(E\otimes G)$ and $M(F)\otimes M(H) \to M(F\otimes H)$ such that*

(1.9) $$\Xi(m)b = m\cdot b \quad\text{and}\quad \Xi(m)^*x = \langle m, x\rangle_{M(F)\otimes M(H)}$$

for each $m\in M(Z)\otimes M(W)$, $b\in F\otimes H$, and $x\in Z\otimes W$.

PROOF. Note that $M(Z)\otimes M(W)$ contains $Z\otimes W$ in the obvious way as a right-Hilbert $(M(E)\otimes M(G)) - (M(F)\otimes M(H))$ sub-bimodule, that $E\otimes G \subseteq M(E)\otimes M(G)$ and $F\otimes H \subseteq M(F)\otimes M(H)$ are essential closed ideals, and that
$$(M(Z)\otimes M(W))\cdot (F\otimes H) = M(Z)\cdot F \otimes M(W)\cdot H \subseteq Z\otimes W.$$
Hence Proposition 1.28 provides an isometric right-Hilbert bimodule homomorphism $\Xi\colon M(Z)\otimes M(W)\to M(Z\otimes W)$ as desired. By (1.5) and (1.6) in the proof of Proposition 1.28, Ξ satisfies (1.9). □

REMARK 1.37. As for balanced tensor products, we will suppress the map Ξ, using the above lemma to identify $M(Z)\otimes M(W)$ with its image in $M(Z\otimes W)$. Note that for $m\in M(Z)$ and $n\in M(W)$ the adjointable map $m\otimes n\colon F\otimes H \to Z\otimes W$ is given on elementary tensors by $(m\otimes n)(f\otimes h) = m\cdot f \otimes n\cdot h$.

PROPOSITION 1.38. *Let $_\varphi\Phi_\psi\colon {_A}X_B \to M(_EZ_F)$ and $_\theta\Psi_\rho\colon {_C}Y_D \to M(_GW_H)$ be right-Hilbert bimodule homomorphisms. There exists a right-Hilbert bimodule homomorphism*
$$\Phi\otimes\Psi\colon {_{A\otimes C}}(X\otimes Y)_{B\otimes D} \to M(_{E\otimes G}(Z\otimes W)_{F\otimes H}),$$
with coefficient maps the usual homomorphisms $\varphi\otimes\theta\colon A\otimes C \to M(E\otimes G)$ and $\psi\otimes\rho\colon B\otimes D \to M(F\otimes H)$, such that

(1.10) $$(\Phi\otimes\Psi)(x\otimes y) = \Phi(x)\otimes\Psi(y).$$

If Φ and Ψ are nondegenerate, then $\Phi\otimes\Psi$ is too.

PROOF. As in the proof of Proposition 1.34, we clearly have a unique linear map $\Phi\odot\Psi\colon X\odot Y \to M(Z)\otimes M(W)$, so for the first part it suffices to check the left module actions and the inner products on the generators: for $a\in A$, $c\in C$, $x,z\in X$, and $y,w\in Y$ we have
$$\begin{aligned}(\Phi\otimes\Psi)\big((a\otimes c)\cdot(x\otimes y)\big) &= (\Phi\otimes\Psi)(a\cdot x \otimes c\cdot y)\\ &= \Phi(a\cdot x)\otimes\Psi(c\cdot y) = \varphi(a)\cdot\Phi(x)\otimes\theta(c)\cdot\Psi(y)\\ &= \big(\varphi(a)\otimes\theta(c)\big)\cdot\big(\Phi(x)\otimes\Psi(y)\big) = (\varphi\otimes\theta)(a\otimes c)\cdot(\Phi\otimes\Psi)(x\otimes y)\end{aligned}$$
and
$$\begin{aligned}\big\langle(\Phi\otimes\Psi)(x\otimes y),(\Phi\otimes\Psi)(z\otimes w)\big\rangle_{M(F\otimes H)} &= \big\langle \Phi(x)\otimes\Psi(y),\Phi(z)\otimes\Psi(w)\big\rangle_{M(F\otimes H)}\\ &= \langle\Phi(x),\Phi(z)\rangle_{M(F)}\otimes\langle\Psi(y),\Psi(w)\rangle_{M(H)} = \psi(\langle x,z\rangle_B)\otimes\rho(\langle y,w\rangle_D)\\ &= (\psi\otimes\rho)\big(\langle x,z\rangle_B\otimes\langle y,w\rangle_D\big) = (\psi\otimes\rho)\big(\langle x\otimes y, z\otimes w\rangle_{B\otimes D}\big).\end{aligned}$$

Now suppose Φ and Ψ are nondegenerate. Then
$$\begin{aligned}\overline{(\Phi\otimes\Psi)(X\otimes Y)\cdot(F\otimes H)} &= \overline{(\Phi(X)\otimes\Psi(Y))\cdot(F\otimes H)}\\ &= \overline{\Phi(X)\cdot F\otimes\Psi(Y)\cdot H} = Z\otimes W,\end{aligned}$$
so $\Phi\otimes\Psi$ is nondegenerate as well. □

1.4. The C-multiplier bimodule $M_C(X \otimes C)$

In this section we introduce the notion of the C-multiplier bimodule of the exterior tensor product $X \otimes C$ of a right-Hilbert bimodule X with a C^*-algebra C. The definition is very similar to the definition of the C-multiplier algebra $M_C(A \otimes C)$ as given in Appendix A, and we shall obtain properties for C-multiplier bimodules similar to those we obtained in Proposition A.5 and Proposition A.6 for C-multiplier bimodules. We start with:

DEFINITION 1.39. Suppose that X is a right-Hilbert $A - B$ bimodule and let C be a C^*-algebra. The *C-multiplier bimodule* $M_C(X \otimes C)$ of $X \otimes C$ is defined as the set
$$M_C(X \otimes C) = \{m \in M(X \otimes C) \mid (1 \otimes C)m \cup m(1 \otimes C) \subseteq X \otimes C\}.$$
The *C-strict topology* on $M_C(X \otimes C)$ is the locally convex topology generated by the seminorms $m \mapsto \|m(1 \otimes c)\|$ and $m \mapsto \|(1 \otimes c)m\|$, $c \in C$.

If $C = C^*(G)$ for some locally compact group G, we shall simply write $M_G(X \otimes C^*(G))$ for $M_{C^*(G)}(X \otimes C^*(G))$ and call it the *G-multiplier bimodule* of $X \otimes C^*(G)$. The following result is the bimodule analogue of Proposition A.5.

LEMMA 1.40. *Suppose that A, B, and C are C^*-algebras and that X is a right-Hilbert $A - B$ bimodule. Then:*

(i) *$M_C(X \otimes C)$ is a closed $A \otimes C - B \otimes C$ sub-bimodule of $M(X \otimes C)$.*
(ii) *$M_C(X \otimes C)$ is a right-Hilbert $M_C(A \otimes C) - M_C(B \otimes C)$ bimodule with respect to the bimodule operations on $M(X \otimes C)$ restricted to $M_C(X \otimes C)$.*
(iii) *The C-strict topology on $M_C(X \otimes C)$ is stronger than the strict topology inherited from the full multiplier bimodule $M(X \otimes C)$.*
(iv) *The right-Hilbert $M_C(A \otimes C) - M_C(B \otimes C)$ bimodule operations on $M_C(X \otimes C)$ are separately C-strictly continuous.*
(v) *$M_C(A \otimes C)$ and $M_C(B \otimes C)$ are the C-strict completions of $A \otimes C$ and $B \otimes C$, respectively, and $M_C(X \otimes C)$ is the C-strict completion of $X \otimes C$.*

PROOF. We omit the straightforward verifications of (i) and (ii). For (iii), suppose that $(m_i)_{i \in I}$ is a net in $M_C(X \otimes C)$ which converges C-strictly to some $m \in M_C(X \otimes C)$. Let $z \in B \otimes C$. Factoring $z = (1 \otimes c)y$ for some $c \in C$, $y \in B \otimes C$ gives
$$\lim_i m_i z = \lim_i (m_i(1 \otimes c))y = m(1 \otimes c)y = mz.$$
A similar argument shows that $zm_i \to zm$ for all $z \in \mathcal{K}(X \otimes C) \cong \mathcal{K}(X) \otimes C$. Thus $m_i \to m$ in the strict topology of $M(X \otimes C)$.

Part (iv) follows directly from (iii) and Proposition 1.27. For the proof of (v), let $(u_i)_{i \in I}$ be a bounded approximate unit for C. Then, if $m \in M_C(X \otimes C)$, we see that $m(1 \otimes u_i)$ is a net in $X \otimes C$ which converges C-strictly to m, so $X \otimes C$ is C-strictly dense in $M_C(X \otimes C)$. Conversely, if $(m_i)_{i \in I}$ is a C-strict Cauchy net in $M_C(X \otimes C)$, it follows from (iii) that it is also a strict Cauchy net in $M(X \otimes C)$. It follows then from Proposition 1.27 that there exists an $m \in M(X \otimes C)$ such that $m_i \to m$ strictly.

We claim that $m_i \to m$ C-strictly and that $m \in M_C(X \otimes C)$. For this we let $\epsilon > 0$ and $c \in C$. We show that there exists an $i_0 \in I$ such that $\|(m_i - m)(1 \otimes c)z\| \leq \epsilon$ for all $z \in B \otimes C$ with $\|z\| \leq 1$ and $i \geq i_0$. For this let $i_0 \in I$ such that

$\|(m_i - m_j)(1 \otimes c)\| \leq \epsilon$ for all $i, j \geq i_0$. Since $m_j(1 \otimes c)z \to m(1 \otimes c)z$ in norm, it follows that

$$\|(m_i - m)(1 \otimes c)z\| = \lim_j \|(m_i - m_j)(1 \otimes c)z\| \leq \epsilon$$

for all $i \geq i_0$. Thus $m_i(1 \otimes c) \to m(1 \otimes c)$ in norm, and a similar argument shows that $(1 \otimes c)m_i \to (1 \otimes c)m$ in norm for all $c \in C$. Thus $m_i \to m$ C-strictly. Finally, since $m_i(1 \otimes c), (1 \otimes c)m_i \in X \otimes C$, it follows from the fact that $X \otimes C$ is norm-closed in $M(X \otimes C)$ that $(1 \otimes c)m, m(1 \otimes c) \in X \otimes C$ for all $c \in C$, hence $m \in M_C(X \otimes C)$. □

REMARK 1.41. It is useful to observe that we always have

$$(1 \otimes M(C)) \cdot M_C(X \otimes C) \cup M_C(X \otimes C) \cdot (1 \otimes M(C)) \subseteq M_C(X \otimes C).$$

We omit the straightforward argument.

PROPOSITION 1.42. *Suppose that A, B, C, D, E, F are C^*-algebras. Let X be a right-Hilbert $A-B$ bimodule and let Y be a right-Hilbert $E-F$ bimodule. Assume further that $_\varphi \Phi_\psi \colon {_A}X_B \to M(_E Y_F)$ is a right-Hilbert bimodule homomorphism, and that $\Psi \colon C \to M(D)$ is a nondegenerate $*$-homomorphism. Let $\Phi \otimes \Psi \colon X \otimes C \to M(Y \otimes D)$ denote the tensor product homomorphism of Proposition 1.38. Then:*

  (i) *There exists a unique bimodule homomorphism*

  $$\overline{\Phi \otimes \Psi} \colon {_{M_C(A \otimes C)}} M_C(X \otimes C)_{M_C(B \otimes C)} \to {_{M(E \otimes D)}} M(Y \otimes D)_{M(F \otimes D)}$$

  which extends $\Phi \otimes \Psi \colon X \otimes C \to M(Y \otimes D)$.
  (ii) *The map $\overline{\Phi \otimes \Psi}$ of (i) is C-strict to strict continuous.*
  (iii) *If $\Phi(X) \subseteq Y$, then $\overline{\Phi \otimes \Psi}((M_C(X \otimes C)) \subseteq M_D(Y \otimes D)$ and $\overline{\Phi \otimes \Psi}$ is C-strict to D-strict continuous.*

REMARK 1.43. In the rest of the paper we shall usually write $\Phi \otimes \Psi$ also for its extension $\overline{\Phi \otimes \Psi}$ to $M_C(X \otimes C)$ if confusion seems unlikely. Note that it already follows from Proposition A.6 that the coefficient maps $\varphi \otimes \Psi \colon A \otimes C \to M(E \otimes D)$ and $\psi \otimes \Psi \colon B \otimes C \to M(F \otimes D)$ have unique extensions to $M_C(A \otimes C)$ and $M_C(B \otimes C)$, respectively, with similar properties as stated for $M_C(X \otimes C)$ above.

PROOF OF PROPOSITION 1.42. The proof follows closely the ideas of the proof of Theorem 1.30, and it is also very similar to the proof of Proposition A.6. We first remark that the map $\Phi \otimes \Psi \colon X \otimes C \to M(Y \otimes D)$ is C-strict to strict continuous. Indeed, if $(x_i)_{i \in I}$ is a net in $X \otimes C$ which converges C-strictly to $x \in X \otimes C$, and if $z \in Y \otimes D$, we can factor $z = (1 \otimes \Psi(c))y$ for some $c \in C$ and $y \in Y \otimes D$ (since Ψ is nondegenerate) to conclude that

$$\Phi \otimes \Psi(x_i)z = \Phi \otimes \Psi(x_i(1 \otimes c))y \to \Phi \otimes \Psi(x(1 \otimes c))y = \Phi \otimes \Psi(x)z,$$

where convergence is in norm. A similar argument shows that $k(\Phi \otimes \Psi(x_i)) \to k(\Phi \otimes \Psi(x))$ in norm for all $k \in \mathcal{K}(Y \otimes D) \cong \mathcal{K}(Y) \otimes D$. Thus there exist unique C-strict to strict continuous linear extensions $\overline{\Phi \otimes \Psi} \colon M_C(X \otimes C) \to M(Y \otimes D)$.

Since all bimodule operations on $_{M_C(A \otimes C)} M_C(X \otimes C)_{M_C(B \otimes C)}$ are separately C-strictly continuous, and since all bimodule operations on $M(Y \otimes D)$ are separately strictly continuous, we conclude that the extensions

$$\overline{\varphi \otimes \Psi} \colon M_C(A \otimes C) \to M(E \otimes D) \quad \text{and} \quad \overline{\psi \otimes \Psi} \colon M_C(B \otimes C) \to M(F \otimes D)$$

are ∗-homomorphisms, and that $\overline{\Phi \otimes \Psi}$ is $\overline{\varphi \otimes \Psi} - \overline{\psi \otimes \Psi}$ compatible. A straightforward argument, similar to the one presented in the proof of Theorem 1.30 and Proposition A.6, shows that $\overline{\varphi \otimes \Psi}\overline{\Phi \otimes \Psi}_{\overline{\psi \otimes \Psi}}$ is indeed the only bimodule extension of $\Phi \otimes \Psi$ to $M_C(X \otimes C)$. This proves (i) and (ii).

For the proof of (iii) assume that $\Phi(X) \subseteq Y$. Then we can copy the arguments given in the proof of part (iv) of Proposition A.6 to conclude that $\Phi \otimes \Psi(X \otimes C) \subseteq M_D(Y \otimes D)$ and that $\Phi \otimes \Psi \colon X \otimes C \to M_D(Y \otimes D)$ is C-strict to D-strict continuous. It then follows that $\Phi \otimes \Psi$ extends to a C-strict to D-strict continuous bimodule homomorphism of $M_C(X \otimes C)$ into $M_D(Y \otimes D)$, and by the uniqueness clause in (i), this must coincide with $\overline{\Phi \otimes \Psi}$. □

It is an important observation that the C-multiplier bimodule $M_C(X \otimes C)$ only depends on X and not on the coefficients.

LEMMA 1.44. *Let $_A X_B$ be a right-Hilbert $A - B$ bimodule and let C be a C^*-algebra. Let $\Phi \colon {_A}X_B \to {_{\mathcal{K}(X)}}X_B$ be the identity map. Then*

$$\overline{\Phi \otimes \mathrm{id}_C} \colon M_C(_A X_B \otimes C) \to M_C(_{\mathcal{K}(X)} X_B \otimes C)$$

is isometric and surjective. Also, if $B_X = \overline{\langle X, X \rangle}_B$ and $\Psi \colon {_A}X_{B_X} \to {_A}X_B$ denotes the identity map, then

$$\overline{\Psi \otimes \mathrm{id}_C} \colon M_C(_A X_{B_X} \otimes C) \to M_C(_A X_B \otimes C)$$

is isometric and surjective. In particular, the range of the $M_C(B \otimes C)$-valued inner product on $M_C(X \otimes C)$ lies in $M_C(B_X \otimes C)$.

PROOF. The result follows directly from the fact that $\Phi \otimes \mathrm{id}_C \colon {_A}X_B \otimes C \to {_{\mathcal{K}(X)}}X_B \otimes C$ and $\Psi \otimes \mathrm{id}_C \colon {_A}X_{B_X} \otimes C \to {_A}X_B \otimes C$ are linear homeomorphisms with respect to the C-strict topologies. □

PROPOSITION 1.45. *Let $\Phi \colon {_A}X_B \to M(_E Y_F)$ and $\Psi \colon C \to M(D)$ be as in Proposition 1.42. If Φ and Ψ are isometric, then so is $\overline{\Phi \otimes \Psi} \colon M_C(X \otimes C) \to M(Y \otimes D)$.*

PROOF. Let $B_X = \overline{\langle X, X \rangle}_B$ and consider the composition

$$_A X_{B_X} \otimes C \longrightarrow {_A}X_B \otimes C \xrightarrow{\Phi \otimes \Psi} M(Y \otimes D).$$

This composition extends to the composition

$$M_C(_A X_{B_X} \otimes C) \xrightarrow{\cong} M_C(_A X_B \otimes C) \xrightarrow{\overline{\Phi \otimes \Psi}} M(Y \otimes D).$$

Thus in order to show that $\overline{\Phi \otimes \Psi}$ is isometric, it is enough to see that the above composition of maps is isometric. But by uniqueness of extensions, this composition equals the extension of $\Phi_1 \otimes \Psi \colon {_A}X_{B_X} \otimes C \to M(Y \otimes D)$, where $\Phi_1 = \Phi$ with right coefficient map restricted to the ideal B_X of B. Since $\Phi \colon X \to M(Y)$ is isometric, it follows that the right coefficient map $\psi \colon B \to M(F)$ restricts to an isometric ∗-homomorphism $\psi_1 \colon B_X \to M(F)$. It follows then from Proposition A.6 that the right coefficient map $\overline{\psi_1 \otimes \Psi} \colon M_C(B_X \otimes C) \to M(Y \otimes D)$ of $\overline{\Phi_1 \otimes \Psi}$ is isometric. Thus $\overline{\Phi_1 \otimes \Psi}$ is isometric, too. □

In the special case where $\Psi = \mathrm{id}_C$ and $\Phi(X) \subseteq Y$ it will be necessary to identify the elements in the image of $\overline{\Phi \otimes \mathrm{id}_C} \colon M_C(X \otimes C) \to M_C(Y \otimes C)$. This can be done as follows:

LEMMA 1.46. *Suppose that* $\Phi\colon {}_AX_B \to M({}_EY_F)$ *is an isometric right-Hilbert bimodule homomorphism with* $\Phi(X) \subseteq Y$. *Then the isometry* $\overline{\Phi \otimes \mathrm{id}_C}\colon M_C(X \otimes C) \to M_C(Y \otimes C)$ *has image*

$$M = \{m \in M_C(Y \otimes C) \mid (1 \otimes C)m \cup m(1 \otimes C) \subseteq \overline{\Phi \otimes \mathrm{id}_C}(X \otimes C)\}.$$

PROOF. It is clear that $\overline{\Phi \otimes \mathrm{id}_C}\big(M_C(X \otimes C)\big) \subseteq M$. So let $m \in M$. Let $(c_i)_{i \in I}$ be a bounded approximate identity of C, and define $(z_i)_{i \in I} \subseteq X \otimes C$ by $z_i = (\Phi \otimes \mathrm{id}_C)^{-1}(m(1 \otimes c_i))$. Since $m(1 \otimes c_i)$ is a C-strict Cauchy net in $Y \otimes C$ which lies in the image of $\Phi \otimes \mathrm{id}_C$, it follows that $(z_i)_{i \in I}$ is a C-strict Cauchy net in $X \otimes C$. Since $M_C(X \otimes C)$ is the C-strict completion of $X \otimes C$, we find an $n \in M_C(X \otimes C)$ with $z_i \to n$ C-strictly. Since $\overline{\Phi \otimes \mathrm{id}_C}\colon M_C(X \otimes C) \to M_C(Y \otimes C)$ is C-strict to C-strict continuous, it follows that

$$\overline{\Phi \otimes \mathrm{id}_C}(n) = \lim_i \overline{\Phi \otimes \mathrm{id}_C}(z_i) = \lim_i m(1 \otimes c_i) = m.$$

□

As a first application of this lemma we get

COROLLARY 1.47. *Suppose that* I *is a closed ideal of* B, *and let us view* $M_C(I \otimes C)$ *as a subalgebra of* $M_C(B \otimes C)$ *via the unique extension of the embedding* $I \otimes C \to B \otimes C$ *as given by Proposition A.6. Then* $M_C(I \otimes C)$ *is an ideal in* $M_C(B \otimes C)$.

PROOF. Let $m \in M_C(I \otimes C)$. Then

$$m(b \otimes c) = m(1 \otimes c)(b \otimes 1) \in (I \otimes C)(B \otimes 1) = I \otimes C$$

for all elementary tensors $b \otimes c \in B \otimes C$, from which it follows that $m(B \otimes C) \subseteq I \otimes C$. Similarly, we get $(B \otimes C)m \subseteq I \otimes C$.

Thus, if $n \in M_C(B \otimes C)$, it follows that

$$mn(1 \otimes C) \subseteq m(B \otimes C) \subseteq I \otimes C,$$

and we also have $(1 \otimes C)mn \subseteq (I \otimes C)n \in I \otimes C$, since $I \otimes C$ is an ideal in $M(B \otimes C)$. Thus, it follows from Lemma 1.46 that $mn \in M_C(I \otimes C)$. For symmetric reasons, this also implies that $nm \in M_C(I \otimes C)$. □

1.5. Linking algebras

We collect here a few facts we will need concerning linking algebras of partial imprimitivity bimodules; the primary references are [4], [33], and [54], where the linking algebras were studied for the special case of imprimitivity bimodules. Linking algebras of Hilbert C^*-bimodules appear in [5].

The *linking algebra* of a partial $A - B$ imprimitivity bimodule X is

$$L(X) = \left\{ \begin{pmatrix} a & x \\ \widetilde{z} & b \end{pmatrix} \middle| a \in A,\ b \in B,\ x, z \in X \right\},$$

with the usual linear structure and $*$-algebra operations

(1.11) $$\begin{pmatrix} a & x \\ \widetilde{z} & b \end{pmatrix} \begin{pmatrix} a' & x' \\ \widetilde{z'} & b' \end{pmatrix} = \begin{pmatrix} aa' + {}_A\langle x, z' \rangle & a \cdot x' + x \cdot b' \\ \widetilde{z} \cdot a' + b \cdot \widetilde{z'} & \langle z, x' \rangle_B + bb' \end{pmatrix}$$

and

(1.12) $$\begin{pmatrix} a & x \\ \widetilde{z} & b \end{pmatrix}^* = \begin{pmatrix} a^* & z \\ \widetilde{x} & b^* \end{pmatrix}.$$

$L(X)$ acts by adjointable operators on the Hilbert B-module $X \oplus B$ via

$$\begin{pmatrix} a & x \\ \tilde{z} & b \end{pmatrix} \begin{pmatrix} y \\ c \end{pmatrix} = \begin{pmatrix} a \cdot y + x \cdot c \\ \langle z, y \rangle_B + bc \end{pmatrix},$$

giving a homomorphism of $L(X)$ into the C^*-algebra $\mathcal{L}_B(X \oplus B)$. By considering the action on vectors of the form $(z, 0)$ or $(0, b)$, we can see that this homomorphism is injective on X, $\widetilde{X}$ and B. Similarly, we can define a right-action on the left Hilbert A-module $A \oplus X$, giving a $*$-homomorphism of $L(X)$ into $\mathcal{L}_A(A \oplus X)$ which is injective on A, X, and $\widetilde{X}$. Thus, defining a norm on $L(X)$ by the maximum of the respective operator norms, we obtain a complete C^*-norm on $L(X)$ (see [**4**] and [**54**, pages 50–51] — where the constructions have been done for imprimitivity bimodules, in which case $L(X)$ embeds injectively into $\mathcal{L}_B(X \otimes B)$). Since any $*$-algebra has at most one complete C^*-norm, this unambiguously makes $L(X)$ into a C^*-algebra. The linking algebra of $_AX_B$ contains copies of A, B, X and the conjugate module $\widetilde{X}$. We can recover these copies by observing that the matrices

$$p = p_{L(X)} = \begin{pmatrix} 1_{M(A)} & 0 \\ 0 & 0 \end{pmatrix} \quad \text{and} \quad q = q_{L(X)} = \begin{pmatrix} 0 & 0 \\ 0 & 1_{M(B)} \end{pmatrix}$$

define double centralizers of $L(X)$, so that $p, q \in M(L(X))$, and noting that, for example,

$$A = \begin{pmatrix} A & 0 \\ 0 & 0 \end{pmatrix} = pL(X)p;$$

more formally, the inclusion $a \mapsto \begin{pmatrix} a & 0 \\ 0 & 0 \end{pmatrix}$ is an isomorphism of A onto the corner $pL(X)p$.

The above construction has an obvious inverse:

PROPOSITION 1.48. *Suppose that L is a C^*-algebra and $p, q \in M(L)$ are complementary projections* (i.e., $p + q = 1$). *Then pLq is a partial $pLp - qLq$ imprimitivity bimodule with operations given by multiplication and involution on L in the canonical way* (e.g., *the qLq-valued inner product on pLq is given by $\langle a, b \rangle_{qLq} = a^*b$*). *Moreover, pLq is a right-partial $pLp - qLq$ imprimitivity bimodule if and only if q is full* (i.e., $\overline{LqL} = L$), *and pLq is a left-partial $pLp - qLq$ imprimitivity bimodule if and only if p is full. In particular, pLq is a $pLp - qLq$ imprimitivity bimodule if and only if both p and q are full.*

PROOF. It is straightforward to check that pLq is a partial $pLp - qLq$ imprimitivity bimodule with respect to the canonical operations. If q is full, then we get

$$_{pLp}\overline{\langle pLq, pLq \rangle} = \overline{pLq(pLq)^*} = \overline{pLqLp} = pLp,$$

which implies that pLq is a right-partial $pLp - qLq$ imprimitivity bimodule. For the converse, we first observe that for any projection $q \in M(L)$ one has the equations $\overline{LqLq} = Lq$ and (hence) $\overline{qLqL} = qL$. This follows from the fact that Lq is a Hilbert qLq-module, which implies that qLq acts nondegenerately on Lq by multiplication. Thus, if $\overline{pLqLp} = {_{pLp}\overline{\langle pLq, pLq \rangle}} = pLp$, then a short computation reveals that

$$\overline{LqL} = \overline{pLqLp} + \overline{pLqLq} + \overline{qLqLp} + \overline{qLqLq} = pLp + pLq + qLp + qLq = L,$$

so q is a full projection. Applying the same arguments to p completes the proof. □

In practice it is crucial to be able to recognize under what conditions a given algebra L with given complementary projections $p,q \in M(L)$ is isomorphic to the linking algebra of a given partial imprimitivity bimodule:

PROPOSITION 1.49. *Let $_AX_B$ be a partial imprimitivity bimodule. Suppose that $(L, p, q, {_\varphi}\Phi_\psi)$ consists of a C^*-algebra L, complementary projections p, q in $M(L)$, and a partial imprimitivity bimodule isomorphism ${_\varphi}\Phi_\psi\colon {_A}X_B \to {_{pLp}}(pLq)_{qLq}$ (with the obvious meaning). Then*

$$\theta\begin{pmatrix} a & x \\ \tilde{z} & b \end{pmatrix} = \varphi(a) + \Phi(x) + \Phi(z)^* + \psi(b)$$

defines an isomorphism $\theta\colon L(X) \xrightarrow{\cong} L$.

PROOF. Routine calculations show that θ is a homomorphism. It is injective because each of the components is; it is surjective because every $d \in L$ can be decomposed as $d = pdp + pdq + qdp + qdq$, and because the maps φ, Φ, and ψ are surjective. $\square$

REMARK 1.50. We shall often write $L = L(X)$ to summarize an application of this proposition; this means that we think there are obvious candidates for the projections p, q and the maps φ, Φ, and ψ, and that these candidates satisfy the hypotheses of Proposition 1.49. Thus, for example, if C is any other C^*-algebra, we write $L(X) \otimes C = L(X \otimes C)$ to mean that the quadruple

$$\bigl(L(X) \otimes C, p_{L(X)} \otimes 1_{M(C)}, q_{L(X)} \otimes 1_{M(C)}, {_\varphi}\Phi_\psi \otimes \mathrm{id}\bigr)$$

satisfies the hypotheses of Proposition 1.49 for the external tensor product $X \otimes C$, which is a partial $(A \otimes C)$–$(B \otimes C)$ imprimitivity bimodule. This leads to statements like $A \otimes C = p(L(X) \otimes C)p$, by which we mean that the obvious map $\varphi \otimes \mathrm{id}$ is an isomorphism of $A \otimes C$ onto the given corner.

For any right-partial $A - B$ imprimitivity bimodule X, we saw in Remark 1.15 that $M(X)$ is a partial $M(A) - M(B)$ imprimitivity bimodule, so we can also form the linking algebra $L(M(X)) = \begin{pmatrix} M(A) & M(X) \\ \widetilde{M(X)} & M(B) \end{pmatrix}$. It follows from the properties of $M(X)$ as listed in Proposition 1.27 that $L(X)$ is a closed ideal in $L(M(X))$. Thus, by the universal properties of the multiplier algebras, there exists a unique algebra homomorphism $\Phi\colon L(M(X)) \to M(L(X))$ which extends the identity on $L(X)$. In fact, this map is always an isomorphism:

PROPOSITION 1.51. *Suppose that X is a right-partial $A - B$ imprimitivity bimodule. Then the canonical homomorphism $\Phi\colon L(M(X)) \to M(L(X))$ is an isomorphism of C^*-algebras. Moreover, Φ is a homeomorphism with respect to the topology on $L(M(X)) = \begin{pmatrix} M(A) & M(X) \\ M(\widetilde{X}) & M(B) \end{pmatrix}$ which is the product of the strict topologies on the corners and the strict topology on $M(L(X))$.*

PROOF. Using the separate strict continuity of all bimodule pairings of $_AX_B$ (compare with Proposition 1.27) it is straightforward to check that the product of the strict topologies on the corners of $L(X)$, viewed as subspaces of the corners of $L(M(X))$, coincides with the strict topology of $L(X)$, viewed as a subalgebra of $M(L(X))$. It follows that the identity map on $L(X)$ extends to a product-strict to strict linear homeomorphism $\bar{\Phi}$ between the corresponding completions $L(M(X))$ (see Proposition 1.27) and $M(L(X))$. Since all algebra operations are separately continuous with respect to these topologies (which in the case of $L(M(X))$ also follows from Proposition 1.27), it follows that $\bar{\Phi}$ is an algebra homomorphism which extends the identity on $L(X)$. Thus $\bar{\Phi} = \Phi$ by the uniqueness of Φ. □

Just as we are going to simplify notation by writing $L = L(X)$ when we believe it is obvious what the projections p, q and the bimodule isomorphism $_\varphi\Phi_\psi$ are meant to be, we shall write things like $qM(L)q = D$ when we think there is an obvious candidate for q and for the embedding of D in $M(L)$. Thus, for example, the identification $L(X) \otimes C = L(X \otimes C)$ induces an identification $M(B \otimes C) = qM(L(X \otimes C))q$ which means, strictly speaking, that the isomorphism $\psi \otimes \mathrm{id}$ of $B \otimes C$ onto the corner $(q \otimes 1)(L(X \otimes C))(q \otimes 1)$ induces, via Proposition 1.51, an isomorphism of $M(B \otimes C)$ onto $(q \otimes 1)M(L(X \otimes C))(q \otimes 1)$.

We close this section with a slight extension of [**20**, Remark (2) on p. 307]:

LEMMA 1.52. *Assume that $_AX_B$ and $_CY_D$ are right-partial imprimitivity bimodules. Let $_\varphi\Phi_\psi\colon X \to M(Y)$ be a partial imprimitivity bimodule homomorphism. Then (using the identification $M(L(Y)) = L(M(Y))$ given in Proposition 1.51) the formula*
$$\Psi\left(\begin{pmatrix} a & x \\ \widetilde{y} & b \end{pmatrix}\right) = \begin{pmatrix} \varphi(a) & \Phi(x) \\ \widetilde{\Phi(y)} & \psi(b) \end{pmatrix}$$
defines a $*$-homomorphism $\Psi\colon L(X) \to M(L(Y))$, and Ψ is nondegenerate if Φ is nondegenerate. The extension $\bar{\Psi}\colon M(L(X)) \to M(L(Y))$ is given by the extensions of the corner maps.

Conversely, assume that $\Psi\colon L(X) \to M(L(Y))$ is a nondegenerate $*$-homomorphism such that $\Psi(p_X l) = p_Y \Psi(l)$ and $\Psi(q_X l) = q_Y \Psi(l)$, where p_X, q_X and p_Y, q_Y denote the respective corner projections. Then Ψ determines a partial imprimitivity bimodule homomorphism $_\varphi\Phi_\psi\colon X \to M(Y)$ by defining
$$\varphi(p_X l p_X) = p_Y \Psi(l) p_Y, \quad \Phi(p_X l q_X) = p_Y \Psi(l) q_Y, \quad \text{and} \quad \psi(q_X l q_X) = q_Y \Psi(l) q_Y$$
for $l \in L(X)$, and Φ is nondegenerate if Ψ is nondegenerate.

PROOF. It follows immediately from the algebraic properties of the linking algebras that the above procedure gives a correspondence between partial imprimitivity bimodule homomorphisms of X into $M(Y)$ and $*$-homomorphisms of $L(X)$ into $M(L(Y))$. So we only have to check that this correspondence preserves nondegeneracy.

If $_\varphi\Phi_\psi$ is nondegenerate, then
$$\Psi(L(X))L(Y) = \begin{pmatrix} \varphi(A)C +\, _A\langle \Phi(X), Y \rangle & \Phi(A)Y + \Phi(X)D \\ \widetilde{\Phi(X)}C + \psi(B)\widetilde{Y} & \langle \Phi(X), Y \rangle_B + \psi(B)D \end{pmatrix}.$$

Since φ and ψ are nondegenerate, it follows that $\varphi(A)C = C$, $\varphi(A)Y = Y$, $Y\psi(B) = Y$, and $\psi(B)D = D$, which implies that $\Psi(L(X))L(Y) = L(Y)$.

Now assume conversely that $\Psi(L(X)) \to M(L(Y))$ is nondegenerate. Using the general equation $pL(X)pL(X) = pL(X)$ (see the proof of Lemma 1.48) we get:
$$\varphi(A)C = p\Psi(L(X))pL(Y)p = p\Psi(L(X))p\Psi(L(X))L(Y)p$$
$$= \Psi(pL(X)pL(X))L(Y)p = \Psi(pL(X))L(Y)p = p\Psi(L(X))L(Y)p$$
$$= pL(Y)p = C.$$

(Here we've omitted the subscripts on the p's and q's.) Hence φ is nondegenerate, and a similar computation shows that ψ is nondegenerate. Finally, using the fullness of q_X (since the left inner product on X is full), we get

$$\overline{\Phi(X)D} = \overline{p\Psi(L(X))qL(Y)q} = \overline{p\Psi(L(X))q\Psi(L(X))L(Y)q}$$
$$= \overline{p\Psi(L(X)qL(X))L(Y)q} = \overline{p\Psi(L(X))L(Y)q} = Y.$$

This finishes the proof. □

REMARK 1.53. (1) Note that if $_AX_B$ and $_CY_D$ are right-partial imprimitivity bimodules, then it follows from Lemma 1.18 that the nondegenerate right-Hilbert bimodule homomorphisms of X into $M(Y)$ are automatically nondegenerate partial imprimitivity bimodule homomorphisms (and *vice versa*). Thus the above lemma gives a one-to-one correspondence between the nondegenerate right-Hilbert bimodule homomorphisms of X into $M(Y)$ and the nondegenerate $*$-homomorphisms of $L(X)$ into $M(L(Y))$.

(2) The proof of the lemma implies the interesting observation that a partial imprimitivity bimodule homomorphism $_\varphi\Phi_\psi \colon {_A}X_B \to M(_CY_D)$ between right-partial imprimitivity bimodules is nondegenerate if (and only if) the coefficient homomorphisms $\varphi \colon A \to M(C)$ and $\psi \colon B \to M(D)$ are nondegenerate, since this was all we needed for the proof of the nondegeneracy of the corresponding homomorphism of the linking algebra. For imprimitivity bimodules this was already observed in Remark 1.19. As remarked there, for this to be true it is necessary that $_\varphi\Phi_\psi$ preserve the left inner products.

CHAPTER 2

The Categories

In this chapter we show that there exists a category $\mathcal{C}$ in which the objects are C^*-algebras, and the morphisms from A to B are the isomorphism classes of right-Hilbert $A - B$ bimodules. For any locally compact group G, there are also equivariant categories $\mathcal{A}(G)$, $\mathcal{C}(G)$, and $\mathcal{AC}(G)$ which combine, respectively, actions of G, coactions of G, and both, with the structure of $\mathcal{C}$. Note that our category differs from the one considered in [**17**] by allowing isomorphism classes of non-full right-Hilbert modules to be morphisms in $\mathcal{C}$.

2.1. C^*-Algebras

Let X and Y be right-Hilbert $A - B$ bimodules. Recall from the preceding chapter that a *right-Hilbert $A - B$ bimodule isomorphism* of X onto Y is a bijective right-Hilbert bimodule homomorphism $\Phi\colon X \to Y$ whose coefficient maps are id_A and id_B. From now on we abuse the terminology and simply say X and Y are *isomorphic* if there exists a right-Hilbert $A - B$ bimodule homomorphism of X onto Y. It is not hard to check that this notion of isomorphism is an equivalence relation on the class of right-Hilbert $A - B$ bimodules.

REMARK 2.1. When we began writing this, the idea of a category in which the morphisms come from bimodules seemed new. In the intervening years, however, this category and close relatives of it have been independently discovered by several others. (See, for instance, [**36, 37**] and [**59, 60, 61**].) It also turns out that this category has been at least implicitly in the air for quite some time; for example, [**4**] briefly mentions the category of C^*-algebras and (isomorphism classes of) imprimitivity bimodules.

THEOREM 2.2. *There is a category $\mathcal{C}$ in which the objects are C^*-algebras, and in which the morphisms from A to B are the isomorphism classes of right-Hilbert $A - B$ bimodules. The composition of $[X]\colon A \to B$ with $[Y]\colon B \to C$ is the isomorphism class of the balanced tensor product $X \otimes_B Y$; the identity morphism on A is the isomorphism class of the standard right-Hilbert bimodule ${}_A A_A$.*

PROOF. We first note that the composition of morphisms is well-defined: suppose $[X] = [X']\colon A \to B$ and $[Y] = [Y']\colon B \to C$, so that we have right-Hilbert $A - B$ bimodule isomorphisms
$$\Phi\colon X \to X' \quad \text{and} \quad \Psi\colon Y \to Y'.$$
Then the tensor product homomorphism $\Phi \otimes_B \Psi$ maps $X \otimes_B Y$ into $X' \otimes_B Y'$ and has inverse $\Phi^{-1} \otimes_B \Psi^{-1}$, so gives an isomorphism $X \otimes_B Y \cong X' \otimes_B Y'$.

Next we establish that composition of morphisms in $\mathcal{C}$ is associative; by the above, it suffices to show that $X \otimes_B (Y \otimes_C Z)$ and $(X \otimes_B Y) \otimes_C Z$ are isomorphic for any right-Hilbert bimodules ${}_A X_B$, ${}_B Y_C$, and ${}_C Z_D$. But straightforward calculations show that the usual linear isomorphism of $X \odot (Y \odot Z)$ onto $(X \odot Y) \odot Z$ respects the module actions and right inner products, so extends to the desired isomorphism.

Finally, note that $A \otimes_A X \cong X$ and $Y \otimes_A A \cong Y$ for any right-Hilbert bimodules ${}_A X_B$ and ${}_B Y_A$. Hence the identity morphism from A to A is given by the isomorphism class of the standard bimodule ${}_A A_A$. □

REMARK 2.3. In any category, $\mathrm{mor}(A, B)$ is required to be a *set* (not merely a class) for each pair of objects A and B. This will fail in $\mathcal{C}$ unless we limit the size of the bimodules involved. We can do this by considering only C^*-algebras and Hilbert modules with dense subsets whose cardinalities do not exceed a fixed large cardinal. For example, we could consider only separable C^*-algebras and bimodules. Alternatively, for each A, B we could restrict attention to right-Hilbert $A - B$ bimodules with cardinality dominated by the larger of the cardinalities of A and B (which would accommodate all the bimodules which occur in the usual imprimitivity theorems, for example). In practice, these issues should never present a real problem, and we shall ignore them.

Recall that in any category, a morphism $f \colon A \to B$ is called an *equivalence* if there exists $g \colon B \to A$ such that $f \circ g = \mathrm{id}_B$ and $g \circ f = \mathrm{id}_A$. The equivalences in $\mathcal{C}$ are exactly the (isomorphism classes of) imprimitivity bimodules. In one direction, if X is an $A - B$ imprimitivity bimodule, then the isomorphism class of the conjugate module $\widetilde{X}$ is in $\mathrm{mor}(B, A)$, and satisfies $[X] \circ [\widetilde{X}] = [B]$ and $[\widetilde{X}] \circ [X] = [A]$. To see the converse requires substantially more work. Schweizer independently discovered the following result in [**59**, Proposition 2.3]; our proof is considerably different.[1]

LEMMA 2.4. *Let ${}_A X_B$ and ${}_B Y_A$ be right-Hilbert bimodules such that*
$$X \otimes_B Y \cong A \quad \text{and} \quad Y \otimes_A X \cong B.$$
Then X is an $A - B$ imprimitivity bimodule, and $Y \cong \widetilde{X}$.

PROOF. We first note that $Y \otimes_A X \cong B$ implies that X is full, since it follows from the definition of the inner product on $Y \otimes_A X$ that its range is always contained in the range of the inner product on X. A similar argument shows that Y is full, too. It therefore suffices to show that the canonical homomorphism $\varphi \colon A \to \mathcal{L}_B(X)$ is an isomorphism of A onto $\mathcal{K}_B(X)$, for this shows X is an $A - B$ imprimitivity bimodule, and then the second statement of the lemma will follow from uniqueness of inverses in a category. We first show that φ is faithful. Since $X \otimes_B Y \cong A$, we know that the canonical homomorphism of A into $\mathcal{L}_A(X \otimes_B Y)$ is faithful (because $A \to \mathcal{L}_A(A)$ is). This implies φ is faithful, for if $a \cdot x = 0$ for all $x \in X$, then $a \cdot (x \otimes y) = 0$ for all $x \in X, y \in Y$, so $a \cdot z = 0$ for all $z \in X \otimes_B Y$, hence $a = 0$.

To see that $\varphi(A) = \mathcal{K}_B(X)$, note that the latter coincides with $\overline{XX^*}$, using the canonical identification of X with $\mathcal{K}_B(B, X)$. Our strategy is to show that there is a right-Hilbert bimodule homomorphism
$$\Phi \colon {}_B Y_A \to \mathcal{L}_B(X, B) = M\bigl({}_B(\mathcal{K}_B(X, B))_{\mathcal{K}(X)}\bigr)$$
such that the right coefficient map of Φ is φ, and $\Phi(Y)^* = X$. This will do the job, since we will then have
$$\overline{XX^*} = \overline{\Phi(Y)^*\Phi(Y)} = \varphi(A).$$

---

[1]It is also different from the representation-theoretic proof used in [**17**], where this theorem was proved for full right-Hilbert bimodules.

Let $\Psi\colon {}_B(Y\otimes_A X)_B \xrightarrow{\cong} {}_BB_B$, and define $\Phi\colon Y \to \mathcal{L}_B(X,B)$ by
$$\Phi(y)x = \Psi(y\otimes x).$$

We have
$$\Phi(b\cdot y)x = \Psi(b\cdot y\otimes x) = b\Psi(y\otimes x) = b\Phi(y)x \quad \text{for all } b\in B,\ x\in X,\ y\in Y,$$
and
$$\begin{aligned}
\langle\langle\Phi(y_1),\Phi(y_2)\rangle_{\mathcal{K}(X)}\,x_1,x_2\rangle_B &= \langle\Phi(y_1)^*\Phi(y_2)x_1,x_2\rangle_B = \langle\Phi(y_2)x_1,\Phi(y_1)x_2\rangle_B\\
&= \langle\Psi(y_2\otimes x_1),\Psi(y_1\otimes x_2)\rangle_B = \langle y_2\otimes x_1, y_1\otimes x_2\rangle_B\\
&= \langle\langle y_1,y_2\rangle_A\cdot x_1, x_2\rangle_B = \langle\varphi(\langle y_1,y_2\rangle_A)x_1, x_2\rangle_B,
\end{aligned}$$
for all $x_i \in X$ and $y_i \in Y$, which implies that Φ is an id_B–φ compatible right-Hilbert bimodule homomorphism.

Finally,
$$X = A\cdot X = \overline{\Phi(Y)^*\Phi(Y)X} = \overline{\Phi(Y)^*\Psi(Y\otimes_A X)} = \overline{\Phi(Y)^*B} = \overline{\Phi(B\cdot Y)^*} = \Phi(Y)^*,$$
where B is identified with $\mathcal{K}_B(B_B)$ where appropriate. $\square$

2.2. Group actions

DEFINITION 2.5. Let G be a locally compact group, let α and β be actions of G on C^*-algebras A and B, and let X be a right-Hilbert A–B bimodule. An α–β *compatible right-Hilbert bimodule action* of G on X is a homomorphism γ of G into the group of invertible linear maps on X such that
  (i) $\gamma_s(a\cdot x) = \alpha_s(a)\cdot\gamma_s(x)$
  (ii) $\gamma_s(x\cdot b) = \gamma_s(x)\cdot\beta_s(b)$
  (iii) $\langle\gamma_s(x),\gamma_s(y)\rangle_B = \beta_s(\langle x,y\rangle_B)$
for each $s\in G$, $a\in A$, $x,y\in X$, and $b\in B$; and such that each map $s\mapsto\gamma_s(x)$ is continuous from G into X. We call α and β the *coefficient actions* of γ.

REMARK 2.6. (1) Note that each γ_s is in particular a right-Hilbert bimodule homomorphism of X onto itself with coefficient maps α_s and β_s; thus condition (ii) is implied by condition (iii) and the linearity of β_s by Remark 1.17.

(2) If X is an A–B imprimitivity bimodule, and if γ is an α–β compatible right-Hilbert bimodule action on X, then γ is automatically an imprimitivity bimodule action in the sense of [**11**]: calculating as in the proof of Lemma 1.18, for each $s\in G$ and $x,y,z\in X$ we have
$$\begin{aligned}
\alpha_s({}_A\langle x,y\rangle)\cdot\gamma_s(z) &= \gamma_s({}_A\langle x,y\rangle\cdot z)\\
&= \gamma_s(x\cdot\langle y,z\rangle_B)\\
&= \gamma_s(x)\cdot\langle\gamma_s(y),\gamma_s(z)\rangle_B\\
&= {}_A\langle\gamma_s(x),\gamma_s(y)\rangle\cdot\gamma_s(z),
\end{aligned}$$
which shows that $\alpha_s({}_A\langle x,y\rangle) = {}_A\langle\gamma_s(x),\gamma_s(y)\rangle$.

By Remark 1.19, we have a sort of converse: when X is an imprimitivity bimodule we can replace (i) in Definition 2.5 by
  (i′) ${}_A\langle\gamma_s(x),\gamma_s(y)\rangle = \alpha_s({}_A\langle x,y\rangle)$.

(3) We should point out that group actions on Hilbert bimodules as introduced above are well known in the literature. In particular they play an important rôle in the construction of Kasparov's equivariant KK-Theory for C^*-algebras (see also [**30**]).

Given right-Hilbert bimodule actions ${}_{(A,\alpha)}(X,\gamma)_{(B,\beta)}$ and ${}_{(B,\beta)}(Y,\rho)_{(C,\epsilon)}$, it is easy to check that the automorphisms of $X \otimes_B Y$ defined by

$$(\gamma \otimes_B \rho)_s = \gamma_s \otimes_B \rho_s$$

for each $s \in G$ give rise to an $\alpha - \epsilon$ compatible action $\gamma \otimes_B \rho$ of G on $X \otimes_B Y$.

DEFINITION 2.7. Let γ and ρ be $\alpha - \beta$ compatible actions of G on right-Hilbert $A - B$ bimodules X and Y. An isomorphism Φ of X onto Y is $\gamma - \rho$ *equivariant*, or *intertwines* γ and ρ, if

$$\Phi \circ \gamma_s = \rho_s \circ \Phi$$

for all $s \in G$. We say γ and ρ are *isomorphic*, or X and Y are *equivariantly isomorphic*, if such a Φ exists.

It is straightforward to check that this notion of isomorphism is an equivalence relation on the class of right-Hilbert $A - B$ bimodules with $\alpha - \beta$ compatible actions of G.

THEOREM 2.8. *Let G be a locally compact group. There is a category $\mathcal{A}(G)$ in which the objects are C^*-algebras with actions of G, and in which the morphisms from (A,α) to (B,β) are the equivariant isomorphism classes of right-Hilbert $A - B$ bimodules with $\alpha - \beta$ compatible actions of G. The composition of $[X,\gamma]\colon (A,\alpha) \to (B,\beta)$ with $[Y,\rho]\colon (B,\beta) \to (C,\nu)$, is the isomorphism class of the tensor product action $(X \otimes_B Y, \gamma \otimes_B \rho)$; the identity morphism on (A,α) is the isomorphism class of the $\alpha - \alpha$ compatible right-Hilbert bimodule action (A,α) itself.*

PROOF. We first note that the composition of morphisms is well-defined: suppose $[X,\gamma] = [X',\gamma']\colon (A,\alpha) \to (B,\beta)$ and $[Y,\rho] = [Y',\rho']\colon (B,\beta) \to (C,\nu)$, so that we have equivariant right-Hilbert bimodule isomorphisms

$$\Phi\colon (X,\gamma) \to (X',\gamma') \quad \text{and} \quad \Psi\colon (Y,\rho) \to (Y',\rho').$$

Then straightforward calculations show that the isomorphism $\Phi \otimes_B \Psi \colon X \otimes_B Y \to X' \otimes_B Y'$ from the proof of Theorem 2.2 satisfies

$$(\Phi \otimes_B \Psi) \circ (\gamma \otimes_B \rho)_s = (\gamma' \otimes_B \rho')_s \circ (\Phi \otimes_B \Psi)$$

for all $s \in G$, hence gives an isomorphism between the $\alpha - \nu$ compatible actions $\gamma \otimes_B \rho$ and $\gamma' \otimes_B \rho'$.

Next we establish that composition of morphisms in $\mathcal{A}(G)$ is associative; by the above, it suffices to show that the actions $\gamma \otimes_B (\rho \otimes_C \sigma)$ and $(\gamma \otimes_B \rho) \otimes_C \sigma$ are isomorphic for any right-Hilbert bimodule actions $({}_A X_B, \gamma)$, $({}_B Y_C, \rho)$, and $({}_C Z_D, \sigma)$. Again, straightforward calculations show that the isomorphism $X \otimes_B (Y \otimes_C Z) \cong (X \otimes_B Y) \otimes_C Z$ from the proof of Theorem 2.2 intertwines the actions as desired.

Finally, note that for any action $({}_A X_B, \gamma)$, with left coefficient action α, the canonical isomorphism $A \otimes_A X \cong X$ intertwines the actions $\alpha \otimes_A \gamma$ and γ. Similarly, for any action $({}_B Y_A, \rho)$ with right coefficient action α, the canonical isomorphism $Y \otimes_A A \cong Y$ intertwines $\rho \otimes_A \alpha$ and ρ. Hence the identity morphism on (A,α) is the isomorphism class of the $\alpha - \alpha$ compatible right-Hilbert bimodule action (A,α). □

REMARK 2.9. The equivalences in $\mathcal{A}(G)$ are exactly the (isomorphism classes of) imprimitivity bimodule actions of G in the sense of [11] or [12]. In one direction, if $_{(A,\alpha)}(X,\gamma)_{(B,\beta)}$ is an imprimitivity bimodule action, then the canonical isomorphisms $X \otimes_B \widetilde{X} \cong A$ and $\widetilde{X} \otimes_A X \cong B$ are easily seen to be $\gamma \otimes_B \widetilde{\gamma} - \alpha$ and $\widetilde{\gamma} \otimes_A \gamma - \beta$ equivariant, respectively. In the other direction, if $(_A X_B, \gamma)$ and $(_B Y_A, \rho)$ are $\alpha - \beta$ and $\beta - \alpha$ compatible right-Hilbert bimodule actions with

$$\gamma \otimes_B \rho \cong \alpha \quad \text{and} \quad \rho \otimes_A \gamma \cong \beta,$$

then X is in particular an $A - B$ imprimitivity bimodule by Lemma 2.4, so γ and ρ are imprimitivity bimodule actions by Remark 2.6.

2.3. Group coactions

In this section we are going to construct a category $\mathcal{C}(G)$, where all actions and morphisms are equipped with coactions of the group G. The necessary background on coactions of groups on C^*-algebras is given in Appendix A.

For any right-Hilbert $A - B$ bimodule X and any locally compact group G, we may take the exterior tensor product of X with the right-Hilbert $C^*(G) - C^*(G)$ module $C^*(G)$, as in Definition 1.35, and get a right-Hilbert $(A \otimes C^*(G)) - (B \otimes C^*(G))$ bimodule $X \otimes C^*(G)$.

DEFINITION 2.10. Let G be a locally compact group, let δ and ϵ be coactions of G on C^*-algebras A and B, and let X be a right-Hilbert $A - B$ bimodule. A $\delta - \epsilon$ *compatible right-Hilbert bimodule coaction* of G on X is a nondegenerate[2] right-Hilbert bimodule homomorphism $\zeta \colon X \to M(X \otimes C^*(G))$, with coefficient maps δ and ϵ (called the *coefficient coactions* of ζ), such that

(i) $(1_{M(A)} \otimes C^*(G))\zeta(X) \subseteq X \otimes C^*(G)$, and
(ii) $(\zeta \otimes \mathrm{id}_G) \circ \zeta = (\mathrm{id}_X \otimes \delta_G) \circ \zeta$ (the *coaction identity*).

Moreover, ζ is called *nondegenerate* if δ and ϵ are nondegenerate coactions and

$$\overline{(1 \otimes C^*(G))\zeta(X)} = X \otimes C^*(G).$$

REMARK 2.11. (1) A condition (as in [20])

(iii) $\zeta(X)(1_{M(B)} \otimes C^*(G)) \subseteq X \otimes C^*(G)$,

symmetric to (i), would be redundant: it follows from the analogous property for the C^*-coaction ϵ, because

$$\zeta(X)(1 \otimes C^*(G)) = \zeta(XB)(1 \otimes C^*(G)) = \zeta(X)\epsilon(B)(1 \otimes C^*(G))$$
$$\subseteq \zeta(X)(B \otimes C^*(G)) \subseteq X \otimes C^*(G).$$

It follows from this that condition (i) of Definition 2.10 is equivalent to the requirement that $\zeta(X) \subseteq M_G(X \otimes C^*(G))$ (see Definition 1.39).

(2) If ζ is nondegenerate, then it automatically satisfies

$$\overline{\zeta(X)(1 \otimes C^*(G))} = X \otimes C^*(G).$$

This is easily checked by using the right module homomorphism property and non-degeneracy of ϵ.

(3) If X is an $A - B$ imprimitivity bimodule, and if ζ is a $\delta - \epsilon$ compatible right-Hilbert bimodule coaction on X, then ζ is automatically an imprimitivity

---

[2]Notice that we have incorporated nondegeneracy of ζ *as a bimodule homomorphism* into our definition, whereas Ng's definition [40, Definition 2.10] does not.

bimodule coaction in the sense of [**20**, Definition 3.1] (*cf.* [**2**, 2.2],[**6**, 2.15], and [**40**, Definition 3.3]). We only need to check that

$$\delta(_A\langle x,y\rangle) = {}_{M(A\otimes C^*(G))}\langle \zeta(x),\zeta(y)\rangle$$

for all $x,y \in X$; but this is immediate from Lemma 1.18 (in fact, for this to be true we only need to assume that X is a right-partial imprimitivity bimodule). Moreover, if X is an imprimitivity bimodule, then condition (i) in Definition 2.10 is redundant, nondegeneracy of ζ as a bimodule homomorphism follows automatically from nondegeneracy of the coefficient homomorphisms, and nondegeneracy of ζ as a coaction follows automatically from nondegeneracy of the coefficient coactions.

The following lemma will be fundamental in all computations with bimodule coactions.

LEMMA 2.12. *Let $_AX_B$ and $_BY_C$ be right-Hilbert bimodules, and let D be a C^*-algebra. There exists a right-Hilbert $A \otimes D - C \otimes D$ bimodule isomorphism*

$$\Theta \colon (X \otimes D) \otimes_{B \otimes D} (Y \otimes D) \to (X \otimes_B Y) \otimes D$$

such that

(2.1) $$\Theta\big((x \otimes d) \otimes_{B \otimes D} (y \otimes e)\big) = (x \otimes_B y) \otimes de$$

for $x \in X$, $y \in Y$, and $d,e \in D$.[3]

Moreover, the unique extension of Θ to the multiplier bimodules satisfies[4]

$$\Theta\big((x \otimes m) \otimes (y \otimes n)\big) = (x \otimes y) \otimes mn \quad \text{for } x \in X, y \in Y, m,n \in M(D).$$

PROOF. Equation (2.1) clearly determines a map $\Theta \colon (X \odot D) \odot (Y \odot D) \to (X \otimes_B Y) \otimes D$. The following computation implies that Θ preserves inner products:

$$\begin{aligned}
\langle \Theta\big((x \otimes d) \otimes (y \otimes e)\big), \Theta\big((z \otimes f) \otimes (w \otimes g)\big) \rangle_{C \otimes D} \\
= \langle (x \otimes y) \otimes de, (z \otimes w) \otimes fg \rangle_{C \otimes D} \\
= \langle x \otimes y, z \otimes w \rangle_C \otimes (de)^* fg \\
= \langle y, \langle x, z\rangle_B \cdot w \rangle_C \otimes e^* d^* fg \\
= \langle y \otimes e, \langle x, z\rangle_B \cdot w \otimes d^* fg \rangle_{C \otimes D} \\
= \langle y \otimes e, (\langle x, z\rangle_B \otimes (d^* f)) \cdot (w \otimes g) \rangle_{C \otimes D} \\
= \langle y \otimes e, \langle x \otimes d, z \otimes f\rangle_{B \otimes D} \cdot (w \otimes g) \rangle_{C \otimes D} \\
= \langle (x \otimes d) \otimes (y \otimes e), (z \otimes f) \otimes (w \otimes g) \rangle_{C \otimes D}.
\end{aligned}$$

Since Θ clearly has dense range in $(X \otimes_B Y) \otimes D$, it therefore extends to an isometry, which we continue to denote by Θ, of $(X \otimes D) \otimes_{B \otimes D} (Y \otimes D)$ onto $(X \otimes_B Y) \otimes D$. Straightforward calculations with elementary tensors verify that Θ intertwines the left actions, so that Θ is indeed a right-Hilbert $A \otimes D - C \otimes D$ bimodule isomorphism.

---

[3]More generally, one can prove that $(_AX_B \otimes {}_DZ_E) \otimes_{B \otimes E} (_BY_C \otimes {}_EW_F) \cong (X \otimes_B Y) \otimes (Z \otimes_E W)$ as right-Hilbert $A \otimes D - C \otimes F$ bimodules.

[4]It is fairly obvious that the following identity can be generalized to allow x and y to be multipliers as well, but we only need the specific fact we record here.

2.3. GROUP COACTIONS

For the other part, take $c \in C$ and $d \in D$ and compute:

$$\begin{aligned}
\Theta\big((x \otimes m) \otimes (y \otimes n)\big) \cdot (c \otimes d) &= \Theta\big((x \otimes m) \otimes (y \otimes n) \cdot (c \otimes d)\big) \\
&= \Theta\big((x \otimes m) \otimes (y \cdot c \otimes nd)\big) \\
&= \Theta\big((x \otimes m) \otimes (y \cdot c \otimes ef)\big) \quad \text{(for some } e, f \in L\text{)} \\
&= \Theta\big((x \otimes me) \otimes (y \cdot c \otimes f)\big) \\
&= (x \otimes y \cdot c) \otimes mef \\
&= (x \otimes y) \cdot (c \otimes mnd) \\
&= \big((x \otimes y) \otimes mn\big) \cdot (c \otimes d).
\end{aligned}$$

$\square$

The following construction of the balanced tensor product of coactions should be compared to [**2**, Proposition 2.10].

PROPOSITION 2.13. *If* $_{(A,\delta)}(X,\zeta)_{(B,\epsilon)}$ *and* $_{(B,\epsilon)}(Y,\eta)_{(C,\vartheta)}$ *are right-Hilbert bimodule coactions of G, then*

$$\zeta \,\natural_B\, \eta = \Theta \circ (\zeta \otimes_B \eta)$$

defines a $\delta - \vartheta$ compatible coaction $\zeta \,\natural_B\, \eta$ of G on $X \otimes_B Y$, where Θ is the isomorphism of Lemma 2.12. Moreover, if ζ and η are nondegenerate, then so is $\zeta \,\natural_B\, \eta$.

PROOF. By Proposition 1.34 and Lemma 2.12, Θ and $\zeta \otimes_B \eta$ are nondegenerate right-Hilbert bimodule homomorphisms, so their composition (which clearly has the desired coefficient maps) is too.

To show that $(1_{M(A)} \otimes C^*(G)) \cdot (\zeta \,\natural_B\, \eta)(X \otimes_B Y) \subseteq (X \otimes_B Y) \otimes C^*(G)$, we first claim that

$$(2.2) \qquad (1_{M(A)} \otimes c) \cdot (\zeta(x) \otimes \eta(y)) \in (X \otimes C^*(G)) \otimes_{B \otimes C^*(G)} (Y \otimes C^*(G))$$

for all $c \in C^*(G)$, $x \in X$, and $y \in Y$. For, since ζ is a coaction on X, $(1 \otimes c) \cdot \zeta(x) \in X \otimes C^*(G)$, so we may write $(1 \otimes c) \cdot \zeta(x) = z \cdot (1 \otimes d)$ for some $z \in X \otimes C^*(G)$ and $d \in C^*(G)$. Since η is a coaction on Y we have $(1 \otimes d) \cdot \eta(y) \in Y \otimes C^*(G)$. Thus

$$\begin{aligned}
(1 \otimes c) \cdot (\zeta(x) \otimes \eta(y)) &= (1 \otimes c) \cdot \zeta(x) \otimes \eta(y) = z \cdot (1 \otimes d) \otimes \eta(y) \\
&= z \otimes (1 \otimes d) \cdot \eta(y) \in (X \otimes C^*(G)) \otimes_{B \otimes C^*(G)} (Y \otimes C^*(G)),
\end{aligned}$$

which gives (2.2).

Next, note that

$$(2.3) \qquad (1_{M(A)} \otimes c) \cdot \Theta(w) = \Theta((1_{M(A)} \otimes c) \cdot w)$$

for all $c \in C^*(G)$ and $w \in M((X \otimes C^*(G)) \otimes_{B \otimes C^*(G)} (Y \otimes C^*(G)))$, because the unique extension (Proposition 1.30) of Θ to the multiplier bimodule has left coefficient map $\mathrm{id}_{M(A)}$.

Now combine (2.2) and (2.3) to get

$$\begin{aligned}
(1_{M(A)} \otimes c) \cdot (\zeta \,\natural_B\, \eta)(x \otimes y) &= (1 \otimes c) \cdot \Theta\big(\zeta(x) \otimes \eta(y)\big) \\
&= \Theta\big((1 \otimes c) \cdot (\zeta(x) \otimes \eta(y))\big) \\
&\in \Theta\big((X \otimes C^*(G)) \otimes_{B \otimes C^*(G)} (Y \otimes C^*(G))\big) \\
&= (X \otimes_B Y) \otimes C^*(G),
\end{aligned}$$

so that $(1_{M(A)} \otimes C^*(G)) \cdot (\zeta \sharp_B \eta)(X \otimes_B Y) \subseteq (X \otimes_B Y) \otimes C^*(G)$ by density and continuity.

If ζ and η are nondegenerate, the coefficient homomorphisms of $\zeta \sharp_B \eta$ are certainly nondegenerate, and using the above considerations we can compute

$$\overline{(1 \otimes C^*(G))(\zeta \sharp_B \eta)(X \otimes_B Y)} = \overline{\Theta\big((1 \otimes C^*(G)(\zeta(X) \otimes_{M(B \otimes C^*(G))} \eta(Y))\big)}$$
$$= \Theta\big(\overline{(X \otimes C^*(G)) \otimes_{M(B \otimes C^*(G))} \eta(Y)}\big)$$
$$= \Theta\big(\overline{(X \otimes C^*(G)) \otimes_{B \otimes C^*(G)} (1 \otimes C^*(G))\eta(Y)}\big)$$
$$= \Theta\big((X \otimes C^*(G)) \otimes_{B \otimes C^*(G)} (Y \otimes C^*(G))\big)$$
$$= (X \otimes_B Y) \otimes C^*(G).$$

It only remains to check the coaction identity:

$$((\zeta \sharp_B \eta) \otimes \mathrm{id}) \circ (\zeta \sharp_B \eta) = (\Theta \circ (\zeta \otimes_B \eta) \otimes \mathrm{id}) \circ \Theta \circ (\zeta \otimes_B \eta)$$
$$= (\Theta \otimes \mathrm{id}) \circ ((\zeta \otimes_B \eta) \otimes \mathrm{id}) \circ \Theta \circ (\zeta \otimes_B \eta)$$
$$\stackrel{(1)}{=} (\Theta \otimes \mathrm{id}) \circ \Theta \circ ((\zeta \otimes \mathrm{id}) \otimes_{B \otimes C^*(G)} (\eta \otimes \mathrm{id})) \circ (\zeta \otimes_B \eta)$$
$$\stackrel{(2)}{=} (\Theta \otimes \mathrm{id}) \circ \Theta \circ ((\mathrm{id} \otimes \delta_G) \otimes_{B \otimes C^*(G)} (\mathrm{id} \otimes \delta_G)) \circ (\zeta \otimes_B \eta)$$
$$\stackrel{(3)}{=} (\mathrm{id} \otimes_G \delta) \circ \Theta \circ (\zeta \otimes_B \eta)$$
$$= (\mathrm{id} \otimes_G \delta) \circ (\zeta \sharp_B \eta),$$

where the equality at (1) is justified by computing, for $x \in X$, $y \in Y$, and $c, d \in C^*(G)$,

$$((\zeta \otimes_B \eta) \otimes \mathrm{id}) \circ \Theta\big((x \otimes c) \otimes (y \otimes d)\big) = ((\zeta \otimes_B \eta) \otimes \mathrm{id})\big((x \otimes y) \otimes cd\big)$$
$$= (\zeta \otimes_B \eta)(x \otimes y) \otimes cd$$
$$= (\zeta(x) \otimes \eta(y)) \otimes cd$$
$$= \Theta\big((\zeta(x) \otimes c) \otimes (\eta(y) \otimes d)\big)$$
$$= \Theta\big((\zeta \otimes \mathrm{id})(x \otimes c) \otimes (\eta \otimes \mathrm{id})(y \otimes d)\big)$$
$$= \Theta \circ \big((\zeta \otimes \mathrm{id}) \otimes_{B \otimes C^*(G)} (\eta \otimes \mathrm{id})\big)\big((x \otimes c) \otimes (y \otimes d)\big),$$

and then appealing to linearity, strict density, and strict continuity; the equality at (2) is justified by

$$((\zeta \otimes \mathrm{id}) \otimes_{B \otimes C^*(G)} (\eta \otimes \mathrm{id})) \circ (\zeta \otimes_B \eta) = (\zeta \otimes \mathrm{id}) \circ \zeta \otimes_B (\eta \otimes \mathrm{id}) \circ \eta$$
$$= (\mathrm{id} \otimes \delta_G) \circ \zeta \otimes_B (\mathrm{id} \otimes \delta_G) \circ \eta$$
$$= ((\mathrm{id} \otimes \delta_G) \otimes_{B \otimes C^*(G)} (\mathrm{id} \otimes \delta_G)) \circ (\zeta \otimes_B \eta);$$

and finally the equality at (3) is justified by computing, for $x \in X$, $y \in Y$, and $s, t \in G$ (viewed as elements of $M(C^*(G))$ via the canonical embedding $G \to$

$UM(C^*(G)))$,

$$(\Theta \otimes \mathrm{id}) \circ \Theta \circ \big((\mathrm{id}\otimes\delta_G) \otimes_{B\otimes C^*(G)} (\mathrm{id}\otimes\delta_G)\big)\big((x \otimes s) \otimes (y \otimes t)\big)$$
$$= (\Theta \otimes \mathrm{id}) \circ \Theta\big((\mathrm{id}\otimes\delta_G)(x \otimes s) \otimes (\mathrm{id}\otimes\delta_G)(y \otimes t)\big)$$
$$= (\Theta \otimes \mathrm{id}) \circ \Theta\big((x \otimes s \otimes s) \otimes (y \otimes t \otimes t)\big)$$
$$= (\Theta \otimes \mathrm{id})\big(((x \otimes s) \otimes (y \otimes t)) \otimes st\big)$$
$$= (x \otimes y) \otimes st \otimes st$$
$$= (\mathrm{id}\otimes\delta_G)\big((x \otimes y) \otimes st\big)$$
$$= (\mathrm{id}\otimes\delta_G) \circ \Theta\big((x \otimes s) \otimes (y \otimes t)\big),$$

and then appealing to linearity, strict density, and strict continuity. □

DEFINITION 2.14. Let ζ and η be $\delta-\epsilon$ compatible coactions on right-Hilbert $A-B$ bimodules X and Y. An isomorphism Φ of X onto Y is $\zeta-\eta$ equivariant, or *intertwines* ζ and η, if

$$\eta \circ \Phi = (\Phi \otimes \mathrm{id}) \circ \zeta.$$

We say ζ and η are *isomorphic*, or X and Y are *equivariantly isomorphic*, if such a Φ exists.

It is straightforward to check that this notion of isomorphism is an equivalence relation on the class of right-Hilbert $A-B$ bimodules with $\delta-\epsilon$ compatible coactions of G.

THEOREM 2.15. *Let G be a locally compact group. There is a category $\mathcal{C}(G)$ in which the objects are C^*-algebras with nondegenerate normal coactions of G (see Definition A.50 for the meaning of normal coaction), and in which the morphisms from (A, δ) to (B, ϵ) are the equivariant isomorphism classes of nondegenerate right-Hilbert $A-B$ bimodules with $\delta-\epsilon$ compatible coactions of G. The composition of $[X, \zeta]\colon (A,\delta) \to (B,\epsilon)$ with $[Y,\eta]\colon (B,\epsilon) \to (C,\vartheta)$ is the isomorphism class of the tensor product coaction $\zeta \sharp_B \eta$ on $X \otimes_B Y$; the identity morphism on (A,δ) is the isomorphism class of the $\delta-\delta$ compatible right-Hilbert bimodule coaction (A,δ) itself.*

PROOF. We first note that the composition of morphisms is well-defined: suppose $[X,\zeta] = [X',\zeta']\colon (A,\delta) \to (B,\epsilon)$ and $[Y,\eta] = [Y',\eta']\colon (B,\epsilon) \to (C,\vartheta)$, so that we have equivariant right-Hilbert bimodule isomorphisms

$$\Phi\colon (X,\zeta) \to (X',\zeta') \quad \text{and} \quad \Psi\colon (Y,\eta) \to (Y',\eta').$$

Then the isomorphism $\Phi \otimes_B \Psi\colon X\otimes_B Y \to X'\otimes_B Y'$ from the proof of Theorem 2.2 satisfies

$$(\zeta' \sharp_B \eta') \circ (\Phi \otimes_B \Psi) = \Theta \circ (\zeta' \otimes_B \eta') \circ (\Phi \otimes_B \Psi)$$
$$= \Theta \circ \big((\zeta' \circ \Phi) \otimes_B (\eta' \circ \Psi)\big)$$
$$= \Theta \circ \big((\Phi \otimes \mathrm{id}) \circ \zeta \otimes_B (\Psi \otimes \mathrm{id}) \circ \eta\big)$$
$$= \Theta \circ \big((\Phi \otimes \mathrm{id}) \otimes_{B\otimes B^*(G)} (\Psi \otimes \mathrm{id})\big) \circ (\zeta \otimes_B \eta)$$
$$= \big((\Phi \otimes_B \Psi) \otimes \mathrm{id}\big) \circ \Theta \circ (\zeta \otimes_B \eta)$$
$$= \big((\Phi \otimes_B \Psi) \otimes \mathrm{id}\big) \circ (\zeta \sharp_B \eta).$$

Next we establish that composition of morphisms in $\mathcal{C}(G)$ is associative; by the above it suffices to show that the coactions $\zeta \,\natural_B\, (\eta \,\natural_C\, \tau)$ and $(\zeta \,\natural_B\, \eta) \,\natural_C\, \tau$ are isomorphic for any tensorable right-Hilbert bimodule coactions $({}_A X_B, \zeta)$, $({}_B Y_C, \eta)$, and $({}_C Z_D, \tau)$. Let $\Phi \colon X \otimes_B (Y \otimes_C Z) \to (X \otimes_B Y) \otimes_C Z$ be the isomorphism from the proof of Theorem 2.2; we need to show that

$$((\zeta \,\natural_B\, \eta) \,\natural_C\, \tau) \circ \Phi = (\Phi \otimes \mathrm{id}) \circ (\zeta \,\natural_B\, (\eta \,\natural_C\, \tau)).$$

We check this for an elementary tensor $x \otimes (y \otimes z)$:

$$((\zeta \,\natural_B\, \eta) \,\natural_C\, \tau) \circ \Phi(x \otimes (y \otimes z)) = \Theta\big(\Theta(\zeta(x) \otimes \eta(y)) \otimes \tau(z)\big)$$
$$\stackrel{(*)}{=} (\Phi \otimes \mathrm{id}) \circ \Theta\big(\zeta(x) \otimes \Theta(\eta(y) \otimes \tau(z))\big)$$
$$= (\Phi \otimes \mathrm{id}) \circ \big(\zeta \,\natural_B\, (\eta \,\natural_C\, \tau)\big)(x \otimes (y \otimes z)),$$

where the equality at $(*)$ is justified by replacing $\zeta(x)$ by an elementary tensor $x' \otimes c \in X \otimes C^*(G)$, replacing $\eta(y)$ by $y' \otimes d$ and $\tau(z)$ by $z' \otimes e$, and then appealing to linearity, strict density, and strict continuity:

$$\Theta\big(\Theta((x' \otimes c) \otimes (y' \otimes d)) \otimes (z' \otimes e)\big) = \Theta\big(((x' \otimes y') \otimes cd) \otimes (z' \otimes e)\big)$$
$$= ((x' \otimes y') \otimes z') \otimes cde$$
$$= \Phi(x' \otimes (y' \otimes z')) \otimes cde$$
$$= (\Phi \otimes \mathrm{id})\big((x' \otimes (y' \otimes z')) \otimes cde\big)$$
$$= (\Phi \otimes \mathrm{id}) \circ \Theta\big((x' \otimes c) \otimes ((y' \otimes z') \otimes de)\big)$$
$$= (\Phi \otimes \mathrm{id}) \circ \Theta\big((x' \otimes c) \otimes \Theta((y' \otimes d) \otimes (z' \otimes e))\big).$$

Finally, note that for any coaction $({}_A X_B, \zeta)$ of G, with left coefficient coaction δ, the canonical isomorphism $\Phi \colon A \otimes_A X \to X$ intertwines the coactions $\delta \,\natural_A\, \zeta$ and ζ. To see this, take $a \in A$ and $x \in X$, and compute:

$$\zeta \circ \Phi(a \otimes x) = \zeta(a \cdot x) = \delta(a) \cdot \zeta(x) \stackrel{(*)}{=} (\Phi \otimes \mathrm{id}) \circ \Theta\big(\delta(a) \otimes \zeta(x)\big) = (\Phi \otimes \mathrm{id}) \circ (\delta \,\natural_A\, \zeta)(a \otimes x),$$

where the equality at $(*)$ is justified by replacing $\delta(a)$ by an elementary tensor $b \otimes c \in A \otimes C^*(G)$, replacing $\zeta(x)$ by $y \otimes d$, and then appealing to linearity, strict density, and strict continuity:

$$(b \otimes c) \cdot (y \otimes d) = b \cdot y \otimes cd = \Phi(b \otimes y) \otimes cd = (\Phi \otimes \mathrm{id})\big((b \otimes y) \otimes cd\big)$$
$$= (\Phi \otimes \mathrm{id}) \circ \Theta\big((b \otimes c) \otimes (y \otimes d)\big).$$

This gives $\zeta \circ \Phi = (\Phi \otimes \mathrm{id}) \circ (\delta \,\natural_A\, \zeta)$. A similar argument shows that for any coaction $({}_B X_A, \zeta)$ with right coefficient coaction δ, the canonical isomorphism $X \otimes A \xrightarrow{\cong} X$ intertwines $\zeta \,\natural_A\, \delta$ and ζ; hence the identity morphism on (A, δ) is the isomorphism class of the $\delta - \delta$ compatible right-Hilbert bimodule coaction (A, δ). $\square$

REMARK 2.16. Notice that we could consider a larger category where the C^*-coactions on the objects are not required to be nondegenerate and normal, in which our category $\mathcal{C}(G)$ would sit as a subcategory. For our purposes, we require the specific category $\mathcal{C}(G)$, so that the Mansfield imprimitivity theorem applies without any further hypotheses (see Appendix B).

REMARK 2.17. The equivalences in $\mathcal{C}(G)$ are exactly the (isomorphism classes of) imprimitivity bimodule coactions. In one direction, suppose $({}_A X_B, \zeta)$ is an imprimitivity bimodule coaction of G, with left coefficient coaction δ. Then the map
$$\widetilde{x} \mapsto \widetilde{\zeta(x)} \colon \widetilde{X} \to M\big((X \otimes C^*(G))\widetilde{}\big)$$
is a nondegenerate homomorphism, and it is routine to check that the map $\Psi \colon (X \otimes C^*(G))\widetilde{} \to \widetilde{X} \otimes C^*(G)$ defined by
$$\Psi(\widetilde{x \otimes c}) = \widetilde{x} \otimes c^*$$
is an isomorphism; hence we can define a nondegenerate homomorphism $\widetilde{\zeta} \colon \widetilde{X} \to M(\widetilde{X} \otimes C^*(G))$ by
$$\widetilde{\zeta}(\widetilde{x}) = \Psi\big(\widetilde{\zeta(x)}\big).$$
Routine computations show that $\widetilde{\zeta}$ is in fact a bimodule coaction, which is nondegenerate if ζ is nondegenerate.

We claim that the canonical isomorphism $\Phi \colon X \otimes_B \widetilde{X} \to A$ intertwines $\zeta \,\sharp_B\, \widetilde{\zeta}$ and δ. To see this, take $x, y \in X$, and compute:
$$(\Phi \otimes \mathrm{id}) \circ (\zeta \,\sharp_B\, \widetilde{\zeta})(x \otimes \widetilde{y}) = (\Phi \otimes \mathrm{id}) \circ \Theta\Big(\zeta(x) \otimes \Psi\big(\widetilde{\zeta(y)}\big)\Big)$$
$$\stackrel{(*)}{=} {}_{A \otimes C^*(G)}\langle \zeta(x), \zeta(y) \rangle = \delta\big({}_A\langle x, y\rangle\big) = \delta \circ \Phi(x \otimes \widetilde{y}),$$
where the equality at $(*)$ is justified by replacing $\zeta(x)$ by an elementary tensor $z \otimes c \in X \otimes C^*(G)$, replacing $\zeta(y)$ by $w \otimes d$, and then appealing to linearity, strict density, and strict continuity:
$$(\Phi \otimes \mathrm{id}) \circ \Theta\Big((z \otimes c) \otimes \Psi\big(\widetilde{w \otimes d}\big)\Big) = (\Phi \otimes \mathrm{id}) \circ \Theta\big((z \otimes c) \otimes (\widetilde{w} \otimes d^*)\big)$$
$$= (\Phi \otimes \mathrm{id})\big((z \otimes \widetilde{w}) \otimes cd^*\big) = \Phi(z \otimes \widetilde{w}) \otimes cd^*$$
$$= {}_A\langle z, w\rangle \otimes cd^* = {}_{A \otimes C^*(G)}\langle z \otimes c, w \otimes d\rangle.$$
A similar argument shows that the canonical homomorphism $\Psi \colon \widetilde{X} \otimes_A X \to B$ intertwines $\widetilde{\zeta} \,\sharp_A\, \zeta$ and ϵ.

In the other direction, if $({}_A X_B, \zeta)$ and $({}_B Y_A, \eta)$ are $\delta - \epsilon$ and $\epsilon - \delta$ compatible right-Hilbert bimodule coactions of G with
$$\zeta \,\sharp_B\, \eta \cong \delta \quad \text{and} \quad \eta \,\sharp_A\, \zeta \cong \epsilon,$$
then in particular X is an $A - B$ imprimitivity bimodule by Lemma 2.4, so ζ and η are imprimitivity bimodule coactions by Remark 2.11.

2.4. Actions and coactions

DEFINITION 2.18. Let γ and ρ be $\alpha - \beta$ compatible actions, and let ζ and η be $\delta - \epsilon$ compatible coactions, of G on right-Hilbert $A - B$ bimodules X and Y. We say the triples (X, γ, ζ) and (Y, ρ, η) are *isomorphic*, or X and Y are *equivariantly isomorphic*, if there exists a right-Hilbert $A - B$ bimodule isomorphism of X onto Y which is both $\gamma - \rho$ equivariant and $\zeta - \eta$ equivariant.

Needless to say, this notion of isomorphism is an equivalence relation on the class of right-Hilbert $A - B$ bimodules with $\alpha - \beta$ compatible actions and $\delta - \epsilon$ compatible coactions of G.

THEOREM 2.19. *Let G be a locally compact group. There is a category $\mathcal{AC}(G)$ in which the objects are C^*-algebras with actions and nondegenerate normal coactions of G, and in which the morphisms from (A, α, δ) to (B, β, ϵ) are the equivariant isomorphism classes of right-Hilbert $A - B$ bimodules with $\alpha - \beta$ compatible actions and nondegenerate $\delta - \epsilon$ compatible coactions of G. The composition of $[X, \gamma, \zeta] \colon (A, \alpha, \delta) \to (B, \beta, \epsilon)$ with $[Y, \rho, \eta] \colon (B, \beta, \epsilon) \to (C, \nu, \vartheta)$ is the isomorphism class of $(X \otimes_B Y, \gamma \otimes_B \rho, \zeta \,\natural_B\, \eta)$; the identity morphism on (A, α, δ) is the isomorphism class of (A, α, δ) itself.*

PROOF. Since the proofs of Theorems 2.8 and 2.15 both use the same isomorphisms $X \otimes_B Y \cong Z \otimes_B W$, $X \otimes_B (Y \otimes_C Z) \cong (X \otimes_B Y) \otimes_C Z$, $A \otimes_A X \cong X$, and $Y \otimes_A A \cong Y$, these proofs combine to show that composition of morphisms is well-defined and associative, and that there are identity morphisms, in $\mathcal{AC}(G)$. □

REMARK 2.20. It might seem that this theorem could be proven by identifying $\mathcal{AC}(G)$ with the "diagonal" subcategory of $\mathcal{A}(G) \times \mathcal{C}(G)$ consisting of the objects $((A, \alpha), (B, \delta))$ which satisfy $A = B$ and the morphisms $([X, \gamma], [Y, \zeta])$ which satisfy $X \cong Y$. However, this approach would fail, because the map $[X, \gamma, \zeta] \mapsto ([X, \gamma], [X, \zeta])$ may not be injective: on the right side there can be different action-equivariant and coaction-equivariant isomorphisms, while on the left side there has to be one isomorphism which is both action- and coaction-equivariant.

2.5. Actions and coactions on linking algebras

One of the basic tools in this work is to exploit the relation between actions and coactions on (right-) partial imprimitivity bimodules and actions and coactions on the corresponding linking algebras. Recall from Section 1.5 in Chapter 1 that for any partial $A - B$ imprimitivity bimodule X, we can form the linking algebra

$$L(X) = \begin{pmatrix} A & X \\ \widetilde{X} & B \end{pmatrix}$$

with multiplication and involution as given in (1.11). As usual, we denote the corner projections $\begin{pmatrix} 1 & 0 \\ 0 & 0 \end{pmatrix}$ and $\begin{pmatrix} 0 & 0 \\ 0 & 1 \end{pmatrix}$ by p and q, respectively. For actions, we get:

LEMMA 2.21. *Suppose that X is a partial $A - B$ imprimitivity bimodule. Let α and β be actions of G on A and B, respectively, and let γ be an $\alpha - \beta$ compatible right-Hilbert bimodule action of G on X. Then there exists an action $\nu \colon G \to \operatorname{Aut} L(X)$ given by*

$$\nu_s \left(\begin{pmatrix} a & x \\ y & b \end{pmatrix} \right) = \begin{pmatrix} \alpha_s(a) & \gamma_s(x) \\ \widetilde{\gamma_s(y)} & \beta_s(b) \end{pmatrix}.$$

Conversely, if $\nu \colon G \to \operatorname{Aut} L(X)$ is an action such that $\nu_s(p) = p$ and (hence) $\nu_s(q) = q$, we obtain actions α, β, and γ on A, B, and X, respectively, such that

$$\alpha_s(plp) = p\nu_s(l)p, \quad \beta_s(qlq) = q\nu_s(l)q, \quad \text{and} \quad \gamma_s(plq) = p\nu_s(l)q,$$

for all $s \in G$ and $l \in L(X)$.

PROOF. Note first that it follows from Lemma 1.18 that a right-Hilbert bimodule action on $_AX_B$ is automatically a partial imprimitivity bimodule action (and vice versa). Using this, the proof follows directly from the algebraic properties in $L(X)$, so we omit further details. □

We now consider coactions. For imprimitivity bimodules, the following result can be found in [**21**]:

LEMMA 2.22. *Suppose that $_AX_B$ is a right-partial imprimitivity bimodule and assume that ζ is coaction of G on X with coefficient coactions δ and ϵ of G on A and B, respectively. Then there is a unique coaction $\nu\colon L(X) \to M(L(X) \otimes C^*(G))$ such that (after identifying $L(X) \otimes C^*(G)$ with $L(X \otimes C^*(G))$ as in Remark 1.50)*

$$\nu((\begin{smallmatrix} a & x \\ y & b \end{smallmatrix})) = \begin{pmatrix} \delta(a) & \zeta(x) \\ \widetilde{\zeta(y)} & \epsilon(b) \end{pmatrix}$$

for all $a \in A$, $x, y \in X$, and $b \in B$. Moreover, ν will be nondegenerate if and only if ζ is, and ν will be normal if and only if ϵ is.

Conversely, if $\nu\colon L(X) \to M(L(X) \otimes C^*(G))$ is a coaction such that $\nu(p) = p \otimes 1$ and $\nu(q) = q \otimes 1$, then ν compresses to give coactions δ and ϵ on the corners A and B and a $\delta - \epsilon$ compatible coaction ζ on X; these coactions will be nondegenerate if and only if ν is, and δ and ϵ will be normal if and only if ν is.

PROOF. Let ζ be as in the first part of the lemma. Then it follows from Lemma 1.52 (see also Remark 1.53) that $\nu\colon L(X) \to M(L(X \otimes C^*(G)) = M(L(X) \otimes C^*(G)))$ is a nondegenerate homomorphism. A straightforward computation then shows that it is a coaction. Since

$$(1 \otimes C^*(G))\nu(L(X)) = \begin{pmatrix} (1 \otimes C^*(G))\delta(A) & (1 \otimes C^*(G))\zeta(X) \\ (1 \otimes C^*(G))\widetilde{\zeta(X)} & (1 \otimes C^*(G))\epsilon(B) \end{pmatrix}$$

it follows that ν is nondegenerate (as a coaction) if and only $_\delta\zeta_\epsilon$ is nondegenerate. Similarly, since

$$(\mathrm{id}_{L(X)} \otimes \lambda) \circ \nu = \begin{pmatrix} (\mathrm{id}_A \otimes \lambda) \circ \delta & (\mathrm{id}_X \otimes \lambda) \circ \zeta \\ (\mathrm{id}_{\widetilde{X}} \otimes \lambda) \circ \widetilde{\zeta} & (\mathrm{id}_B \otimes \lambda) \circ \epsilon \end{pmatrix},$$

and since, by the Rieffel correspondence, all corner maps are injective if and only if the lower-right-hand-corner map is injective, it follows that ν is normal if and only if ϵ is normal.

Conversely, if $\nu\colon L(X) \to M(L(X) \otimes C^*(G))$ is a coaction, it follows from Lemma 1.52 that it compresses to a nondegenerate right-Hilbert bimodule homomorphism $_\delta\zeta_\epsilon$, and it is then easy to check that ζ is a $\delta - \epsilon$ compatible right-Hilbert bimodule homomorphism. The final assertion then follows from the first part of the proof. $\square$

2.6. Standard factorization of morphisms

In this section we want to show that every morphism in our categories $\mathcal{C}$, $\mathcal{A}(G)$, $\mathcal{C}(G)$, and $\mathcal{AC}(G)$ can be factored as a product of a nondegenerate standard morphism and a right-partial equivalence (see the definitions below). We start by defining what we mean by a standard morphism. This is easy in the category $\mathcal{C}$: recall from Example 1.3 that a standard right-Hilbert $A - B$ bimodule is one of the form $X = \varphi(A)B$, where $\varphi\colon A \to M(B)$ is a $*$-homomorphism. We shall call the corresponding morphism $[X]\colon A \to B$ in our category $\mathcal{C}$ the *standard morphism associated to* φ. The following observation allows us to extend this notion to the category $\mathcal{A}(G)$.

LEMMA 2.23. *Suppose that (A,α) and (B,β) are actions and assume that $\varphi\colon A \to M(B)$ is a (possibly degenerate) $\alpha - \beta$ equivariant $*$-homomorphism. Let $X = \varphi(A)B \subseteq B$, and for each $s \in G$ let $\beta_X(s)$ denote the restriction of β_s to X. Then β_X is an $\alpha - \beta$ compatible action of G on X.*

We omit the straightforward proof. The coaction analogue of the above result is a bit more complicated.

LEMMA 2.24. *Suppose that (A,δ) and (B,ϵ) are nondegenerate coactions of G, and assume that $\varphi\colon A \to M(B)$ is a (possibly degenerate) $\delta - \epsilon$ equivariant $*$-homomorphism. Then the restriction of ϵ to $X = \varphi(A)B \subseteq B$ determines a $\delta - \epsilon$ compatible coaction ϵ_X on the standard $A - B$ bimodule X. Moreover, if δ and ϵ are nondegenerate, then so is ϵ_X.*

PROOF. Let $\iota\colon X \hookrightarrow B$ denote the inclusion map. Then ι is a right-Hilbert bimodule homomorphism from ${}_A X_B$ to $M({}_B B_B)$. Thus it follows from Proposition 1.45 that the inclusion $\iota \otimes \mathrm{id}_G \colon X \otimes C^*(G) \to M_G(B \otimes C^*(G))$ extends to an inclusion of the G-multiplier bimodules $M_G(X \otimes C^*(G)) \subseteq M_G(B \otimes C^*(G))$. We show that $\epsilon(X) \subseteq M_G(X \otimes C^*(G))$. Using Lemma 1.46, it is enough to show that

$$(1 \otimes C^*(G))\epsilon(X) \cup \epsilon(X)(1 \otimes C^*(G)) \subseteq X \otimes C^*(G).$$

But this follows from

$$\begin{aligned}\epsilon(X)(1 \otimes C^*(G)) &= \epsilon(\varphi(A)B)(1 \otimes C^*(G)) \\ &= (\varphi \otimes \mathrm{id}_G)(\delta(A))(\epsilon(B)(1 \otimes C^*(G))) \\ &\subseteq (\varphi \otimes \mathrm{id}_G)(\delta(A))(B \otimes C^*(G)) \\ &= (\varphi \otimes \mathrm{id}_G)(\delta(A))((1 \otimes C^*(G))(B \otimes C^*(G))) \\ &= (\varphi \otimes \mathrm{id}_G)(\delta(A)(1 \otimes C^*(G)))(B \otimes C^*(G)) \\ &\subseteq \varphi \otimes \mathrm{id}_G \left(A \otimes C^*(G)\right)(B \otimes C^*(G)) \\ &= X \otimes C^*(G),\end{aligned}$$

and a similar computation which shows that $(1 \otimes C^*(G))\epsilon(X) \subseteq X \otimes C^*(G)$. Note that if δ and ϵ are both nondegenerate, then taking closures of all terms in the above computation allows us to replace all inclusions by equal signs.

It now makes sense to define the restriction $\epsilon_X \colon X \to M_G(X \otimes C^*(G))$ of ϵ to X. Then ϵ_X is a right-Hilbert bimodule homomorphism with coefficient maps δ and ϵ. To see that it is nondegenerate (as a bimodule homomorphism), we compute as above

$$\begin{aligned}\epsilon_X(X)(B \otimes C^*(G)) &= (\varphi \otimes \mathrm{id}_G) \circ \delta(A)(B \otimes C^*(G)) \\ &\supseteq (\varphi \otimes \mathrm{id}_G)(\delta(A)(A \otimes C^*(G)))(B \otimes C^*(G)) \\ &= (\varphi \otimes \mathrm{id}_G)(A \otimes C^*(G))(B \otimes C^*(G)) \\ &= X \otimes C^*(G).\end{aligned}$$

The coaction identity for ϵ_X follows directly from the coaction identity for ϵ. If δ and ϵ are nondegenerate, then we already saw above that

$$\overline{(1 \otimes C^*(G))\epsilon_X(X)} = \overline{(1 \otimes C^*(G))\epsilon(X)} = X \otimes C^*(G),$$

so ϵ_X is a nondegenerate coaction. $\square$

DEFINITION 2.25. Suppose that $\varphi\colon A\to M(B)$ is a $*$-homomorphism and let $X=\varphi(A)B$ be the standard right-Hilbert $A-B$ bimodule associated to φ.
  (i) If $\varphi\colon A\to M(B)$ is $\alpha-\beta$ equivariant for the actions α and β of G on A and B, respectively, and if β_X is the restriction of β to X as in Lemma 2.23, then we say that (X,β_X) is the *standard action* associated to φ.
  (ii) If $\varphi\colon A\to M(B)$ is $\delta-\epsilon$ equivariant for the nondegenerate coactions δ and ϵ of G on A and B, respectively, and if $\epsilon_X\colon X\to M_G(X\otimes C^*(G))$ is as in Lemma 2.24, then (X,ϵ_X) is called the *standard coaction* associated to φ.
  (iii) If $\varphi\colon A\to M(B)$ is nondegenerate, then we say that the above-defined standard morphisms, actions, or coactions, are *nondegenerate*. (Note that in this case we have $X=B$.)

We now introduce the notion of partial equivalences.

DEFINITION 2.26. Suppose that ${}_A X_B$ is right-partial imprimitivity bimodule. Then $[X]\colon A\to B$ is called a *right-partial equivalence* between A and B in the category $\mathcal{C}$. Similarly, morphisms $[(X,\gamma)]$, $[(X,\zeta)]$ and $[(X,\gamma,\zeta)]$ in the categories $\mathcal{A}(G), \mathcal{C}(G)$ and $\mathcal{AC}(G)$, respectively, are called *right-partial equivalences*, if the underlying module X is a right-partial imprimitivity bimodule.

Suppose now that ${}_A X_B$ is any right-Hilbert $A-B$ bimodule. If we put $K=\mathcal{K}(X)$ and let $\kappa\colon A\to M(K)$ be the associated nondegenerate homomorphism, we obtain a nondegenerate standard bimodule ${}_A K_K$, and ${}_K X_B$ becomes a right-partial $K-B$ imprimitivity bimodule. It is trivial to check that the map

$$k\otimes x\mapsto k\cdot x$$

extends to a right-Hilbert $A-B$ bimodule isomorphism

$$_A(K\otimes_K X)_B \cong {}_A X_B,$$

so we see that any morphism in the category $\mathcal{C}$ can be factored as the composition of a nondegenerate standard morphism, and a right-partial equivalence. It is easy to extend this observation to morphisms in the category $\mathcal{A}(G)$:

PROPOSITION 2.27. Suppose that ${}_{(A,\alpha)}(X,\gamma)_{(B,\beta)}$ is an action, and let $K=\mathcal{K}(X)$. Then:
  (i) There exists a unique action μ on K such that γ is $\mu-\beta$ compatible.
  (ii) The canonical nondegenerate homomorphism $\kappa\colon A\to M(K)$ is $\alpha-\mu$ equivariant; thus $({}_A K_K,\mu)$ is a nondegenerate standard action with left coefficient action α.
  (iii) The map $k\otimes x\mapsto k\cdot x$ implements an equivariant isomorphism between $({}_A(K\otimes_K X)_B,\mu\otimes_K\gamma)$ and $({}_A X_B,\gamma)$.

In particular, every morphism in $\mathcal{A}(G)$ can be factored as the composition of a nondegenerate standard morphism, and a right partial equivalence.

PROOF. Since by definition $\gamma_s(x)\cdot\langle\gamma_s(y),\gamma_s(z)\rangle_B=\gamma_s(x\cdot\langle y,z\rangle_B)$ for all $s\in G$ and $x,y,z\in X$ it follows from [11, page 292] that $\mu_s({}_K\langle x,y\rangle)={}_K\langle\gamma_s(x),\gamma_s(y)\rangle$ determines a unique strongly continuous action μ of G on K such that γ is $\mu-\beta$ compatible. This gives (i).

For each $a\in A$, $s\in G$, and $x\in X$ we have

$$\varphi(\alpha_s(a))\gamma_s(x)=\alpha_s(a)\cdot\gamma_s(x)=\gamma_s(a\cdot x)=\gamma_s(\varphi(a)x)=\mu_s(\varphi(a))\gamma_s(x).$$

Thus $\varphi\colon A \to M(K)$ is $\alpha - \mu$ equivariant and so $({}_A K_K, \mu)$ is a standard action.

For (iii) just check that $k \otimes x \mapsto k \cdot x$ is $(\mu \otimes_K \gamma) - \gamma$ equivariant, which is trivial. $\square$

We have to work a bit more to get the analogous result for morphisms in $\mathcal{C}(G)$. Recall from Remark 1.19 that if ${}_\varphi\Phi_\psi\colon {}_A X_B \to M({}_C Y_D)$ is any nondegenerate right-Hilbert bimodule homomorphism then there exists a map $\mu\colon \mathcal{K}_B(X) \to M(\mathcal{K}_D(Y))$ such that ${}_\mu\Phi_\psi$ is a nondegenerate partial imprimitivity bimodule homomorphism. The following easy lemma is quite fundamental.

LEMMA 2.28. *With the above notation*:
(i) *If $\kappa_A\colon A \to M(\mathcal{K}_B(X))$ and $\kappa_C\colon C \to M(\mathcal{K}_D(Y))$ are the canonical nondegenerate homomorphisms associated with the left module actions on X and Y, then the following diagram commutes*:

$$\begin{array}{ccc} A & \xrightarrow{\kappa_A} & M(\mathcal{K}_B(X)) \\ {\scriptstyle \varphi}\downarrow & & \downarrow{\scriptstyle \mu} \\ M(C) & \xrightarrow{\kappa_C} & M(\mathcal{K}_D(Y)). \end{array}$$

(ii) $\ker \mu$ *is the ideal induced from* $\ker \psi$ *via* X.

PROOF. For (i) let $a \in A$, $x \in X$, and $d \in D$, and compute
$$\mu(\kappa_A(a)) \cdot \Phi(x)d = \Phi(\kappa_A(a)x)d = \Phi(a \cdot x)d = \varphi(a) \cdot \Phi(x)d = \kappa_C(\varphi(a))\Phi(x)d.$$
Since $\overline{\Phi(X) \cdot D} = Y$ by the nondegeneracy of Φ, it follows that $\mu(\kappa_A(a))$ and $\kappa_C(\varphi(a))$ are identical in $\mathcal{L}(Y)$. The assertion (ii) follows from Lemma 1.20. $\square$

LEMMA 2.29. *Let ${}_A X_B$ and ${}_C Y_D$ be right-Hilbert bimodules. The equation*
$$\Phi(k \otimes l)(x \otimes y) = k(x) \otimes l(y)$$
determines a C^*-algebra isomorphism Φ of $\mathcal{K}_B(X) \otimes \mathcal{K}_D(Y)$ onto $\mathcal{K}_{B \otimes D}(X \otimes Y)$ such that
$$\Phi \circ (\kappa_A \otimes \kappa_C) = \kappa_{A \otimes C},$$
where the maps $\kappa_A\colon A \to M(\mathcal{K}_B(X))$, $\kappa_C\colon C \to M(\mathcal{K}_D(Y))$, and $\kappa_{A \otimes C}\colon A \otimes C \to M(\mathcal{K}_{B \otimes D}(X \otimes Y))$ are the nondegenerate homomorphisms associated with X, Y, and $X \otimes Y$.

PROOF. The first part of the lemma is [**33**, pages 35–37]. For the second part, simply compute:
$$\Phi \circ (\kappa_A \otimes \kappa_C)(a \otimes c)(x \otimes y) = \Phi(\kappa_A(a) \otimes \kappa_C(c))(x \otimes y)$$
$$= \kappa_A(a)x \otimes \kappa_C(c)y = a \cdot x \otimes c \cdot y = (a \otimes c) \cdot (x \otimes y) = \kappa_{A \otimes C}(a \otimes c)(x \otimes y).$$
$\square$

PROPOSITION 2.30. *Suppose that ${}_{(A,\delta)}(X,\zeta)_{(B,\epsilon)}$ is a coaction, and let $K = \mathcal{K}_B(X)$. Then*:
(i) *There exists a unique coaction μ on K such that ζ is $\mu - \epsilon$ compatible.*
(ii) *The canonical nondegenerate homomorphism $\kappa_A\colon A \to M(K)$ is $\delta - \mu$ equivariant; thus $({}_A K_K, \mu)$ is a nondegenerate standard coaction with left coefficient coaction δ.*

(iii) *The map $k \otimes x \mapsto k \cdot x$ induces an equivariant isomorphism between $({}_A(K \otimes_K X)_B, \mu \natural_K \zeta)$ and $({}_A X_B, \zeta)$.*

Moreover, μ is nondegenerate if ζ is, and μ is normal if ϵ is.

PROOF. Since $\zeta\colon X \to M(X \otimes C^*(G))$ is a nondegenerate bimodule homomorphism, by Remark 1.19 there exists a map $\nu\colon K \to M(\mathcal{K}_{B \otimes C^*(G)}(X \otimes C^*(G)))$ such that ${}_\nu \zeta_\epsilon$ is a nondegenerate right-Hilbert bimodule homomorphism. We compose ν with the isomorphism $\Phi^{-1}\colon M(\mathcal{K}_{B \otimes C^*(G)}(X \otimes C^*(G))) \to M(K \otimes C^*(G))$ provided by Lemma 2.29 to get a nondegenerate homomorphism $\mu\colon K \to M(K \otimes C^*(G))$ such that ${}_\mu \zeta_\epsilon$ is a nondegenerate right-Hilbert bimodule homomorphism. Since ϵ is injective and since, by Lemma 2.29, $\ker \mu = \ker \nu$ is the ideal induced from $\ker \epsilon$ via X, it follows that μ is injective, too. Thus to show that μ is a coaction on K it only remains to show that

(a) $(1 \otimes z)\mu(k) \in K \otimes C^*(G)$ for all $z \in C^*(G)$, $k \in K$, and
(b) $(\mu \otimes \mathrm{id}_G) \circ \mu = (\mathrm{id}_K \otimes \delta_G) \circ \mu$.

But (a) follows from the simple calculation

$$(1 \otimes z)\mu({}_K\langle x, y\rangle) = (1 \otimes z){}_{M(K \otimes C^*(G))}\langle \zeta(x), \zeta(y)\rangle$$
$$= {}_{M(K \otimes C^*(G))}\langle (1 \otimes z) \cdot \zeta(x), \zeta(y)\rangle,$$

which is in $K \otimes C^*(G)$ since $(1 \otimes z) \cdot \zeta(x) \in X \otimes C^*(G)$ and

$${}_{M(K \otimes C^*(G))}\langle X \otimes C^*(G), M(X \otimes C^*(G))\rangle \subseteq K \otimes C^*(G).$$

In order to prove (b) we just compute

$$(\mu \otimes \mathrm{id}_G) \circ \mu({}_K\langle x, y\rangle) = (\mu \otimes \mathrm{id}_G)\big({}_{M(K \otimes C^*(G))}\langle \zeta(x), \zeta(y)\rangle\big)$$
$$= {}_{M(K \otimes C^*(G) \otimes C^*(G))}\langle (\zeta \otimes \mathrm{id}_G) \circ \zeta(x), (\zeta \otimes \mathrm{id}_G) \circ \zeta(y)\rangle$$
$$= {}_{M(K \otimes C^*(G) \otimes C^*(G))}\langle (\mathrm{id}_X \otimes \delta_G) \circ \zeta(x), (\mathrm{id}_X \otimes \delta_G) \circ \zeta(y)\rangle$$
$$= (\mathrm{id}_K \otimes \delta_G) \circ \mu({}_K\langle x, y\rangle)$$

for all $x, y \in X$. Then (b) follows from the density of ${}_K\langle X, X\rangle$ in K. Since the left coefficient map on $\mathcal{K}(X)$ of any right-Hilbert bimodule homomorphism $\Phi\colon {}_{\mathcal{K}(X)} X_B \to M({}_C Y_D)$ is uniquely determined by Φ (since it is compatible with the $\mathcal{K}(X)$-valued inner product on X), this proves (i).

It follows from Lemma 2.28 that

$$\mu \circ \kappa_A = \Phi^{-1} \circ \kappa_{A \otimes C^*(G)} \circ \delta,$$

where $\kappa_{A \otimes C^*(G)}\colon A \otimes C^*(G) \to M(\mathcal{K}_{B \otimes C^*(G)}(X \otimes C^*(G)))$ is the canonical nondegenerate homomorphism. Thus the second part of Lemma 2.29 gives

$$\mu \circ \kappa_A = (\kappa_A \otimes \mathrm{id}) \circ \delta,$$

which establishes (ii).

For (iii), we need to show that

$$\zeta \circ \Phi = (\Phi \otimes \mathrm{id}) \circ (\mu \natural_K \zeta),$$

where $\Phi\colon K \otimes_K X \to X$ is the canonical isomorphism. But this was verified in the proof of Theorem 2.15, where we showed that there are identities in $\mathcal{C}(G)$.

Now assume that ζ (and therefore also δ and ϵ) is a nondegenerate coaction. Then

$$\overline{(1 \otimes C^*(G))\mu(K)} = \overline{(1 \otimes C^*(G))_{M(K \otimes C^*(G))}\langle \zeta(X), \zeta(X)\rangle}$$
$$= {}_{K \otimes C^*(G)}\overline{\langle (1 \otimes C^*(G))\zeta(X), \zeta(X)\rangle}$$
$$= {}_{K \otimes C^*(G)}\overline{\langle X \otimes C^*(G), \zeta(X)\rangle}$$
$$= {}_{K \otimes C^*(G)}\overline{\langle (X \otimes C^*(G))(1 \otimes C^*(G)), \zeta(X)\rangle}$$
$$= {}_{K \otimes C^*(G)}\overline{\langle X \otimes C^*(G), \zeta(X)(1 \otimes C^*(G))\rangle}$$
$$= {}_{K \otimes C^*(G)}\overline{\langle X \otimes C^*(G), X \otimes C^*(G)\rangle}$$
$$= \mathcal{K}(X) \otimes C^*(G).$$

Finally, if ϵ is normal, the normality of μ follows from the fact that

$$(\mathrm{id}_X \otimes \lambda) \circ \zeta \colon {}_K X_B \to M\big({}_{K \otimes \mathcal{K}(L^2(G))}(X \otimes \mathcal{K}(L^2(G)))_{B \otimes \mathcal{K}(L^2(G))}\big)$$

is a right-partial imprimitivity bimodule homomorphism. Thus injectivity of the right coefficient map $(\mathrm{id}_B \otimes \lambda) \circ \epsilon$ implies injectivity of the left coefficient map $(\mathrm{id}_K \otimes \lambda) \circ \mu$. □

REMARK 2.31. Combining Proposition 2.27 with Proposition 2.30 gives a similar factorization result for morphisms in $\mathcal{AC}(G)$. We omit the obvious details.

In some situations, it becomes necessary to further factor the right-partial equivalence part of the morphisms in $\mathcal{C}$, $\mathcal{A}(G)$, $\mathcal{C}(G)$, or $\mathcal{AC}(G)$, into an equivalence and a standard morphism coming from an inclusion of an ideal. To be more precise, if ${}_A X_B$ is a right-Hilbert $A - B$ bimodule, then X can be regarded as an ${}_A X_{B_X}$ bimodule in the canonical way, where $B_X = \overline{\langle X, X\rangle}_B$. It is then clear that the map of ${}_A(X \otimes_{B_X} B_X)_B$ into ${}_A X_B$ determined by $x \otimes b \mapsto x \cdot b$ is a right-Hilbert bimodule isomorphism. If ${}_A X_B$ is a right-partial $A - B$ imprimitivity bimodule, then ${}_A X_{B_X}$ is an imprimitivity bimodule, and the above isomorphism gives a factorization

$$[{}_A X_B] = [{}_{B_X} B_B] \circ [{}_A X_{B_X}]$$

of $[{}_A X_B]$ as the product of the equivalence $[{}_A X_{B_X}]$ and the (probably degenerate) standard morphism $[{}_{B_X} B_B]$ associated to the inclusion of the ideal B_X into B.

More generally, if $({}_A X_B, \gamma)$ is a right-Hilbert bimodule action, then the above isomorphism is easily seen to be $\gamma - \gamma \otimes_{B_X} \beta_X$ equivariant, where β_X denotes the restriction of β to B_X. Thus we get a similar factorization for right-partial imprimitivity bimodule actions. As usual, the case of coactions is a bit more complicated, so we do it in a lemma. Interestingly, it seems to be necessary to assume nondegeneracy of all coactions to do this step.

LEMMA 2.32. *Assume that* ${}_{(A,\delta)}(X, \zeta)_{(B,\epsilon)}$ *is a nondegenerate coaction and let* $B_X = \overline{\langle X, X\rangle}_B$. *Then:*

(i) *There exists a unique nondegenerate coaction* $\epsilon_X \colon B_X \to M(B_X \otimes C^*(G))$ *such that* $({}_A X_{B_X}, \zeta)$ *becomes a nondegenerate* $\delta - \epsilon_X$ *compatible coaction on* ${}_A X_{B_X}$.

(ii) *The inclusion* $B_X \hookrightarrow B$ *is* $\epsilon_X - \epsilon$ *equivariant, and*

$$_{(A,\delta)}(X, \zeta)_{(B,\epsilon)} \cong {}_{(A,\delta)}(X \otimes_{B_X} B_X, \zeta \sharp_{B_X} \epsilon_X)_{(B,\epsilon)}$$

via the canonical isomorphism ${}_A(X \otimes_{B_X} B_X)_B \cong {}_A X_B$.

(iii) *If ϵ is normal then so is ϵ_X.*

PROOF. Of course, we want to define ϵ_X as the restriction of ϵ to B_X. For this to make sense, we have to show, similarly to the proof of Lemma 2.24, that $\epsilon(B_X) \subseteq M_G(B_X \otimes C^*(G))$, where we view $M_G(B_X \otimes C^*(G))$ as an ideal of $M_G(B \otimes C^*(G))$ via the canonical inclusion (see Corollary 1.47). But this follows from

$$\epsilon(B_X) = \epsilon(\langle X, X \rangle_B) \subseteq \langle M_G(X \otimes C^*(G)), M_G(X \otimes C^*(G))\rangle_{M_G(B \otimes C^*(G))},$$

and the fact that the range of the $M_G(B \otimes C^*(G))$-valued inner product on $M_G(X \otimes C^*(G))$ lies in $M_G(B_X \otimes C^*(G))$ (see Lemma 1.44). The coaction identity of ϵ_X follows directly from the coaction identity for ϵ. Thus, to see that ϵ_X is a nondegenerate coaction, it only remains to check that $\overline{\epsilon_X(B_X)(1 \otimes C^*(G))} = B_X \otimes C^*(G)$. For this we compute

$$\begin{aligned}\overline{\epsilon_X(B_X)(1 \otimes C^*(G))} &= \overline{\langle \zeta(X), \zeta(X)\rangle_{M(B \otimes C^*(G))}(1 \otimes C^*(G))}\\ &= \overline{\langle \zeta(X), \zeta(X)(1 \otimes C^*(G))\rangle_{B \otimes C^*(G)}}\\ &= \overline{\langle \zeta(X), X \otimes C^*(G)\rangle_{B \otimes C^*(G)}}\\ &= \overline{\langle (1 \otimes C^*(G))\zeta(X), X \otimes C^*(G)\rangle_{B \otimes C^*(G)}}\\ &= \overline{\langle X \otimes C^*(G), X \otimes C^*(G)\rangle_{B \otimes C^*(G)}}\\ &= B_X \otimes C^*(G).\end{aligned}$$

By Lemma 1.44 we can canonically identify $M_G({}_A X_B \otimes C^*(G))$ with $M_G({}_A X_{B_X} \otimes C^*(G))$, which now shows that ζ can be viewed as a $\delta - \epsilon_X$ compatible coaction on ${}_A X_{B_X}$. This proves (i).

It is clear from the construction of ϵ_X that the inclusion $B_X \hookrightarrow B$ is $\epsilon_X - \epsilon$ equivariant. For (ii), it only remains to show that the canonical isomorphism $X \otimes_{B_X} B_X \cong X$ is $\zeta \sharp_{B_X} \epsilon_X - \zeta$ equivariant. But this follows from the last paragraph in the proof of Theorem 2.15.

Finally, using the identifications made in Corollary 1.47, $(\mathrm{id}_{B_X} \otimes \lambda) \circ \epsilon_X$ is precisely the restriction to B_X of $(\mathrm{id}_B \otimes \lambda) \circ \epsilon$. Thus, if ϵ is normal (see Remark A.51), then ϵ_X is too.

Finally, if ϵ is normal, and if $\iota \colon B_X \hookrightarrow B$ is the inclusion map, it follows from Proposition A.6 that $\iota \otimes \mathrm{id}_{\mathcal{K}}$, with $\mathcal{K} = \mathcal{K}(L^2(G))$, extends to an injective inclusion of $M_{\mathcal{K}}(B_X \otimes \mathcal{K})$ into $M_{\mathcal{K}}(B \otimes \mathcal{K})$, and then we get

$$(\iota \otimes \mathrm{id}_{\mathcal{K}}) \circ (\mathrm{id}_{B_X} \otimes \lambda) \circ \epsilon_X = (\mathrm{id}_B \otimes \lambda) \circ \epsilon \circ \iota,$$

where $\lambda \colon C^*(G) \to \mathcal{B}(L^2(G)) = M(\mathcal{K})$ is the regular representation. If ϵ is normal, the right side of this equality is injective, which then implies that $(\mathrm{id}_{B_X} \otimes \lambda) \circ \epsilon_X$ is injective, too. But (see Proposition A.37 and Remark A.51) this implies that ϵ_X is normal. □

2.7. Morphisms and induced representations

In this section we give an outline of the relationship between our categories $\mathcal{C}$, $\mathcal{A}(G)$, $\mathcal{C}(G)$, and $\mathcal{AC}(G)$, and (covariant) representations on Hilbert space. As always in this paper, we assume that the inner product on a Hilbert space $\mathcal{H}$ is linear in the second and conjugate linear in the first variable. Thus, if A is a C^*-algebra, it follows from this convention that a nondegenerate representation $\pi \colon A \to \mathcal{B}(\mathcal{H})$ gives $\mathcal{H}$ the structure of a right-Hilbert $A - \mathbb{C}$ bimodule. Moreover, two such modules $\mathcal{H}_1$ and $\mathcal{H}_2$ are isomorphic as right-Hilbert $A - \mathbb{C}$ bimodules if and only if there exists a unitary $U\colon \mathcal{H}_1 \to \mathcal{H}_2$ which intertwines the left A-actions, *i.e.*, if and only if the corresponding nondegenerate representations of A on $\mathcal{H}_1$ and $\mathcal{H}_2$ are unitarily equivalent. It follows that the unitary equivalence classes of nondegenerate Hilbert space representations of A are precisely the morphisms from A to $\mathbb{C}$ in the category $\mathcal{C}$. Note also that $\mathcal{H} = \{0\}$ is the only right-Hilbert $A - \mathbb{C}$ bimodule which is not full.

DEFINITION 2.33. Let A be a C^*-algebra. We denote by $\operatorname{Rep}(A)$ the class $\operatorname{mor}(A, \mathbb{C})$ of morphisms from A to $\mathbb{C}$ in the category $\mathcal{C}$. Similarly, if (A, α) is an action of a locally compact group G on A, then we put

$$\operatorname{Rep}(A, \alpha) = \operatorname{mor}\big((A, \alpha), (\mathbb{C}, \operatorname{id})\big)$$

in $\mathcal{A}(G)$, and if (A, δ) is a coaction of G, we put

$$\operatorname{Rep}(A, \delta) = \operatorname{mor}\big((A, \delta), (\mathbb{C}, \operatorname{id} \otimes 1)\big)$$

in $\mathcal{C}(G)$. (Here $\operatorname{id} \otimes 1\colon \mathbb{C} \to M(\mathbb{C} \otimes C^*(G))$ denotes the trivial coaction of G on $\mathbb{C}$.) We similarly define $\operatorname{Rep}(A, \alpha, \delta)$ for our mixed category $\mathcal{AC}(G)$.

We already saw above that $\operatorname{Rep}(A)$ coincides with the class of all unitary equivalence classes of nondegenerate Hilbert-space representations of the C^*-algebra A (including the zero-representation $\mathcal{H} = \{0\}$). We shall later check that $\operatorname{Rep}(A, \alpha)$ and $\operatorname{Rep}(A, \delta)$ are also precisely the unitary equivalence classes of nondegenerate covariant representations of (A, α) and (A, δ), respectively (also including the zero-representation).

But before we do this, we note that the definition of $\operatorname{Rep}(A)$ as $\operatorname{mor}(A, \mathbb{C})$ in the category $\mathcal{C}$ directly gives us a procedure to use any right-Hilbert $B - A$ bimodule ${}_B X_A$ to induce representations from A to B: we simply compose the morphisms in $\operatorname{Rep}(A)$ with the morphism $[X]$ to obtain the map

$$\operatorname{Rep}(A) \to \operatorname{Rep}(B)\colon [\mathcal{H}] \mapsto [\mathcal{H}] \circ [X] = [X \otimes_A \mathcal{H}],$$

which is a bijection if $[X]$ is an equivalence in $\mathcal{C}$, *i.e.*, if X is a $B - A$ imprimitivity bimodule. Exactly the same arguments work in the equivariant cases: if (X, γ) is a $\beta - \alpha$ compatible right-Hilbert bimodule action, we get an induction map

$$\operatorname{Rep}(A, \alpha) \to \operatorname{Rep}(B, \beta)\colon [(\mathcal{H}, U)] \mapsto [(\mathcal{H}, U)] \circ [(X, \gamma)],$$

where composition is in $\mathcal{A}(G)$, and if (X, ζ) is an $\epsilon - \delta$ compatible right-Hilbert bimodule coaction we get a map

$$\operatorname{Rep}(A, \delta) \to \operatorname{Rep}(B, \epsilon)\colon [(\mathcal{H}, \eta)] \mapsto [(\mathcal{H}, \eta)] \circ [(X, \zeta)],$$

where composition is in $\mathcal{C}(G)$; the maps are bijections if the respective morphisms are equivalences in the appropriate categories. Of course, similar observations can be made for $\mathcal{AC}(G)$.

We now turn to identifying the spaces $\operatorname{Rep}(A, \alpha)$ and $\operatorname{Rep}(A, \delta)$ with the equivalence classes of the covariant representations of (A, α) and (A, δ), respectively. We start with the easy case of an action:

LEMMA 2.34. *Assume that (π, U) is a covariant representation of (A, α) on a Hilbert space $\mathcal{H}_\pi$. Then $(\mathcal{H}_\pi, U)$ is an $\alpha - \operatorname{id}$ compatible right-Hilbert bimodule action, and the assignment*

$$[(\pi, U)] \to [(\mathcal{H}_\pi, U)]$$

2.7. MORPHISMS AND INDUCED REPRESENTATIONS

is a one-to-one correspondence between the unitary equivalence classes of nondegenerate covariant representations of (A, α) (including the zero-representation) and $\operatorname{Rep}(A, \alpha)$.

PROOF. Note that in our definition of covariant representations (π, U), we already assume that $\pi \colon A \to \mathcal{B}(\mathcal{H}_\pi)$ is nondegenerate (see Section A.2 of Appendix A). Thus it follows from the definition of covariance and the definition of right-Hilbert bimodule actions (see Definition 2.5) that $(\pi, U) \mapsto (\mathcal{H}_\pi, U)$ is actually a one-to-one correspondence between the nondegenerate covariant Hilbert-space representations of (A, α) and the $\alpha - \mathrm{id}$ compatible right-Hilbert $A - \mathbb{C}$ bimodule actions. It is then straightforward to check that two covariant representations are unitarily equivalent if and only if the corresponding bimodule actions are isomorphic. $\square$

As usual, the coaction case requires a bit more work. Recall from Definition A.32 that a covariant representation of a coaction (A, δ) on a Hilbert space $\mathcal{H}_\pi$ consists of a pair (π, μ) such that $\pi \colon A \to \mathcal{B}(\mathcal{H}_\pi)$ and $\mu \colon C_0(G) \to \mathcal{B}(\mathcal{H}_\pi)$ are nondegenerate representations satisfying

$$(\pi \otimes \mathrm{id}_G) \circ \delta(a) = (\mu \otimes \mathrm{id}_G)(w_G)(\pi(a) \otimes 1)(\mu \otimes \mathrm{id}_G)(w_G)^*,$$

where $w_G \in C^b(G, M^\beta(C^*(G))) = M(C_0(G) \otimes C^*(G))$ denotes the canonical embedding $s \mapsto u(s)$ of G into $UM(C^*(G))$.

LEMMA 2.35. *Assume that (π, μ) is a nondegenerate covariant representation of the coaction (A, δ). Then the map $\zeta_\mu \colon \mathcal{H}_\pi \to M(\mathcal{H}_\pi \otimes C^*(G))$ defined by*

$$\zeta_\mu(x) = (\mu \otimes \mathrm{id}_G)(w_G)(x \otimes 1)$$

is a $\delta - \mathrm{id} \otimes 1$ compatible coaction on $\mathcal{H}_\pi$.

PROOF. We first check that ζ_μ is indeed a right-Hilbert bimodule coaction in the sense of Definition 2.10. To see that $\zeta_\mu(\mathcal{H}_\pi) \subseteq M(\mathcal{H}_\pi \otimes C^*(G))$, note that $(\mu \otimes \mathrm{id}_G)(w_G) \in M(\mathcal{K}(\mathcal{H}_\pi) \otimes C^*(G)) = \mathcal{L}_{\mathbb{C} \otimes C^*(G)}(\mathcal{H}_\pi \otimes C^*(G))$. Moreover, if $x \in \mathcal{H}_\pi$, then $x \otimes 1 \in M(\mathcal{H}_\pi \otimes C^*(G)) = \mathcal{L}_{\mathbb{C} \otimes C^*(G)}(\mathbb{C} \otimes C^*(G), \mathcal{H}_\pi \otimes C^*(G))$. Thus it is clear that $\zeta_\mu(x) = (\mu \otimes \mathrm{id}_G)(w_G)(x \otimes 1) \in \mathcal{L}_{\mathbb{C} \otimes C^*(G)}(\mathbb{C} \otimes C^*(G), \mathcal{H}_\pi \otimes C^*(G)) = M(\mathcal{H}_\pi \otimes C^*(G))$.

Next we check that $\zeta_\mu \colon \mathcal{H}_\pi \to M(\mathcal{H}_\pi \otimes C^*(G))$ is a bimodule homomorphism with coefficient maps δ and $\mathrm{id} \otimes 1$. Indeed, if we write $a \cdot x$ for $\pi(a)x$, the covariance condition for (δ, μ) implies that

$$\zeta_\mu(a \cdot x) = (\mu \otimes \mathrm{id}_G)(w_G)(a \cdot x \otimes 1) = (\mu \otimes \mathrm{id}_G)(w_G)(\pi(a) \otimes 1)(x \otimes 1)$$
$$= (\pi \otimes \mathrm{id}_G) \circ \delta(a)\big((\mu \otimes \mathrm{id}_G)(w_G)(x \otimes 1)\big) = \delta(a) \cdot \zeta_\mu(x)$$

for all $a \in A$ and $x \in \mathcal{H}_\pi$. On the other side, viewing bimodule elements as adjointable operators, we have

$$\langle \zeta_\mu(x), \zeta_\mu(y) \rangle_{\mathbb{C} \otimes C^*(G)} = (x \otimes 1)^* (\mu \otimes \mathrm{id}_G)(w_G)^* (\mu \otimes \mathrm{id}_G)(w_G)(y \otimes 1)$$
$$= x^*y \otimes 1 = \langle x, y \rangle_\mathbb{C} \otimes 1 = (\mathrm{id} \otimes 1)(\langle x, y \rangle_\mathbb{C})$$

for all $x, y \in \mathcal{H}_\pi$.

Since $(\mu \otimes \mathrm{id}_G)(w_G)$ is actually a unitary operator on $\mathcal{H}_\pi \otimes C^*(G)$, we get

$$\overline{\zeta_\mu(\mathcal{H}_\pi)(\mathbb{C} \otimes C^*(G))} = (\mu \otimes \mathrm{id}_G)(w_G)\overline{((H_\pi \otimes 1)(1 \otimes C^*(G)))}$$
$$= (\mu \otimes \mathrm{id}_G)(w_G)(\mathcal{H}_\pi \otimes C^*(G))$$
$$= \mathcal{H}_\pi \otimes C^*(G),$$

which proves that ζ_μ is nondegenerate.

We now check that $(1 \otimes C^*(G))\zeta_\mu(\mathcal{H}_\pi) \subseteq \mathcal{H}_\pi \otimes C^*(G)$. If $z \in C_c(G) \subseteq C^*(G)$, then $(1 \otimes z)(w_G) \in M(C_0(G) \otimes C^*(G))$ is given by the continuous function $s \mapsto zu(s)$ of G into $C^*(G)$. Since $u(s)$ is the canonical image of s in $UM(C^*(G))$, it follows from the formulas given in Remark A.8 (where $u(s)$ is denoted $i_G(s)$) that $zu(s) \in C_c(G) \subseteq C^*(G)$ and $(zu(s))(t) = \Delta(s^{-1})z(ts^{-1})$ for $t \in G$. It follows that the function $s \mapsto zu(s)$ lies in $C_c(G, C^*(G))$ for all $z \in C_c(G) \subseteq C^*(G)$; hence we see that $(1 \otimes C^*(G))w_G \subseteq C_0(G) \otimes C^*(G)$. From this we obtain

$$(1 \otimes C^*(G))\zeta_\mu(\mathcal{H}_\pi) = (1 \otimes C^*(G))(\mu \otimes \mathrm{id}_G)(w_G)(\mathcal{H}_\pi \otimes 1)$$
$$= (\mu \otimes \mathrm{id}_G)\big((1 \otimes C^*(G))w_G\big)(\mathcal{H}_\pi \otimes 1)$$
$$\subseteq (\mu \otimes \mathrm{id}_G)(C_0(G) \otimes C^*(G))(\mathcal{H}_\pi \otimes 1)$$
$$= (\mu \otimes \mathrm{id}_G)(C_0(G) \otimes C^*(G))(1 \otimes C^*(G))(\mathcal{H}_\pi \otimes 1)$$
$$\subseteq (\mu \otimes \mathrm{id}_G)(C_0(G) \otimes C^*(G))(\mathcal{H}_\pi \otimes C^*(G))$$
$$\subseteq \mathcal{H}_\pi \otimes C^*(G).$$

So it only remains to check the coaction identity for ζ_μ. For this recall from Proposition A.34 that if $w = (\mu \otimes \mathrm{id}_G)(w_G)$, then we have $w_{12}w_{13} = (\mu \otimes \delta_G)(w_G)$, with notation as in that proposition. Using this equation we compute

$$(\zeta_\mu \otimes \mathrm{id}_G) \circ \zeta_\mu(x) = (\zeta_\mu \otimes \mathrm{id}_G)\big(w(x \otimes 1)\big)$$
$$= w_{12}w_{13}(x \otimes 1 \otimes 1)$$
$$= (\mu \otimes \delta_G)(w_G)(x \otimes 1 \otimes 1)$$
$$= (\mathrm{id}_{\mathcal{H}_\pi} \otimes \delta_G)\big((\mu \otimes \mathrm{id}_G)(w_G)(x \otimes 1)\big)$$
$$= (\mathrm{id}_{\mathcal{H}_\pi} \otimes \delta_G) \circ \zeta_\mu(x)$$

for all $x \in X$. This completes the proof of the lemma. $\square$

Thus, every covariant representation (π, μ) of (A, δ) determines a $\delta - \mathrm{id} \otimes 1$ equivariant coaction on $\mathcal{H}_\pi$. Conversely, we get:

LEMMA 2.36. *Suppose that $(\mathcal{H}, \zeta)$ is a $\delta - \mathrm{id} \otimes 1$ right-Hilbert $A - \mathbb{C}$ bimodule coaction. Let $\pi \colon A \to \mathcal{B}(\mathcal{H})$ be the nondegenerate representation coming from the action of A on $\mathcal{H}$. Then there exists a unique nondegenerate representation $\mu \colon C_0(G) \to \mathcal{B}(\mathcal{H})$ such that (π, μ) is a covariant representation of (A, δ) on $\mathcal{H}$ and such that $\zeta = \zeta_\mu$ with ζ_μ as in Lemma 2.35.*

PROOF. We may assume without loss of generality that $\mathcal{H} \neq \{0\}$. We first show that there exists a unitary $w \in U(\mathcal{H} \otimes C^*(G))$ such that $\zeta(x) = w(x \otimes 1)$ for all $x \in \mathcal{H}$. For this we first define a map $w \colon \mathcal{H} \odot C^*(G) \to \mathcal{H} \otimes C^*(G)$ by

$$w(x \otimes u) = \zeta(x)(1 \otimes u).$$

Note that $\zeta(x)(1 \otimes u) \in \mathcal{H} \otimes C^*(G)$, since $\zeta(\mathcal{H}) \subseteq M_G(\mathcal{H} \otimes C^*(G))$. Moreover, w preserves the $\mathbb{C} \otimes C^*(G)$-valued inner products: for $x, y \in \mathcal{H}$ and $u, v \in C^*(G)$ we have

$$\langle w(x \otimes u), w(y \otimes v)\rangle_{\mathbb{C} \otimes C^*(G)} = \langle \zeta(x)(1 \otimes u), \zeta(y)(1 \otimes v)\rangle_{\mathbb{C} \otimes C^*(G)}$$
$$= (1 \otimes u^*) \langle \zeta(x), \zeta(y)\rangle_{M(\mathbb{C} \otimes C^*(G))} (1 \otimes v)$$
$$= (1 \otimes u^*)\big((\mathrm{id}_{\mathbb{C}} \otimes 1)(\langle x, y\rangle_{\mathbb{C}})\big)(1 \otimes v)$$
$$= (1 \otimes u^*)\big(\langle x, y\rangle_{\mathbb{C}} \otimes 1\big)(1 \otimes v)$$
$$= \langle x, y\rangle_{\mathbb{C}} \otimes u^* v$$
$$= \langle x \otimes u, y \otimes v\rangle_{\mathbb{C} \otimes C^*(G)}.$$

It follows that w extends to an isometry $w \colon \mathcal{H} \otimes C^*(G) \to \mathcal{H} \otimes C^*(G)$, which is surjective since $w(\mathcal{H} \otimes C^*(G)) \supseteq \zeta(\mathcal{H})(1 \otimes C^*(G)) = \zeta(\mathcal{H})(\mathbb{C} \otimes C^*(G)) = \mathcal{H} \otimes C^*(G)$ by nondegeneracy of ζ.

Now, if $(c_i)_i$ is a bounded approximate identity of $C^*(G)$, we get

$$\zeta(x) = \lim_i \zeta(x)(1 \otimes c_i) = \lim_i w(x \otimes c_i) = w(x \otimes 1)$$

(note that both $\zeta(x)$ and $w(x \otimes 1)$ are in $M_G(\mathcal{H} \otimes C^*(G))$ — compare with the proof of Lemma 1.40). Thus we have $\zeta(x) = w(x \otimes 1)$ for all $x \in \mathcal{H}$. Using the coaction identity for ζ, it follows that

$$w_{12} w_{13}(x \otimes 1 \otimes 1) = (\zeta \otimes \mathrm{id}_G) \circ \zeta(x)$$
$$= (\mathrm{id}_{\mathcal{H}} \otimes \delta_G) \circ \zeta(x) = (\mathrm{id}_{\mathcal{K}(\mathcal{H})} \otimes \delta_G)(w)(x \otimes 1 \otimes 1)$$

for all $x \in \mathcal{H}$, from which we get the equation $w_{12} w_{13} = (\mathrm{id}_{\mathcal{K}(\mathcal{H})} \otimes \delta_G)(w)$. It follows then from Remark A.35 that there exists a unique nondegenerate homomorphism $\mu \colon C_0(G) \to \mathcal{B}(\mathcal{H}_\pi)$ such that $w = (\mu \otimes \mathrm{id}_G)(w_G)$. It clearly follows from our constructions that μ is then uniquely determined by the property that $\zeta = \zeta_\mu$.

It only remains to show that (π, μ) is a covariant homomorphism of (A, δ), where $\pi \colon A \to \mathcal{B}(\mathcal{H})$ comes from the left action of A on X. For this we just compute

$$\big((\pi \otimes \mathrm{id}_G) \circ \delta(a)\big)(w(x \otimes 1)) = \big((\pi \otimes \mathrm{id}_G) \circ \delta(a)\big)\zeta(x)$$
$$= \zeta(\pi(a)x) = w(\pi(a)x \otimes 1)\big(w(\pi(a) \otimes 1)w^*\big)w(x \otimes 1),$$

from which it follows that $(\pi \otimes \mathrm{id}_G) \circ \delta(a) = w(\pi(a) \otimes 1)w^*$ for all $a \in A$. Since $w = (\mu \otimes \mathrm{id}_G)(w_G)$, this just means that (π, μ) is covariant. $\square$

We now combine the above results to get:

PROPOSITION 2.37. *Let (A, δ) be a coaction. For a covariant representation (π, μ) of (A, δ) let $(\mathcal{H}_\pi, \zeta_\mu)$ be the corresponding $\delta - \mathrm{id} \otimes 1$ compatible right-Hilbert $A - \mathbb{C}$ bimodule coaction as in Lemma 2.35. Then*

$$[(\pi, \mu)] \mapsto [(\mathcal{H}_\pi, \zeta_\mu)]$$

is a one-to-one correspondence between the unitary equivalence classes of covariant Hilbert-space representations of (A, δ) and $\mathrm{Rep}(A, \delta)$.

PROOF. Having Lemmas 2.35 and 2.36 at hand, we only have to show that two representations (π, μ) and (ρ, ν) of (A, δ) are unitarily equivalent if and only if $(\mathcal{H}_\pi, \zeta_\mu)$ and $(\mathcal{H}_\rho, \zeta_\nu)$ are isomorphic right-Hilbert bimodule coactions. So let

$U \colon \mathcal{H}_\pi \to \mathcal{H}_\rho$ be a unitary such that $U\pi(a) = \rho(a)U$ for all $a \in A$ and $U\mu(f) = \nu(f)U$ for all $f \in C_0(G)$. It then follows that

$$(U \otimes 1)(\mu \otimes \mathrm{id}_G)(w_G) = (\nu \otimes \mathrm{id}_G)(w_G)(U \otimes 1).$$

This implies that

$$\zeta_\nu(Ux) = (\nu \otimes \mathrm{id}_G)(w_G)(Ux \otimes 1) = (\nu \otimes \mathrm{id}_G)(w_G)(U \otimes 1)(x \otimes 1)$$
$$= (U \otimes 1)(\mu \otimes \mathrm{id}_G)(w_G)(x \otimes 1) = (U \otimes 1)\zeta_\mu(x),$$

which means that U is a $\zeta_\mu - \zeta_\nu$ compatible isomorphism between $\mathcal{H}_\pi$ and $\mathcal{H}_\rho$.

Conversely, if $V \colon \mathcal{H}_\pi \to \mathcal{H}_\rho$ is such a compatible isomorphism, then we have $V\pi(a) = \rho(a)V$ for all $a \in A$, and a short computation as above shows that

$$(V \otimes 1)(\mu \otimes \mathrm{id}_G)(w_G) = (\nu \otimes \mathrm{id}_G)(w_G)(V \otimes 1).$$

Using slice maps S_f for f in the Fourier algebra $A(G)$, it follows from Proposition A.34 that

$$V\mu(f) = S_f\big((V \otimes 1)(\mu \otimes \mathrm{id}_G)(w_G)\big) = S_f\big((\nu \otimes \mathrm{id}_G)(w_G)(V \otimes 1)\big) = \nu(f)V.$$

Since $A(G)$ is norm-dense in $C_0(G)$, we see that V is a unitary which intertwines (π, μ) and (ρ, ν). □

We close this chapter with a short remark on the additive structure of our categories. Recall that if X and Y are two right-Hilbert $A - B$ bimodules, then we can equip the direct sum $X \oplus Y$ with the structure of a right-Hilbert $A - B$ bimodule by defining

$$\langle x_1 + y_1, x_2 + y_2 \rangle_B = \langle x_1, x_2 \rangle_B + \langle y_1, y_2 \rangle_B,$$
$$a \cdot (x+y) = a \cdot x + a \cdot y, \quad \text{and} \quad (x+y) \cdot b = x \cdot b + y \cdot b,$$

for $x, x_1, x_2 \in X, y, y_1, y_2 \in Y$, $a \in A$ and $b \in B$. It is not hard to check that $[X] + [Y] = [X \oplus Y]$ defines an additive structure on the morphisms $\mathrm{mor}(A, B)$ in our category $\mathcal{C}$, so that the zero-morphism (i.e., $X = \{0\}$) serves as an additive identity. It is also straightforward to check that the distributive laws

$$[Z] \circ ([X] + [Y]) = [Z] \circ [X] + [Z] \circ [Y] \quad \text{and} \quad ([X] + [Y]) \circ [Z] = [X] \circ [Z] + [Y] \circ [Z]$$

hold in $\mathcal{C}$. Of course we can define similar additive structures on the categories $\mathcal{A}(G)$, $\mathcal{C}(G)$, and $\mathcal{AC}(G)$ — we omit further details on this. One can even define infinite direct sums of right-Hilbert bimodules, and then we get a notion of infinite sums of morphisms from A to B: if $\{X_i\}_{i \in I}$ is a family of right-Hilbert $A - B$ bimodules, then, as usual, one defines $\oplus_{i \in I} X_i$ as the completion of the vector space

$$\{(x_i)_{i \in I} \mid x_i \neq 0 \text{ for only finitely many } i \in I\}$$

with respect to the B-valued inner product $\langle (x_i), (x_i') \rangle_B = \sum_{i \in I} \langle x_i, x_i' \rangle_B$ and the obvious left and right actions of A and B. We then put $\sum_{i \in I} [X_i] = [\oplus_{i \in I} X_i]$.

Note that these additive structures on $\mathcal{C}$, $\mathcal{A}(G)$, $\mathcal{C}(G)$, and $\mathcal{AC}(G)$, are something which is not available in the category of C^*-algebras with $*$-homomorphisms as morphisms, since in general the sum of two $*$-homomorphisms $\varphi_1, \varphi_2 \colon A \to B$ is not a $*$-homomorphism any more.

CHAPTER 3

The Functors

In this chapter we show that many fundamental C^*-algebraic constructions— including restricting, inflating, decomposing, and taking crossed products by actions and coactions—have right-Hilbert bimodule counterparts, and that these constructions give functors among the various categories $\mathcal{A}(G)$, $\mathcal{C}(G)$, and $\mathcal{AC}(G)$.

3.1. Crossed products

3.1.1. Actions. In Theorem 3.7 below we will define a functor from $\mathcal{A}(G)$ to $\mathcal{C}(G)$ with object map $(A, \alpha) \mapsto (A \times_r G, \hat{\alpha})$. Because the objects in $\mathcal{A}(G)$ are actions on C^*-algebras, rather than isomorphism classes of such, we need to choose a single version of the reduced crossed product and stick with it. We use Definition A.13: the action α can also be regarded as a nondegenerate homomorphism $\alpha\colon A \to C_b(G, A) \subseteq M(A \otimes C_0(G))$ via $\alpha(a)(s) = \alpha_{s^{-1}}(a)$, and we define

$$A \times_{\alpha, r} G = (i_A^r \times i_G^r)(A \times_\alpha G) \subseteq M(A \otimes \mathcal{K}(L^2(G))),$$

where (i_A^r, i_G^r) is the covariant homomorphism $\big((\mathrm{id}_A \otimes M) \circ \alpha, 1 \otimes \lambda\big)$.

The dual coaction of G on $A \times_{\alpha, r} G$ is defined (see Example A.26 in Appendix A) by

$$\hat{\alpha}(i_A^r(a) i_G^r(f)) = (i_A^r(a) \otimes 1)(i_G^r \otimes u)(f) \qquad \text{for } a \in A, f \in C_c(G),$$

where $u\colon G \to UM(C^*(G))$ is the canonical inclusion. Recall that we require all coactions in $\mathcal{C}(G)$ to be normal and nondegenerate; every dual coaction $\hat{\alpha}$ has these properties (see Example A.26 and Proposition A.61).

We find it convenient to do many of our calculations with vector-valued C_c-functions. To facilitate this, by a slight abuse of notation we *identify* $C_c(G, A)$ with its image $(i_A^r \times i_G^r)(C_c(G, A))$ in $A \times_{\alpha, r} G$. The calculations will sometimes involve integrals of functions with values in locally convex spaces, rather than just C^*-algebras. Because we (almost always) only need to integrate continuous functions of compact support, the standard theorems about vector-valued integration apply, and we have chosen to use this theory to avoid having to insert C^*-algebra elements and linear functionals at every turn; the technical foundations we require are laid out in Appendix C. As an example of how such integrals arise, notice that the embedding of $C_c(G, A)$ in $A \times_{\alpha, r} G$ is described in terms of the canonical embeddings (i_A^r, i_G^r) by the strictly convergent integral

$$f = \int_G i_A^r(f(s)) i_G^r(s) \, ds.$$

As a point of notation, we usually identify G with its canonical image in $UM(C^*(G))$, so that, for example, we can say that the dual coaction of G on $A \times_{\alpha, r} G$ is given

on $C_c(G, A)$ by the strictly convergent integral
$$\hat{\alpha}(f) = \int_G i_A^r(f(s))i_G^r(s) \otimes s \, ds.$$

We now show how to regard $\hat{\alpha}(f)$ as a C_c-function. To prepare for this, we require some technical flexibility involving tensor products. If B is a C^*-algebra, we (often without comment) make extensive use of the canonical embedding of the algebraic tensor product $C_c(G, A) \odot B$ into $C_c(G, A \otimes B)$ given by $(f \otimes b)(s) = f(s) \otimes b$. (We occasionally extend this convention to other function spaces, such as C_b, the bounded continuous functions.) Note that this embedding extends to the canonical isomorphism of $(A \times_{\alpha, r} G) \otimes B$ onto $(A \otimes B) \times_{\alpha \otimes \mathrm{id}, r} G$ from Lemma A.20. We employ this trick many times throughout this work when dealing with coactions in terms of C_c-functions. The idea is to treat the extra copy of $C^*(G)$ as a freely moving object.

This allows us to embed $C_c(G, A \otimes B)$ in $(A \times_{\alpha, r} G) \otimes B$, and $C_c(G, M^\beta(A \otimes B))$ in $M((A \times_{\alpha, r} G) \otimes B)$; for the latter we use the embedding of $C_c(G, M^\beta(A))$ into $M(A \times_{\alpha, r} G)$ provided by Corollary C.7. (Recall that we write $M^\beta(A)$ for $M(A)$ equipped with the strict topology.) Note that for $f \in C_c(G, A)$ and $b \in B$, we have

$$f \otimes b = \int_G i_A^r(f(s))i_G^r(s) \, ds \otimes b = \int_G i_A^r(f(s))i_G^r(s) \otimes b \, ds$$
$$= \int_G (i_A^r(f(s)) \otimes b)(i_G^r(s) \otimes 1) \, ds = \int_G (i_A^r \otimes \mathrm{id})\big((f \otimes b)(s)\big)(i_G^r(s) \otimes 1) \, ds,$$

so that, by linearity, density and continuity, if $f \in C_c(G, M^\beta(A \otimes B))$ then as an element of $M((A \times_{\alpha, r} G) \otimes B)$ we have (by the same abuse of notation as above)

$$f = \int_G (i_B^r \otimes \mathrm{id})(f(s))(i_G^r(s) \otimes 1) \, ds.$$

LEMMA 3.1. *Let (A, G, α) be an action. For each $f \in C_c(G, A)$, we have*
$$\hat{\alpha}(f) \in C_c(G, M^\beta(A \otimes C^*(G))) \subseteq M\big((A \times_r G) \otimes C^*(G)\big),$$
with $\hat{\alpha}(f)(s) = f(s) \otimes s$ for $s \in G$.

PROOF. Compute:
$$\hat{\alpha}(f) = \hat{\alpha}\left(\int_G i_A^r(f(s))i_G^r(s) \, ds\right) = \int_G i_A^r(f(s))i_G^r(s) \otimes s \, ds$$
$$= \int_G (i_A^r \otimes \mathrm{id})(f(s) \otimes s)(i_G^r(s) \otimes 1) \, ds.$$

Thus, $\hat{\alpha}(f)$ agrees with the element $s \mapsto f(s) \otimes s$ of $C_c(G, M^\beta(A \otimes C^*(G)))$. □

Functoriality requires that we also give a construction of crossed products of right-Hilbert bimodules. Such constructions have appeared in several places in the literature (*e.g.*, see [12, 11, 30, 2, 17] for several constructions of full and reduced crossed products by Hilbert modules). In order to have a construction which is best suited for our needs, and for completeness, we shall give our own construction of reduced crossed products by a right-Hilbert bimodule actions. Of course we closely follow the ideas presented in the literature cited above.

Let $_{(A,\alpha)}(X, \gamma)_{(B,\beta)}$ be a right-Hilbert bimodule action. Consider the dense subalgebras $C_c(G, A)$ and $C_c(G, B)$ of $A \times_{\alpha, r} G$ and $B \times_{\beta, r} G$. We want to equip

$C_c(G,X)$ with a pre-right-Hilbert $C_c(G,A) - C_c(G,B)$ bimodule structure with operations

(3.1)
$$f \cdot x(s) = \int_G f(t) \cdot \gamma_t(x(t^{-1}s)) \, dt,$$
$$x \cdot g(s) = \int_G x(t) \cdot \beta_t(g(t^{-1}s)) \, dt, \quad \text{and}$$
$$\langle x, y \rangle_{B \times_\beta G}(s) = \int_G \beta_{t^{-1}}\big(\langle x(t), y(ts) \rangle_B\big) \, dt$$

for $f \in C_c(G,A)$, $x,y \in C_c(G,X)$, and $g \in C_c(G,B)$.

PROPOSITION 3.2. *With the above operations, $C_c(G,X)$ completes to give a right-Hilbert $A \times_{\alpha,r} G - B \times_{\beta,r} G$ bimodule. Moreover, if $B_X = \overline{\langle X,X \rangle}_B$, then $\overline{\langle X \times_{\gamma,r} G, X \times_{\gamma,r} G \rangle}_{B \times_{\beta,r} G} = B_X \times_{\beta,r} G$.*

We do the proof in two steps. In the first step we consider the right-Hilbert bimodule action $({}_K X_B, {}_\mu \gamma_\beta)$, with $K = \mathcal{K}(X)$, as provided by Proposition 2.27, and use a linking algebra argument to see that the proposition is true in this special situation. We then observe that the homomorphism $\kappa \times_r G \colon A \times_{\alpha,r} G \to M(K \times_{\mu,r} G)$ coming from the $\alpha - \mu$ equivariant homomorphism $\kappa \colon A \to M(K) \cong \mathcal{L}(X)$ induces a left action of $A \times_{\alpha,r} G$ on $X \times_{\gamma,r} G$ which coincides on C_c-functions with the actions as given in (3.1).

If $({}_K X_B, {}_\mu \gamma_\beta)$ is a right-Hilbert imprimitivity bimodule action of G on the partial imprimitivity bimodule ${}_K X_B$, then we obtain a strongly continuous action $\nu \colon G \to \operatorname{Aut} L(X)$ by defining

(3.2)
$$\nu_s\left(\begin{pmatrix} k & x \\ \tilde{y} & b \end{pmatrix}\right) = \begin{pmatrix} \mu_s(k) & \gamma_s(x) \\ \widetilde{\gamma_s(y)} & \beta_s(b) \end{pmatrix}$$

(see Lemma 2.21). The convolution algebra $C_c(G, L(X))$ has a canonical decomposition as two-by-two matrices

(3.3)
$$C_c(G, L(X)) = \begin{pmatrix} C_c(G,K) & C_c(G,X) \\ C_c(G,\widetilde{X}) & C_c(G,B) \end{pmatrix},$$

and it is straightforward to check that the pairings among the corners $C_c(G,K)$, $C_c(G,X)$ and $C_c(G,B)$ given by convolution on $C_c(G, L(X))$ are precisely the ones given by (3.1).

Now let $p = \begin{pmatrix} 1 & 0 \\ 0 & 0 \end{pmatrix}$ and $q = \begin{pmatrix} 0 & 0 \\ 0 & 1 \end{pmatrix}$ denote the corner projections in $M(L(X))$, and let $i_L^r(p)$ and $i_L^r(q)$ denote the images of p and q in $M(L(X) \times_{\nu,r} G)$. It follows from Lemma A.16 that the canonical inclusions of $C_c(G,K)$ and $C_c(G,B)$ into $C_c(G,L(X))$ provided by (3.3) extend to isomorphisms

$$K \times_{\mu,r} G \cong i_L^r(p)\big(L(X) \times_{\nu,r} G\big)i_L^r(p) \quad \text{and} \quad B \times_{\beta,r} G \cong i_L^r(q)\big(L(X) \times_{\nu,r} G\big)i_L^r(q),$$

and we have

$$\overline{C_c(G,X)} = i_L^r(p)\big(L(X) \times_{\nu,r} G\big)i_L^r(q),$$

where we identify $C_c(G,X)$ with the upper right corner of $C_c(G,L(X))$ as in (3.3). In particular, it follows from Lemma 1.48 that $C_c(G,X)$ completes to give a partial $K \times_{\mu,r} G - B \times_{\beta,r} G$ imprimitivity bimodule $X \times_{\gamma,r} G$, with actions and inner products given by (3.1). Using Lemma 1.49 we also see that $L(X) \times_{\nu,r} G$ is then canonically isomorphic to $L(X \times_{\gamma,r} G)$.

We now gather the above observations:

LEMMA 3.3. *Suppose that $({}_K X_B, {}_\mu \gamma_\beta)$ is a right-Hilbert bimodule action of G on the partial imprimitivity bimodule ${}_K X_B$. Then $C_c(G, X)$, equipped with the $C_c(G, K) - C_c(G, B)$ pre-right-Hilbert bimodule structure of (3.1), completes to a partial $K \times_{\mu,r} G - B \times_{\beta,r} G$ imprimitivity bimodule such that the identification (3.3) extends to a canonical isomorphism*

$$L(X) \times_{\nu,r} G \cong L(X \times_{\gamma,r} G),$$

with ν as in (3.2). Moreover, the ranges of the left- and right-valued inner products on $X \times_{\gamma,r} G$ are given by the ideals $K_X \times_{\mu,r} G$ and $B_X \times_{\beta,r} G$, where K_X and B_X denote the ranges of the inner products on X. In particular, $X \times_{\gamma,r} G$ is a $K \times_{\mu,r} G - B \times_{\beta,r} G$ imprimitivity bimodule if and only if X is a $K - B$ imprimitivity bimodule.

PROOF. Everything but the assertion on the ranges of the inner products follows from the above considerations.

For the ranges we first consider the case where $K_X = K$ and $B_X = B$, i.e., where X is a $K - B$ imprimitivity bimodule. It follows then from Lemma 1.48 that p and q are full projections in $L(X)$, which implies that $i_L^r(q)$ and $i_L^r(p)$ are full projections in $M(L(X) \times_{\nu,r} G)$. To see this, compute

$$\begin{aligned} L(X) \times_{\nu,r} G &= \overline{i_G^r(C^*(G)) i_L^r(L(X)) i_G^r(C^*(G))} \\ &= \overline{i_G^r(C^*(G)) i_L^r(L(X) q L(X)) i_G^r(C^*(G))} \\ &= \overline{\big(i_G^r(C^*(G)) i_L^r(L(X))\big) i_L^r(q) \big(i_L^r(L(X)) i_G^r(C^*(G))\big)} \\ &= \overline{\big(L(X) \times_{\nu,r} G\big) i_L^r(q) \big(L(X) \times_{\nu,r} G\big)}. \end{aligned}$$

Thus it follows that $X \times_{\gamma,r} G$ is a $K \times_{\mu,r} G - B \times_{\beta,r} G$ imprimitivity bimodule.

If ${}_K X_B$ is a partial imprimitivity bimodule, then $\begin{pmatrix} K_X & X \\ \widetilde{X} & B_X \end{pmatrix}$, which we denote by $L^{\mathrm{imp}}(X)$, embeds as an ideal in $L(X)$ and therefore $L^{\mathrm{imp}}(X) \times_{\nu,r} G = \begin{pmatrix} K_X \times_{\mu,r} G & X \times_{\gamma,r} G \\ (X \times_{\gamma,r} G)\widetilde{} & B_X \times_{\beta,r} G \end{pmatrix}$ canonically embeds as an ideal in $L(X \times_{\gamma,r} G)$. Since this embedding is the identity on $X \times_{\gamma,r} G$, and since ${}_{K_X} X_{B_X}$ is an imprimitivity bimodule, it follows that the ranges of the inner products on $X \times_{\gamma,r} G$ are precisely $K_X \times_{\mu,r} G$ and $B_X \times_{\beta,r} G$, respectively. $\square$

PROOF OF PROPOSITION 3.2. Let $K = \mathcal{K}(X)$. Then it follows from Proposition 2.27 that there exists an action μ of G on K such that $({}_K X_B, {}_\mu \gamma_\beta)$ is a right-Hilbert bimodule action on the right-partial $K - A$ imprimitivity bimodule ${}_K X_B$, and such that the homomorphism $\kappa \colon A \to M(K) \cong \mathcal{L}_B(X)$ coming from the left action of A on X is $\alpha - \mu$ equivariant. Lemma 3.3 implies that $C_c(G, X)$ completes to a right-partial $K \times_{\mu,r} G - B \times_{\beta,r} G$ imprimitivity bimodule, and it follows from Lemma A.16 that there exists a homomorphism $\kappa \times_r G \colon A \times_{\alpha,r} G \to M(K \times_{\nu,r} G) \cong \mathcal{L}_{B \times_r G}(X \times_{\gamma,r} G)$ given on $C_c(G, A)$ by $f \mapsto \kappa \circ f$. Using this, it is straightforward to check that the resulting left action of $A \times_{\alpha,r} G$ on $X \times_{\gamma,r} G$ coincides on $C_c(G, A)$ with the action given by (3.1). $\square$

As with reduced C^*-crossed products, we identify $C_c(G, X)$ with its image in $X \times_{\gamma,r} G$.

To complete our functor, we need a dual coaction on $X \times_{\gamma,r} G$, and for this we will view $C_c(G, M^\beta(X \otimes C^*(G)))$ as a subspace of $M((X \times_{\gamma,r} G) \otimes C^*(G))$. The following general lemma justifies this embedding, since for any C^*-algebra C, it follows from Lemma C.8 that we have an embedding $C_c(G, M^\beta(X \otimes C)) \hookrightarrow M((X \otimes C) \times_{\gamma \otimes \mathrm{id}, r} G)$.

LEMMA 3.4. *For any action* $_{(A,\alpha)}(X, \gamma)_{(B,\beta)}$ *and for any C^*-algebra C the canonical embedding of $C_c(G, X) \odot C$ in $C_c(G, X \otimes C)$ extends to an isomorphism*

$$\Phi\colon {}_{(A \times_\alpha,r G) \otimes C}((X \times_{\gamma,r} G) \otimes C)_{(B \times_\beta,r G) \otimes C}$$
$$\xrightarrow{\cong} {}_{(A \otimes C) \times_{\alpha \otimes \mathrm{id}, r} G} \left((X \otimes C) \times_{\gamma \otimes \mathrm{id}, r} G\right)_{(B \otimes C) \times_{\beta \otimes \mathrm{id}, r} G}.$$

PROOF. We know that the embedding of $C_c(G, A) \odot C$ into $C_c(G, A \otimes C)$ determines an isomorphism of $(A \times_{\gamma,r} G) \otimes C$ with $(A \otimes C) \times_{\gamma \otimes \mathrm{id}, r} G$, and similarly for B. A routine calculation now shows that the embedding of $C_c(G, X) \odot C \subseteq (X \times_{\gamma,r} G) \otimes C$ into $C_c(G, X \otimes C) \subseteq (X \otimes C) \times_{\gamma \otimes \mathrm{id}, r} G$ is compatible with these coefficient maps and all bimodule operations. Since the image of $C_c(G, X) \otimes C$ is inductive limit dense in $C_c(G, X \otimes C)$, the result follows. □

PROPOSITION 3.5. *If γ is an $\alpha - \beta$ compatible action of G on a right-Hilbert bimodule $_AX_B$, then there is a unique nondegenerate $\hat\alpha - \hat\beta$ compatible coaction $\hat\gamma$ of G on $X \times_{\gamma,r} G$ such that for each $x \in C_c(G, X)$ we have*

$$\hat\gamma(x) \in C_c(G, M^\beta(X \otimes C^*(G))) \subseteq M((X \times_{\gamma,r} G) \otimes C^*(G)),$$

with $\hat\gamma(x)(s) = x(s) \otimes s$ for $s \in G$.

PROOF. As for the construction of the crossed product $X \times_{\gamma,r} G$, we use a linking algebra argument. For this let $C = C^*(G)$, let $K = \mathcal{K}(X)$, let $\mu\colon G \to \mathrm{Aut}\,K$ denote the action of G on K provided by Proposition 2.27, and let $\nu\colon G \to \mathrm{Aut}\,L(X)$ denote the corresponding action on the linking algebra. It then follows from Lemma 3.3 that $X \times_{\gamma,r} G$ is a right-partial $K \times_{\mu,r} G - B \times_{\beta,r} G$ imprimitivity bimodule and $L(X \times_{\gamma,r} G) \cong L(X) \times_{\nu,r} G$. By Lemma 3.1, the dual coaction $\hat\nu$ maps $C_c(G, L(X)) \subseteq L(X) \times_{\nu,r} G$ into $C_c(G, M^\beta(L(X) \otimes C)) \subseteq M((L(X) \times_{\nu,r} G) \otimes C)$ according to the rule $\hat\nu(f)(s) = f(s) \otimes s$ for $s \in G$. We make the canonical identifications

$$C_c(G, L(X)) = \begin{pmatrix} C_c(G, A) & C_c(G, X) \\ C_c(G, X)^\sim & C_c(G, B) \end{pmatrix}$$

and

$$C_c(G, M^\beta(L(X \otimes C))) = \begin{pmatrix} C_c(G, M^\beta(A \otimes C)) & C_c(G, M^\beta(X \otimes C)) \\ C_c(G, M^\beta(X \otimes C))^\sim & C_c(G, M^\beta(B \otimes C)) \end{pmatrix},$$

where the latter identification is allowed since the strict topology on $M(L(X) \otimes C)$ coincides with the product of the strict topologies of the corners by Proposition 1.51. With this identification, the formula for $\hat\nu$ becomes

$$\hat\nu\left(\begin{pmatrix} g & x \\ \tilde y & l \end{pmatrix}\right)(s) = \begin{pmatrix} g(s) \otimes s & x(s) \otimes s \\ \widetilde{y(s) \otimes s} & l(s) \otimes s \end{pmatrix},$$

for $g \in C_c(G, A)$, $x, y \in C_c(G, X)$, and $l \in C_c(G, B)$. In particular, it follows from this that $\hat\nu(p) = p \otimes 1$ and $\hat\nu(q) = q \otimes 1$, where $p, q \in M(L(X \times_{\gamma,r} G)) = M(L(K) \times_{\nu,r} G)$ denote the usual corner projections. Thus it follows from Lemma 2.22 that $\hat\nu$

compresses to a $\hat{\mu} - \hat{\beta}$ compatible coaction $\hat{\gamma}$ on the upper right corner $X \times_{\gamma, r} G$ which then has to be given by the formula

$$\hat{\gamma}(x)(s) = x(s) \otimes s$$

for $x \in C_c(G, X)$. Since $\hat{\nu}$ is nondegenerate by Example A.27, it follows from Lemma 2.22 that $\hat{\gamma}$ is nondegenerate, too.

We now complete the proof by observing that it follows directly from the formulas for $\hat{\alpha}$ and $\hat{\mu}$ that the homomorphism $\kappa \times_r G : A \times_{\alpha, r} G \to M(K \times_{\mu, r} G)$ is $\hat{\alpha} - \hat{\beta}$ equivariant (where $\kappa : A \to M(K)$ is determined by the left A-action on X). This shows that $\hat{\gamma}$ is a nondegenerate $\hat{\alpha} - \hat{\beta}$ compatible coaction on $X \times_{\gamma, r} G$. $\square$

For later use it is important to see that the crossed product by a standard morphism $({}_A X_B, \gamma)$ associated to an $\alpha - \beta$ equivariant homomorphism $\varphi : A \to M(B)$ is again standard.

PROPOSITION 3.6. *Suppose that $\varphi : A \to M(B)$ is a (possibly degenerate) $\alpha - \beta$ equivariant homomorphism, and let $({}_A X_B, \gamma)$ be the standard right-Hilbert $A - B$ bimodule action associated to φ (see Definition 2.25). Then ${}_{A \times_r G}(X \times_{\gamma, r} G)_{B \times_r G}$ is canonically isomorphic to the standard right-Hilbert $A \times_{\alpha, r} G - B \times_{\beta, r} G$ bimodule $Y = \varphi \times_r G(A \times_{\alpha, r} G)(B \times_{\beta, r} G)$, and this isomorphism is equivariant with respect to $\hat{\gamma}$ and the restriction $\hat{\beta}_Y$ of $\hat{\beta}$ to Y.*

PROOF. Since ${}_A X_B = \varphi(A) B$, we see that

$$\varphi \times_r G(C_c(G, A)) C_c(G, B) \supseteq (C_c(G) \odot \varphi(A))(C_c(G) \odot B) = C_c(G) \odot X$$

is inductive limit dense in $C_c(G, X)$. The result then follows from the fact that the above inclusion of $\varphi \times_r G(C_c(G, A)) C_c(G, B)$ preserves the left and right $C_c(G, A)$- and $C_c(G, B)$-actions and the $C_c(G, B)$-valued inner products. The $\hat{\gamma} - \hat{\beta}_Y$ equivariance follows directly from the formulas for the respective dual coactions as given in Lemma 3.1 and Proposition 3.5. $\square$

THEOREM 3.7. *The object and morphism maps*

$$(A, \alpha) \mapsto (A \times_{\alpha, r} G, \hat{\alpha}) \quad \text{and} \quad [{}_A X_B, \gamma] \mapsto [{}_{A \times_r G}(X \times_{\gamma, r} G)_{B \times_r G}, \hat{\gamma}]$$

define a functor from $\mathcal{A}(G)$ to $\mathcal{C}(G)$.

PROOF. We first show that the map on morphisms is well-defined. Suppose $\Phi : X \to Y$ is an isomorphism of right-Hilbert $A - B$ bimodules which is equivariant for $\alpha - \beta$ compatible actions γ and ρ of G. Then $(\Phi \times_r G)(x)(s) = \Phi(x(s))$ is easily seen to give a bijective map $\Phi \times_r G : C_c(G, X) \to C_c(G, Y)$ which respects the pre-right-Hilbert bimodule structures given by Equation (3.1) and hence extends to a right-Hilbert $(A \times_{\alpha, r} G) - (B \times_{\beta, r} G)$ bimodule isomorphism of $X \times_{\gamma, r} G$ onto $Y \times_{\rho, r} G$. It suffices to check $\hat{\gamma} - \hat{\rho}$ equivariance of $\Phi \times_r G$ for $x \in C_c(G, X)$:

$$((\Phi \times_r G) \otimes \text{id}) \circ \hat{\gamma}(x)(s) = (\Phi \otimes \text{id})(\hat{\gamma}(x)(s)) = \Phi(x(s)) \otimes s$$
$$= (\Phi \times_r G)(x)(s) \otimes s = \hat{\rho} \circ (\Phi \times_r G)(x)(s).$$

If (A, G, α) is an action then it follows from Proposition 3.6 that the bimodule crossed product ${}_A A_A \times_r G$ is equivariantly isomorphic to the standard right-Hilbert bimodule ${}_{A \times_r G}(A \times_r G)_{A \times_r G}$, which shows that the crossed-product functor preserves identities.

3.1. CROSSED PRODUCTS

It only remains to see that the assignment $[X, \gamma] \mapsto [X \times_{\gamma,r} G]$ respects composition of morphisms;[1] that is, if $({}_A X_B, \gamma)$ is $\alpha - \beta$ compatible and $({}_B Y_C, \rho)$ is $\beta - \mu$ compatible, we need to find a right-Hilbert $(A \times_{\alpha,r} G) - (C \times_{\mu,r} G)$ bimodule isomorphism

$$\Upsilon \colon (X \times_{\gamma,r} G) \otimes_{B \times_r G} (Y \times_{\rho,r} G) \xrightarrow{\cong} (X \otimes_B Y) \times_{\gamma \otimes \rho, r} G$$

which is $(\hat{\gamma} \, \natural_{B \times_r G} \, \hat{\rho}) - \widehat{\gamma \otimes \rho}$ equivariant. The rule

$$\Upsilon(x \otimes y)(s) = \int_G x(t) \otimes \rho_t(y(t^{-1}s)) \, dt$$

defines a map from $C_c(G, X) \odot C_c(G, Y)$ to $C_c(G, X \otimes_B Y)$ which is easily checked to preserve the pre-right-Hilbert bimodule structures. In order to see that Υ extends to an isomorphism of the completions, we need only verify that Υ has dense range for the inductive limit topology. For this, let $x_1 \in X$ and $f \in C_c(G, B)$, and define $x \in C_c(G, X)$ by $x(s) = x_1 \cdot f(s)$. Then for $y \in C_c(G, Y)$ we have

$$\Upsilon(x \otimes y)(s) = x_1 \otimes \int_G f(t) \cdot \rho_t(y(t^{-1}s)) \, dt = x_1 \otimes (f \cdot y)(s).$$

Now, we can approximate y by $f \cdot y$ in the inductive limit topology, and taking y of the form $y(s) = y_1 g(s)$ for $y_1 \in Y$ and $g \in C_c(G)$ we can thus approximate the function $s \mapsto (x_1 \otimes y_1) g(s)$. But such functions have inductive limit dense span in $C_c(G, X \otimes_B Y)$.

For the equivariance, it suffices to show that for $x \in C_c(G, X) \subseteq X \times_{\gamma,r} G$ and $y \in C_c(G, Y) \subseteq Y \times_{\rho,r} G$ we have

$$\widehat{\gamma \otimes \rho} \circ \Upsilon(x \otimes y) = (\Upsilon \otimes \mathrm{id}) \circ (\hat{\gamma} \, \natural \, \hat{\rho})(x \otimes y)$$

as elements of $C_c\big(G, M^\beta\big((X \otimes_B Y) \otimes C^*(G)\big)\big) \subseteq M\big(((X \otimes_B Y) \times_{\gamma \otimes \rho, r} G) \otimes C^*(G)\big)$. First note that for $x' \in C_c(G, X)$, $y' \in C_c(G, Y)$, and $u, v, s \in G$ we have

$$(\Upsilon \otimes \mathrm{id}) \circ \Theta\big((x' \otimes u) \otimes (y' \otimes v)\big)(s) = (\Upsilon \otimes \mathrm{id})\big((x' \otimes y') \otimes uv\big)(s)$$
$$= (\Upsilon(x' \otimes y') \otimes uv)(s) = \Upsilon(x' \otimes y')(s) \otimes uv$$
$$= \int_G x'(t) \otimes \rho_t(y'(t^{-1}s)) \, dt \otimes uv$$
$$= \int_G \big(x'(t) \otimes \rho_t(y'(t^{-1}s)))\big) \otimes uv \, dt$$
$$= \int_G \Theta\big((x'(t) \otimes u) \otimes (\rho_t \otimes \mathrm{id})(y'(t^{-1}s) \otimes v)\big) \, dt$$
$$= \int_G \Theta\Big((x' \otimes u)(t) \otimes (\rho_t \otimes \mathrm{id})\big((y' \otimes v)(t^{-1}s)\big)\Big) \, dt.$$

---

[1]This is also proven (using non-categorical language) in [**30**, Lemma 3.10]; we include our own proof for completeness.

By linearity, density, and continuity we conclude that for $x \in C_c(G, X)$, $y \in C_c(G, Y)$, and $s \in G$ we have

$$(\Upsilon \otimes \mathrm{id}) \circ (\hat{\gamma} \,\sharp\, \hat{\rho})(x \otimes y)(s) = (\Upsilon \otimes \mathrm{id}) \circ \Theta\big(\hat{\gamma}(x) \otimes \hat{\rho}(y)\big)(s)$$

$$= \int_G \Theta\Big(\hat{\gamma}(x)(t) \otimes (\rho_t \otimes \mathrm{id})(\hat{\rho}(y)(t^{-1}s))\Big) dt$$

$$= \int_G \Theta\Big((x(t) \otimes t) \otimes (\rho_t(y(t^{-1}s)) \otimes t^{-1}s)\Big) dt$$

$$= \int_G \big(x(t) \otimes \rho_t(y(t^{-1}s))\big) \otimes s \, dt = \int_G x(t) \otimes \rho_t(y(t^{-1}s)) \, dt \otimes s$$

$$= \Upsilon(x \otimes y)(s) \otimes s = \widehat{\gamma \otimes \rho} \circ \Upsilon(x \otimes y)(s).$$

□

REMARK 3.8. For later use, it is worthwhile to observe that the proof of Proposition 3.2 implies that if ${}_A X_B = {}_A(K \otimes_K X)_B$ is the (equivariant) standard factorization of X, as described in Proposition 2.27, then we get the standard factorization

$$_{A \times_{\alpha,r} G}(X \times_{\gamma,r} G)_{B \times_{\beta,r} G} \cong {}_{A \times_{\alpha,r} G}\big((K \times_{\mu,r} G) \otimes_{K \times_{\nu,r} G} (X \times_{\gamma,r} G)\big)_{B \times_{\beta,r} G},$$

which is clearly equivariant with respect to the dual coactions. It follows from Proposition 3.6 below that ${}_{A \times_{\alpha,r} G}(K \times_{\mu,r} G)_{K \times_{\mu,r} G}$ is equivariantly isomorphic to ${}_A K_K \times_{\mu,r} G$, so that the action-crossed-product functor of Theorem 3.7 preserves standard factorizations of morphisms.

3.1.2. Coactions. In Theorem 3.13 below we define a functor from $\mathcal{C}(G)$ to $\mathcal{A}(G)$ with object map

$$(A, \delta) \mapsto (A \times_\delta G, \hat{\delta}).$$

As for actions, we first need to choose once and for all a single version of the crossed product, and we use Definition A.39: the coaction δ is a nondegenerate homomorphism into $M(A \otimes C^*(G))$, and composing with the nondegenerate homomorphism $\mathrm{id} \otimes \lambda \colon A \otimes C^*(G) \to M(A \otimes \mathcal{K}(L^2(G)))$ gives a nondegenerate homomorphism

$$j_A = (\mathrm{id} \otimes \lambda) \circ \delta \colon A \to M(A \otimes \mathcal{K}(L^2(G))).$$

Writing $j_G = 1 \otimes M$, we get a covariant homomorphism (j_A, j_G) of (A, G, δ) to $M(A \otimes \mathcal{K}(L^2(G)))$. (We sometimes write j_G^A for j_G if there is danger of confusion.) Then the crossed product is defined as

$$A \times_\delta G = \overline{j_A(A) j_G(C_0(G))} \subseteq M(A \otimes \mathcal{K}(L^2(G))).$$

The standard theory of coactions (see Appendix A) tells us that $A \times_\delta G$ is a C^*-algebra and that the triple $(A \times_\delta G, j_A, j_G)$ has the universal property that for every covariant homomorphism (π, μ) of $(A, C_0(G))$ into a multiplier algebra $M(B)$ there exists a unique nondegenerate homomorphism $\pi \times \mu \colon A \times_\delta G \to M(B)$ such that $(\pi \times \mu)(j_A(a) j_G(f)) = \pi(a) \mu(f)$ for all $a \in A$ and $f \in C_0(G)$.

The *dual action* $\hat{\delta}$ of G on $A \times_\delta G$ is defined (see Definition A.47) by $\hat{\delta}_s = j_A \times (j_G \circ \rho_s)$, where ρ_s denotes right translation on $C_0(G)$.

If $\varphi \colon A \to M(B)$ is a homomorphism which is equivariant for coactions δ and ϵ of G on A and B, there is an associated homomorphism

$$\varphi \times G = (j_B \circ \varphi) \times j_G$$

3.1. CROSSED PRODUCTS

of $A \times_\delta G$ into $M(B \times_\epsilon G)$, which is nondegenerate if φ is nondegenerate (see Lemma A.46).

Let $\zeta \colon X \to M(X \otimes C^*(G))$ be a $\delta - \epsilon$ compatible nondegenerate coaction on a right-Hilbert A–B bimodule X. We will define the crossed product $X \times_\zeta G$ following [20, Theorem 3.2] (except here we use full coactions and avoid representing things on Hilbert space). Put

$$j_X = (\mathrm{id} \otimes \lambda) \circ \zeta.$$

PROPOSITION 3.9. *Assume that $({}_A X_B, \zeta)$ is a nondegenerate right-Hilbert bimodule coaction. With the above notation, the subspace*

$$X \times_\zeta G = \overline{j_X(X) \cdot j_G^B(C_0(G))}$$

of $M(X \otimes \mathcal{K}(L^2(G)))$ is a right-Hilbert $(A \times_\delta G) - (B \times_\epsilon G)$ bimodule. Indeed, $X \times_\zeta G$ is closed under the bimodule actions of $A \times_\delta G$ and $B \times_\epsilon G$, and the $M(B \otimes \mathcal{K}(L^2(G)))$-valued inner product on $X \times_\zeta G$ actually takes values in $B \times_\epsilon G$. Moreover, if $B_X = \overline{\langle X, X \rangle}_B$, then $\overline{\langle X \times_\zeta G, X \times_\zeta G \rangle}_{B \times_\epsilon G} = B_X \times_{\epsilon_X} G$.

As for actions, we first give the proof for the case of a nondegenerate right-partial imprimitivity bimodule coaction $({}_K X_B, \zeta)$, using the linking algebra approach. So let $L(X) = \begin{pmatrix} K & X \\ \widetilde{X} & B \end{pmatrix}$ denote the linking algebra of X. If we identify $L(X) \otimes C^*(G)$ with $L(X \otimes C^*(G))$ as described in Section 1.5 of Chapter 1, we can use Proposition 1.51 to get a canonical identification $M(L(X) \otimes C^*(G)) \cong L(M(X \otimes C^*(G)))$. Thus, by Lemma 2.22 we obtain a coaction ν of G on $L(X)$ determined by

$$(3.4) \qquad \nu\left(\begin{pmatrix} k & x \\ \widetilde{y} & b \end{pmatrix}\right) = \begin{pmatrix} \mu(k) & \zeta(x) \\ \widetilde{\zeta(y)} & \epsilon(b) \end{pmatrix}.$$

If we similarly identify $M(L(X) \otimes \mathcal{K}(L^2(G)))$ with $L(M(X \otimes \mathcal{K}(L^2(G))))$, we see that the maps $(j_L, j_G^L) \colon (L(X), C_0(G)) \to M(L(X) \otimes \mathcal{K}(L^2(G)))$ can be written as

$$j_L = (\mathrm{id}_L \otimes \lambda) \circ \nu = \begin{pmatrix} (\mathrm{id}_K \otimes \lambda) \circ \mu & (\mathrm{id}_X \otimes \lambda) \circ \zeta \\ (\mathrm{id}_{\widetilde{X}} \otimes \lambda) \circ \widetilde{\zeta} & (\mathrm{id}_B \otimes \lambda) \circ \epsilon \end{pmatrix} = \begin{pmatrix} j_K & j_X \\ \widetilde{j_X} & j_B \end{pmatrix}$$

and

$$j_G^L = 1_{M(L)} \otimes M = \begin{pmatrix} 1_{M(K)} \otimes M & 0 \\ 0 & 1_{M(B)} \otimes M \end{pmatrix} = \begin{pmatrix} j_G^K & 0 \\ 0 & j_G^B \end{pmatrix}.$$

Thus, for the crossed product $L(X) \times_\nu G$, we can write

$$L(X) \times_\nu G = \overline{j_L(L(X)) j_G^L(C_0(G))} = \begin{pmatrix} \overline{j_K(K) j_G^K(C_0(G))} & \overline{j_X(X) j_G^B(C_0(G))} \\ \overline{\widetilde{j_X}(X) j_G^K(C_0(G))} & \overline{j_B(B) j_G^B(C_0(G))} \end{pmatrix}.$$

Note that if $p = \begin{pmatrix} 1 & 0 \\ 0 & 0 \end{pmatrix}$ and $q = \begin{pmatrix} 0 & 0 \\ 0 & 1 \end{pmatrix}$ denote the corner projections in $L(X)$, then $j_L(p)$ and $j_L(q)$ are the corner projections of $L(X) \times_\nu G \cong L(X \times_\zeta G)$. We now gather these results into:

LEMMA 3.10. *Suppose that $({}_K X_B, {}_\mu \zeta_\epsilon)$ is a nondegenerate right-Hilbert bimodule coaction of G on the right-partial imprimitivity bimodule ${}_K X_B$. Then*

$$(3.5) \qquad X \times_\zeta G = \overline{j_X(X) j_G^B(C_0(G))} = \overline{j_G^K(C_0(G)) j_X(X)}$$

is a right-partial $K \times_\mu G$–$B \times_\epsilon G$ imprimitivity bimodule with respect to the bimodule operations as described in Proposition 3.9.

If $_KX_B$ is an imprimitivity bimodule, then so is $_{K\times_\mu G}(X \times_\zeta G)_{B\times_\epsilon G}$ — in general, the range of the $B \times_\epsilon G$-valued inner product on $X \times_\zeta G$ is $B_X \times_{\epsilon_X} G$, where B_X denotes the range of the B-valued inner product on X and ϵ_X denotes the restriction of ϵ to B_X (see Lemma 2.32).

Finally, if $\nu = \begin{pmatrix} \mu & \zeta \\ \tilde\zeta & \epsilon \end{pmatrix}$ is the corresponding coaction on the linking algebra $L(X)$, then $L(X) \times_\nu G$ is canonically isomorphic to $L(X \times_\zeta G)$.

PROOF. All assertions follow directly from the above considerations, except the fullness of the left inner product on $X \times_\zeta G$ and the identification of the range of the right inner product with $B_X \times_{\epsilon_X} G$. (The identity at (3.5) follows from identifying $\overline{j_G^K(C_0(G))j_X(X)}$ as the upper right corner of $\overline{j_G^L(C_0(G))j_L(L(X))}$.)

The fullness of the left inner product follows from the fact that if $q \in M(L(X))$ is a full projection, then $j_L(q)$ is a full projection in $L(X) \times_\nu G$, which follows from

$$\overline{(L(X) \times_\nu G)j_L(q)(L(X) \times_\nu G)} = \overline{j_G^L(C_0(G))j_L(L(X)qL(X))j_G^L(C_0(G))}$$
$$= \overline{j_G^L(C_0(G))j_L(L(X))j_G^L(C_0(G))} = L(X) \times_\nu G.$$

(Compare with the proof of Lemma 3.3.) If $B_X = B$, the same argument applies to give fullness of the right inner product on $X \times_\zeta G$. Thus, in this case $X \times_\zeta G$ is an imprimitivity bimodule.

To get the general formula for the right inner product we do the same trick as in the action case. Let $L^{\mathrm{imp}}(X) = \begin{pmatrix} K & X \\ \tilde X & B_X \end{pmatrix}$. We know from Lemma 2.32 that ζ is also $\mu - \epsilon_X$ compatible, so we can view $L^{\mathrm{imp}}(X)$ as a coaction invariant ideal in $L(X)$, and obtain a corresponding embedding of

$$L^{\mathrm{imp}}(X) \times_\nu G = \begin{pmatrix} K \times_\mu G & X \times_\zeta G \\ (X \times_\zeta G)^\sim & B_X \times_{\epsilon_X} G \end{pmatrix}$$

into $L(X) \times_\nu G = L(X \times_\zeta G)$, which is the identity on the upper right corner $X \times_\zeta G$ and which compresses to the inclusion $B_X \times_{\epsilon_X} G \subseteq B \times_\epsilon G$ in the lower right corner. □

PROOF OF PROPOSITION 3.9. Let $K = \mathcal{K}(X)$. By Proposition 2.30 we see that there is a unique nondegenerate coaction $\mu\colon K \to M(K \otimes C^*(G))$ such that the homomorphism $\kappa\colon A \to M(K)$ coming from the left action of A on X is $\delta - \mu$ equivariant and such that $(_KX_B, \zeta)$ is $\mu - \epsilon$ compatible. By Lemma 3.10 we get an identification $K \times_\mu G \cong \mathcal{K}(X \times_\zeta G)$. Thus the homomorphism

$$\kappa \times G\colon A \times_\delta G \to M(K \times_\mu G) \cong \mathcal{L}_{B\times_\epsilon G}(X \times_\zeta G)$$

determines a left action of $A \times_\delta G$ on $X \times_\zeta G$. Since

$$\kappa \times G = (j_K \circ \kappa) \times j_G^K = (\kappa \otimes \mathrm{id}_\mathcal{K}) \circ (j_A \times j_G^A)$$

(see the proof of Lemma A.46), it follows that this left action coincides with the action described in the proposition. □

We now introduce the dual action of G on $X \times_\zeta G$.

DEFINITION 3.11. Given a $\delta - \epsilon$ compatible nondegenerate coaction ζ of G on a right-Hilbert $A - B$ bimodule X, the *dual action* $\hat\zeta$ of G on the crossed product $X \times_\zeta G$ is defined on generators by

$$\hat\zeta_s(j_X(x) \cdot j_G(f)) = j_X(x) \cdot j_G(\rho_s(f)).$$

(Recall that ρ_s denotes right translation on $C_0(G)$ by s.)

It is straightforward to check that the above formula determines a $\hat\delta - \hat\epsilon$ compatible action on the right-Hilbert $(A \times_\delta G) - (B \times_\epsilon G)$ bimodule $X \times_\zeta G$.

For later use it is necessary to know that coaction-crossed products by standard right-Hilbert modules are standard.

LEMMA 3.12. *Suppose that $\varphi\colon A \to M(B)$ is a (possibly degenerate) homomorphism which is equivariant for nondegenerate coactions δ and ϵ of G, and let $({}_A X_B, \epsilon_X)$ be the standard coaction associated to φ (see Definition 2.25). Then the action $({}_{A\times_\delta G}(X \times_{\epsilon_X} G)_{B\times_\epsilon G}, \hat\epsilon_X)$ is equivariantly isomorphic to the standard action $({}_{A\times_\delta G} Z_{B\times_\epsilon G}, \hat\epsilon_Z)$ associated to $\varphi \times G\colon A \times_\delta G \to M(B \times_\epsilon G)$, where $Z = \varphi \times G(A \times_\delta G)(B \times_\epsilon G)$.*

PROOF. Since ϵ_X is defined as the restriction of ϵ to $X = \varphi(A)B$, it is clear that the inclusion $\iota\colon {}_A X_B \hookrightarrow {}_B B_B$ is $\epsilon_X - \epsilon$ equivariant. It follows from Lemma 1.46 that $\iota \otimes \mathrm{id}_{\mathcal{K}}\colon M_{\mathcal{K}(L^2(G))}(X \otimes \mathcal{K}(L^2(G))) \hookrightarrow M_{\mathcal{K}(L^2(G))}(B \otimes \mathcal{K}(L^2(G)))$ is an isometric inclusion. Thus, for $a \in A$ and $b \in B$, we can compute

$$(\iota \otimes \mathrm{id}_{\mathcal{K}}) \circ j_X(\varphi(a)b) = (\mathrm{id}_B \otimes \lambda) \circ (\iota \otimes \mathrm{id}_G) \circ \epsilon_X(\varphi(a)b)$$
$$= (\mathrm{id}_B \otimes \lambda) \circ \epsilon(\varphi(a)b) = j_B(\varphi(a)b).$$

Restricting to $X \times_{\epsilon_X} G = \overline{j_G(C_0(G)) j_X(\varphi(A)B) j_G(C_0(G))}$, it follows that

$$\iota \otimes \mathrm{id}_{\mathcal{K}}(X \times_{\epsilon_X} G) = \overline{j_G(C_0(G)) j_B(\varphi(A)) j_B(B) j_G(C_0(G))}$$
$$= \overline{\varphi \times G(j_G(C_0(G)) j_A(A))(j_B(B) j_G(C_0(G)))}$$
$$= \varphi \times G(A \times_\delta G)(B \times_\epsilon G).$$

It is clear that this isomorphism is $\widehat{\epsilon_X} - \hat\epsilon_Z$ equivariant and preserves all bimodule operations. □

THEOREM 3.13. *The object and morphism maps*

$$(A, \delta) \mapsto (A \times_\delta G, \hat\delta) \quad \text{and} \quad [{}_A X_B, \zeta] \mapsto [{}_{A\times G}(X \times_\zeta G)_{B\times G}, \hat\zeta]$$

define a functor from $\mathcal{C}(G)$ to $\mathcal{A}(G)$.

PROOF. To see that the morphism map is well-defined, let

$$\Phi\colon {}_{(A,\delta)}(X, \zeta)_{(B,\epsilon)} \xrightarrow{\cong} {}_{(A,\delta)}(Y, \eta)_{(B,\epsilon)}$$

be an equivariant isomorphism. Then the isomorphism $\Phi \otimes \mathrm{id}\colon X \otimes \mathcal{K}(L^2(G)) \to Y \otimes \mathcal{K}(L^2(G))$ takes j_X onto j_Y (after extending to the multiplier bimodule), hence maps $X \times_\zeta G$ onto $Y \times_\eta G$. Obviously, $\Phi \otimes \mathrm{id}$ preserves the right-Hilbert bimodule operations and is equivariant for the dual actions. Thus $[X \times_\zeta G, \hat\zeta] = [Y \times_\eta G, \hat\eta]$.

It is a very special case of Lemma 3.12 that the morphism map preserves identities.

For compositions, if $({}_A X_B, \zeta)$ is $\delta - \epsilon$ compatible and $({}_B Y_C, \eta)$ is $\epsilon - \vartheta$ compatible, we need to find a right-Hilbert $(A \times_\delta G) - (C \times_\vartheta G)$ bimodule isomorphism

$$\Theta\colon (X \times_\zeta G) \otimes_{B\times_\epsilon G} (Y \times_\eta G) \xrightarrow{\cong} (X \otimes_B Y) \times_{\zeta\sharp\eta} G$$

which is $(\hat\zeta \otimes \hat\eta) - \widehat{\zeta \sharp \eta}$ equivariant. The notation Θ is suggestive: by Lemma 2.12 we already have a right-Hilbert $(A \otimes \mathcal{K}(L^2(G))) - (C \otimes \mathcal{K}(L^2(G)))$ bimodule isomorphism

$$\Theta\colon (X \otimes \mathcal{K}(L^2(G))) \otimes_{B \otimes \mathcal{K}(L^2(G))} (Y \otimes \mathcal{K}(L^2(G))) \xrightarrow{\cong} (X \otimes_B Y) \otimes \mathcal{K}(L^2(G)),$$

and we will show that it takes $(X \times_\zeta G) \otimes_{B \times_\epsilon G} (Y \times_\eta G)$ onto $(X \otimes_B Y) \times_{\zeta \sharp \eta} G$ (after extending to multiplier bimodules). For $f, g \in C_0(G)$, $x \in X$, and $y \in Y$ we have

$$\Theta\bigl(j_G^A(f) \cdot j_X(x) \otimes j_Y(y) \cdot j_G^C(g)\bigr) = j_G^A(f) \cdot \Theta\bigl((\mathrm{id} \otimes \lambda) \circ \zeta(x) \otimes (\mathrm{id} \otimes \lambda) \circ \eta(y)\bigr) \cdot j_G^C(g)$$
$$= j_G^A(f) \cdot (\mathrm{id} \otimes \lambda) \circ (\zeta \sharp \eta)(x \otimes y) \cdot j_G^C(g) = j_G^A(f) \cdot j_{X \otimes_B Y}(x \otimes y) \cdot j_G^C(g),$$

and such elements densely span $(X \otimes_B Y) \times_{\zeta \sharp \eta} G$. It follows immediately from the construction that Θ gives an isomorphism $(X \times_\zeta G) \otimes_{B \times_\epsilon G} (Y \times_\eta G) \cong (X \otimes_B Y) \times_{\zeta \sharp \eta} G$, and a quick calculation verifies that Θ intertwines the actions $\hat\zeta \otimes \hat\eta$ and $\widehat{\zeta \sharp \eta}$. □

REMARK 3.14. It follows from the proof of Proposition 3.9 that the right-Hilbert bimodule $_{A \times_\delta G}(X \times_\zeta G)_{B \times_\epsilon G}$ factors as the product

$$_{A \times_\delta G}\bigl((K \times_\nu G) \otimes_{K \times_\mu G} (X \times_\zeta G)\bigr)_{B \times_\epsilon G},$$

where the first factor is the nondegenerate standard bimodule corresponding to the homomorphism $\kappa \times G \colon A \times_\delta G \to M(K \times_\nu G)$, and the second factor is the right-partial imprimitivity bimodule $_{K \times_\mu G}(X \times_\zeta G)_{B \times_\epsilon G}$. This decomposition is clearly equivariant with respect to the dual actions. It follows from Lemma 3.12 that $_{A \times_\delta G}(K \times_\nu G)_{K \times_\mu G}$ is equivariantly isomorphic to $_A K_K \times_\mu G$, so we see that the coaction crossed-product functor of Theorem 3.13 preserves standard decomposition of morphisms. (Compare with Remark 3.8.)

3.2. Restriction and inflation

3.2.1. Actions. If H is a closed subgroup of a locally compact group G, we can restrict any action of G to an action of H, and if N is a closed normal subgroup of G, we can inflate any action of G/N to an action of G (by composing with the quotient map). More generally, suppose we have a continuous homomorphism $\varphi \colon G \to F$ of locally compact groups. Then any action α of F gives rise to an action $\alpha \circ \varphi$ of G. We want to make this (and hence restriction and inflation) into a functor, so we need to handle the morphisms.

LEMMA 3.15. *Let γ be an $\alpha - \beta$ compatible action of F on $_A X_B$, and let $\varphi \colon G \to F$ be a continuous homomorphism. Then $\gamma \circ \varphi$ is an $\alpha \circ \varphi - \beta \circ \varphi$ compatible action of G on $_A X_B$.*

PROOF. Just note that for each $x \in X$, the map $s \mapsto (\gamma \circ \varphi)_s(x) = \gamma_{\varphi(s)}(x)$ is continuous from G into X. □

PROPOSITION 3.16. *Let G and F be locally compact groups, and let $\varphi \colon G \to F$ be a continuous homomorphism. Then the object and morphism maps*

$$(A, \alpha) \mapsto (A, \alpha \circ \varphi) \quad \text{and} \quad [_A X_B, \gamma] \mapsto [_A X_B, \gamma \circ \varphi]$$

define a functor from $\mathcal{A}(F)$ to $\mathcal{A}(G)$.

PROOF. It is obvious that the morphism map is well-defined on isomorphism classes (because any F-equivariant isomorphism is also G-equivariant) and sends identities to identities. For compositions, let $(_A X_B, \gamma)$ and $(_B Y_C, \rho)$ be actions of F. Then just observe that

$$(\gamma \otimes_B \rho) \circ \varphi = (\gamma \circ \varphi) \otimes_B (\rho \circ \varphi)$$

on ${}_A X \otimes_B Y_C$. □

Specializing to restriction and inflation, we immediately get the following corollaries:

COROLLARY 3.17. *For a closed subgroup H of G, the object and morphism maps*
$$(A, G, \alpha) \mapsto (A, H, \alpha|_H) \quad \text{and} \quad [{}_A X_B, G, \gamma] \mapsto [{}_A X_B, H, \gamma|_H]$$
define a functor from $\mathcal{A}(G)$ to $\mathcal{A}(H)$.

COROLLARY 3.18. *For a closed normal subgroup N of G, the object and morphism maps*
$$(A, G/N, \alpha) \mapsto (A, G, \operatorname{Inf} \alpha) \quad \text{and} \quad [{}_A X_B, G/N, \gamma] \mapsto [{}_A X_B, G, \operatorname{Inf} \gamma]$$
define a functor from $\mathcal{A}(G/N)$ to $\mathcal{A}(G)$.

3.2.2. Coactions. If H is a closed subgroup of a locally compact group G, we can inflate any coaction of H to a coaction of G (see Definition A.29), and if N is a closed normal subgroup we can restrict any coaction of G to a coaction of G/N (see Definition A.28). Again, we want to make functors out of restriction $\delta \mapsto \delta|$ and inflation $\delta \mapsto \operatorname{Inf} \delta$, and again there is a more general situation which unifies both: that of a continuous homomorphism $\varphi \colon G \to F$ of locally compact groups. We can integrate φ up to a nondegenerate homomorphism, still denoted by φ, from $C^*(G)$ to $M(C^*(F))$, and this allows us to pass from coactions of G to coactions of F.

LEMMA 3.19. (i) *If δ is a coaction of G on a C^*-algebra A, then $(\operatorname{id} \otimes \varphi) \circ \delta$ is a coaction of F on A. Moreover, if δ is normal or nondegenerate, then so is $(\operatorname{id} \otimes \varphi) \circ \delta$.*

(ii) *If ζ is a coaction of G on a right-Hilbert bimodule X, then $(\operatorname{id} \otimes \varphi) \circ \zeta$ is a coaction of F on X. If ζ is nondegenerate, then so is $(\operatorname{id} \otimes \varphi) \circ \zeta$.*

PROOF. (i) Certainly $(\operatorname{id} \otimes \varphi) \circ \delta$ is a nondegenerate injective homomorphism from A to $M(A \otimes C^*(F))$. Since $\delta(A) \subseteq M_G(A \otimes C^*(G))$ it follows from Proposition A.6 that $(\operatorname{id}_A \otimes \varphi) \circ \delta(A) \subseteq M_F(A \otimes C^*(F))$.

For the coaction identity,
$$(\operatorname{id} \otimes \delta_F) \circ (\operatorname{id} \otimes \varphi) \circ \delta = (\operatorname{id} \otimes \delta_F \circ \varphi) \circ \delta = \big(\operatorname{id} \otimes (\varphi \otimes \varphi) \circ \delta_G\big) \circ \delta$$
$$= (\operatorname{id} \otimes \varphi \otimes \varphi) \circ (\operatorname{id} \otimes \delta_G) \circ \delta = (\operatorname{id} \otimes \varphi \otimes \varphi) \circ (\delta \otimes \operatorname{id}) \circ \delta$$
$$= \big((\operatorname{id} \otimes \varphi) \circ \delta \otimes \varphi\big) \circ \delta = \big((\operatorname{id} \otimes \varphi) \circ \delta \otimes \operatorname{id}\big) \circ (\operatorname{id} \otimes \varphi) \circ \delta.$$

For the normality, it suffices to show that if (π, μ) is a covariant homomorphism of (A, G, δ) then $(\pi, \mu \circ \varphi^*)$ is a covariant homomorphism of $(A, F, (\operatorname{id} \otimes \varphi) \circ \delta)$, where $\varphi^* \colon C_0(F) \to M(C_0(G)) = C_b(G)$ is defined by $\varphi^*(f) = f \circ \varphi$. Normality of $(\operatorname{id} \otimes \varphi) \circ \delta$ then follows from normality of δ by applying this fact to (j_A, j_G). We have

$$\operatorname{Ad}(\mu \circ \varphi^* \otimes \operatorname{id})(w_F)(\pi(a) \otimes 1) = \operatorname{Ad}(\mu \otimes \operatorname{id})\big((\varphi^* \otimes \operatorname{id})(w_F)\big)(\pi(a) \otimes 1)$$
$$= \operatorname{Ad}(\mu \otimes \operatorname{id})\big((\operatorname{id} \otimes \varphi)(w_G)\big)(\pi(a) \otimes 1) = \operatorname{Ad}(\operatorname{id} \otimes \varphi)\big((\mu \otimes \operatorname{id})(w_G)\big)(\pi(a) \otimes 1)$$
$$= (\operatorname{id} \otimes \varphi) \circ \operatorname{Ad}(\mu \otimes \operatorname{id})(w_G)(\pi(a) \otimes 1) = (\operatorname{id} \otimes \varphi) \circ (\pi \otimes \operatorname{id}) \circ \delta(a)$$
$$= (\pi \otimes \operatorname{id}) \circ (\operatorname{id} \otimes \varphi) \circ \delta(a),$$

as desired.

For the nondegeneracy, we have

$$\begin{aligned}
\overline{(\mathrm{id}\otimes\varphi)\circ\delta(A)(1\otimes C^*(F))} &= \overline{(\mathrm{id}\otimes\varphi)\circ\delta(A)\big(1\otimes\varphi(C^*(G))C^*(F)\big)} \\
&= \overline{(\mathrm{id}\otimes\varphi)\circ\delta(A)\big(1\otimes\varphi(C^*(G))\big)(1\otimes C^*(F))} \\
&= \overline{(\mathrm{id}\otimes\varphi)\big(\delta(A)(1\otimes C^*(G))\big)(1\otimes C^*(F))} \\
&= \overline{(\mathrm{id}\otimes\varphi)(A\otimes C^*(G))(1\otimes C^*(F))} \\
&= A\otimes \varphi(C^*(G))C^*(F) = A\otimes C^*(F).
\end{aligned}$$

(ii) A similar argument handles bimodules. $\square$

THEOREM 3.20. *Let F and G be locally compact groups, and let $\varphi\colon G\to F$ be a continuous homomorphism. Then the object and morphism maps*

$$(A,\delta) \mapsto (A,(\mathrm{id}\otimes\varphi)\circ\delta) \quad \text{and} \quad [{}_AX_B,\zeta] \mapsto [{}_AX_B,(\mathrm{id}\otimes\varphi)\circ\zeta]$$

define a functor from $\mathcal{C}(G)$ to $\mathcal{C}(F)$.

PROOF. The morphism map is clearly well-defined and preserves identities. For compositions, let $({}_AX_B,\zeta)$ be $\delta-\epsilon$ compatible and $({}_BY_C,\eta)$ be $\epsilon-\vartheta$ compatible. We must show that the $(\mathrm{id}\otimes\varphi)\circ\delta - (\mathrm{id}\otimes\varphi)\circ\vartheta$ compatible coactions $(\mathrm{id}\otimes\varphi)\circ(\zeta\,\sharp_B\,\eta)$ and $((\mathrm{id}\otimes\varphi)\circ\zeta)\,\sharp_B\,((\mathrm{id}\otimes\varphi)\circ\eta)$ on the right-Hilbert $A-C$ bimodule $X\otimes_B Y$ agree. Because this requires mildly fussy bimodule-multiplier calculations, we go through it in detail. By definition of the tensor product coactions, it suffices to show that the triangle

$$X\otimes_B Y \xrightarrow{\zeta\otimes_B\eta} M\big((X\otimes C^*(G))\otimes_{B\otimes C^*(G)}(Y\otimes C^*(G))\big)$$

with maps $(\mathrm{id}\otimes\varphi)\circ\zeta\otimes_B(\mathrm{id}\otimes\varphi)\circ\eta$ and $(\mathrm{id}\otimes\varphi)\otimes_{B\otimes C^*(G)}(\mathrm{id}\otimes\varphi)$ going down to

$$M\big((X\otimes C^*(F))\otimes_{B\otimes C^*(F)}(Y\otimes C^*(F))\big)$$

and the square

$$\begin{array}{ccc}
(X\otimes C^*(G))\otimes_{B\otimes C^*(G)}(Y\otimes C^*(G)) & \xrightarrow{\Theta_G} & (X\otimes_B Y)\otimes C^*(G) \\
{\scriptstyle (\mathrm{id}\otimes\varphi)\otimes_{B\otimes C^*(G)}(\mathrm{id}\otimes\varphi)}\Big\downarrow & & \Big\downarrow {\scriptstyle \mathrm{id}\otimes\varphi} \\
M\big((X\otimes C^*(F))\otimes_{B\otimes C^*(F)}(Y\otimes C^*(F))\big) & \xrightarrow{\Theta_F} & M\big((X\otimes_B Y)\otimes C^*(F)\big)
\end{array}$$

commute, where Θ_G is the canonical isomorphism of $(X\otimes C^*(G))\otimes_{B\otimes C^*(G)}(Y\otimes C^*(G))$ onto $(X\otimes_B Y)\otimes C^*(G)$ from Lemma 2.12, and similarly for Θ_F. A trivial computation on elementary tensors shows that the triangle commutes.

We show that the square commutes: for $x\in X$, $y\in Y$, and $f,g\in C^*(G)$ we have

$$\begin{aligned}
\Theta_F \circ \big((\mathrm{id}\otimes\varphi)&\otimes_{B\otimes C^*(G)}(\mathrm{id}\otimes\varphi)\big)\big((x\otimes f)\otimes(y\otimes g)\big) \\
&= \Theta_F\big((x\otimes\varphi(f))\otimes(y\otimes\varphi(g))\big) = (x\otimes y)\otimes\varphi(fg) \\
&= (\mathrm{id}\otimes\varphi)\big((x\otimes y)\otimes fg\big) = (\mathrm{id}\otimes\varphi)\circ\Theta_G\big((x\otimes f)\otimes(y\otimes g)\big).
\end{aligned}$$

$\square$

Specializing to restriction and inflation, we immediately get the following corollaries:

COROLLARY 3.21. *For a closed normal subgroup N of G, the object and morphism maps*

$$(A, G, \delta) \mapsto (A, G/N, \delta|_{G/N}) \quad \text{and} \quad [_A X_B, G, \zeta] \mapsto [_A X_B, G/N, \zeta|_{G/N}]$$

define a functor from $\mathcal{C}(G)$ to $\mathcal{C}(G/N)$.

COROLLARY 3.22. *For a closed subgroup H of G, the object and morphism maps*

$$(A, H, \delta) \mapsto (A, G, \operatorname{Inf} \delta) \quad \text{and} \quad [_A X_B, H, \zeta] \mapsto [_A X_B, G, \operatorname{Inf} \zeta]$$

define a functor from $\mathcal{C}(H)$ to $\mathcal{C}(G)$.

3.3. Decomposition

3.3.1. Actions. Let (A, G, α) be an action, and let N be a closed normal subgroup of G. Then there is a canonical action $\tilde{\alpha}$ of G on the restricted crossed product $A \times_{\alpha|} N$, and this is usually called the decomposition action because [**25**, Proposition 1] tells us that $A \times_\alpha G$ can be decomposed into a twisted crossed product of $A \times_{\alpha|} N$ by $\tilde{\alpha}$. We will define a corresponding decomposition action α^{dec} on the reduced crossed product, and show that $(A, \alpha) \mapsto (A \times_{\alpha|,r} N, \alpha^{\mathrm{dec}})$ is the object map of a functor on $\mathcal{A}(G)$.

Decomposition actions use the modular function for conjugation of G on N:

$$\Delta_{G,N}(s) = \Delta_G(s) \Delta_{G/N}(sN)^{-1},$$

which has the property that

$$\int_N f(n)\, dn = \Delta_{G,N}(s) \int_N f(s^{-1}ns)\, dn \quad \text{for } f \in C_c(N),\ s \in G.$$

LEMMA 3.23. *Let N be a closed normal subgroup of G.*

(i) *If α is an action of G on a C^*-algebra A, then there exists a unique action α^{dec} of G on $A \times_{\alpha|,r} N$ given on $C_c(N, A)$ by*

$$\alpha_s^{\mathrm{dec}}(f)(n) = \Delta_{G,N}(s) \alpha_s(f(s^{-1}ns)).$$

(ii) *If γ is an $\alpha - \beta$ compatible action of G on a right-Hilbert bimodule $_A X_B$, then there exists a unique $\alpha^{\mathrm{dec}} - \beta^{\mathrm{dec}}$ compatible action γ^{dec} of G on $X \times_{\gamma|,r} N$ given on $C_c(N, X)$ by*

$$\gamma_s^{\mathrm{dec}}(x)(n) = \Delta_{G,N}(s) \gamma_s(x(s^{-1}ns)).$$

PROOF. For (i), since the right side of the given formula is the one defining the decomposition action $\tilde{\alpha}$ on the full crossed product $A \times_{\alpha|} N$, and since $\tilde{\alpha}$ leaves the kernel of the regular homomorphism $i_A^r \times i_N^r$ invariant [**25**, Lemma 10], the desired action α^{dec} on the reduced crossed product $A \times_{\alpha|,r} N$ exists.

For (ii), the right side of the formula gives an automorphism of the pre-right-Hilbert $C_c(N, A) - C_c(N, B)$ bimodule $C_c(N, X)$ which is continuous for the inductive limit topology and is compatible with the the actions α^{dec} and β^{dec} on the coefficients. It therefore extends to an automorphism γ_s^{dec} of $X \times_{\gamma,r} N$ by Lemma 1.23. □

THEOREM 3.24. *The object and morphism maps*

$$(A, \alpha) \mapsto (A \times_{\alpha|,r} N, \alpha^{\mathrm{dec}}) \quad \text{and} \quad [_A X_B, \gamma] \mapsto [_{A \times_r N}(X \times_{\gamma|,r} N)_{B \times_r N}, \gamma^{\mathrm{dec}}]$$

define a functor from $\mathcal{A}(G)$ to itself.

PROOF. Composing the reduced-crossed-product functor with the restriction functor, and then forgetting about the actions on the image, we get a functor $\mathcal{A}(G) \to \mathcal{C}$; we must trace the decomposition actions through the various isomorphisms. First we show that, given a right-Hilbert $A - B$ bimodule isomorphism $\Phi \colon X \to Y$ which is equivariant for $\alpha - \beta$ compatible actions γ and ρ, the associated isomorphism $\Phi \times_r N \colon X \times_{\gamma|,r} N \to Y \times_{\rho|,r} N$ from Theorem 3.7 is $\gamma^{\mathrm{dec}} - \rho^{\mathrm{dec}}$ equivariant. But this is easily checked for $x \in C_c(N, X)$:

$$(\Phi \times_r N) \circ \gamma_s^{\mathrm{dec}}(x)(n) = \Phi\big(\gamma_s^{\mathrm{dec}}(x(n))\big) = \Phi\big(\Delta_{G,N}(s)\gamma_s(x(s^{-1}ns))\big)$$
$$= \rho_s\big(\Delta_{G,N}(s)\Phi(x(s^{-1}ns))\big) = \rho_s^{\mathrm{dec}} \circ (\Phi \times_r N)(x)(n).$$

For the identities, let (A, α) be an action. From Theorem 3.7 we know that the right-Hilbert bimodule crossed product ${}_A A_A \times_r N$ is isomorphic to the right-Hilbert bimodule ${}_{A \times_r N}(A \times_r N)_{A \times_r N}$. Since the decomposition actions coincide on C_c-functions, the isomorphism is equivariant.

Turning to compositions, let $({}_A X_B, \gamma)$ be $\alpha - \beta$ compatible and $({}_B Y_C, \rho)$ be $\beta - \nu$ compatible. We need to show that the isomorphism Υ of $(X \times_r N) \otimes_{B \times_r N} (Y \times_r N)$ onto $(X \otimes_B Y) \times_r N$ from Theorem 3.7 is $(\gamma^{\mathrm{dec}} \otimes \rho^{\mathrm{dec}}) - (\gamma \otimes \rho)^{\mathrm{dec}}$ equivariant. For $s \in G$, $x \in C_c(N, X)$, $y \in C_c(N, Y)$, and $n \in N$, we have:

$$\Upsilon \circ (\gamma^{\mathrm{dec}} \otimes \rho^{\mathrm{dec}})_s(x \otimes y)(n) = \Upsilon\big(\gamma_s^{\mathrm{dec}}(x) \otimes \rho_s^{\mathrm{dec}}(y)\big)(n)$$
$$= \int_N \gamma_s^{\mathrm{dec}}(x)(k) \otimes \nu_k\big(\rho_s^{\mathrm{dec}}(y)(k^{-1}n)\big)\, dk$$
$$= \int_N \Delta_{G,N}(s)\gamma_s(x(s^{-1}ks)) \otimes \nu_k\big(\Delta_{G,N}(s)\rho_s(y(s^{-1}k^{-1}ns))\big)\, dk$$
$$= \Delta_{G,N}(s) \int_N \gamma_s(x(k)) \otimes \nu_k\big(\rho_s(y(k^{-1}s^{-1}ns))\big)\, dk$$
$$= \Delta_{G,N}(s)(\gamma \otimes \rho)_s \left(\int_N x(k) \otimes \nu_k\big(y(k^{-1}s^{-1}ns)\big)\, dk \right)$$
$$= \Delta_{G,N}(s)(\gamma \otimes \rho)_s\big(\Upsilon(x \otimes y)(s^{-1}ns)\big) = (\gamma \otimes \rho)_s^{\mathrm{dec}} \circ \Upsilon(x \otimes y)(n).$$

$\square$

3.3.2. Coactions. Let (A, G, δ) be a nondegenerate normal coaction, and let N be a closed normal subgroup of G. Recall from Lemma A.49 that the decomposition coaction δ^{dec} of G on $A \times_{\delta|} G/N$ is defined on the generators by

$$\delta^{\mathrm{dec}}(j_A(a) j_{G/N}(f)) = (j_A \otimes \mathrm{id}) \circ \delta(a)(j_{G/N}(f) \otimes 1),$$

and moreover δ^{dec} is also nondegenerate and normal. We will define a functor from $\mathcal{C}(G)$ to itself with object map

$$(A, \delta) \mapsto (A \times_{\delta|} G/N, \delta^{\mathrm{dec}}).$$

For the morphism map, we need decomposition coactions for bimodules.

PROPOSITION 3.25. *If ζ is a nondegenerate $\delta - \epsilon$ compatible coaction of G on ${}_A X_B$, then there is a unique $\delta^{\mathrm{dec}} - \epsilon^{\mathrm{dec}}$ compatible coaction ζ^{dec} of G on $X \times_{\zeta|} G/N$ which is given for $x \in X$ and $f \in C_0(G/N)$ by*

(3.6) $$\zeta^{\mathrm{dec}}(j_X(x) \cdot j_{G/N}^B(f)) = (j_X \otimes \mathrm{id}) \circ \zeta(x) \cdot (j_{G/N}^B(f) \otimes 1).$$

PROOF. Let $K = \mathcal{K}(X)$ and let $\mu\colon K \to M(K \otimes C^*(G))$ denote the unique coaction on K which makes $({}_K X_B, \zeta)$ a $\mu - \epsilon$ compatible coaction (see Proposition 2.30). Moreover, let $L(X) = \begin{pmatrix} K & X \\ \tilde{X} & B \end{pmatrix}$ denote the associated linking algebra and let $\nu| = \begin{pmatrix} \mu| & \zeta| \\ \tilde{\zeta}| & \epsilon| \end{pmatrix}$ be the corresponding coaction of G/N on $L(X)$. By the proof of Lemma 3.10 we get an identification

$$L(X) \times_{\nu|} G/N = L(X \times_{\zeta|} G/N) = \begin{pmatrix} K \times_{\mu|} G/N & X \times_{\zeta|} G/N \\ (X \times_{\zeta|} G/N)\widetilde{} & B \times_{\epsilon|} G/N \end{pmatrix}$$

which identifies the dense subspace $j_L(L(X))j^L_{G/N}(C_0(G/N))$ of $L(X) \times_{\nu|} G/N$ with

$$\begin{pmatrix} j_K(K)j^K_{G/N}(C_0(G/N)) & j_X(X)j^B_{G/N}(C_0(G/N)) \\ \widetilde{j_X(X)}j^K_{G/N}(C_0(G/N)) & j_B(B)j^B_{G/N}(C_0(G/N)) \end{pmatrix}.$$

Using Proposition 1.51 we get a canonical identification

$$M\big((L(X) \times_{\nu|} G/N) \otimes C^*(G)\big)$$
$$\cong \begin{pmatrix} M\big((K \times_{\nu|} G/N) \otimes C^*(G)\big) & M\big((X \times_{\zeta|} G/N) \otimes C^*(G)\big) \\ M\big((X \times_{\zeta|} G/N) \otimes C^*(G)\big)\widetilde{} & M\big((B \times_{\epsilon|} G/N) \otimes C^*(G)\big) \end{pmatrix}.$$

With these identifications, the decomposition coaction ν^{dec} of G on $L(X) \times_{\nu|} G/N$ is then given on the generators

$$j_L\left(\begin{pmatrix} k & x \\ y & b \end{pmatrix}\right) j^L_{G/N}(f) = \begin{pmatrix} j_K(k)j^K_{G/N}(f) & j_X(x)j^B_{G/N}(f) \\ \widetilde{j_X(y)}j^K_{G/N}(f) & j_B(b)j^B_{G/N}(f) \end{pmatrix}$$

by

$$\nu^{\mathrm{dec}}\left(j_L\left(\begin{pmatrix} k & x \\ y & b \end{pmatrix}\right) j^L_{G/N}(f)\right) = (j_L \otimes \mathrm{id}) \circ \nu\left(\begin{pmatrix} k & x \\ y & b \end{pmatrix}\right)(j^L_{G/N}(f) \otimes 1)$$
$$= \begin{pmatrix} (j_K \otimes \mathrm{id}) \circ \mu(k) & (j_X \otimes \mathrm{id}) \circ \zeta(x) \\ (j_X \otimes \mathrm{id}) \circ \zeta(y)\widetilde{} & (j_B \otimes \mathrm{id}) \circ \epsilon(b) \end{pmatrix} \begin{pmatrix} j^K_{G/N}(f) \otimes 1 & 0 \\ 0 & j^B_{G/N}(f) \otimes 1 \end{pmatrix}$$
$$= \begin{pmatrix} (j_K \otimes \mathrm{id}) \circ \mu(k)(j^K_{G/N}(f) \otimes 1) & (j_X \otimes \mathrm{id}) \circ \zeta(x)(j^B_{G/N}(f) \otimes 1) \\ (j_X \otimes \mathrm{id}) \circ \zeta(y)\widetilde{}(j^K_{G/N}(f) \otimes 1) & (j_B \otimes \mathrm{id}) \circ \epsilon(b)(j^B_{G/N}(f) \otimes 1) \end{pmatrix}.$$

Thus we see that ν^{dec} compresses on the corners $K \times_{\nu|} G/N$ and $B \times_{\epsilon|} G/N$ to the coactions μ^{dec} and β^{dec}, and it follows from Lemma 2.22 that ν^{dec} compresses to a nondegenerate $\mu^{\mathrm{dec}} - \beta^{\mathrm{dec}}$ compatible coaction on $X \times_{\zeta|} G/N$, which is then given on the generators by (3.6).

To see that ζ^{dec} is also $\delta^{\mathrm{dec}} - \epsilon^{\mathrm{dec}}$ compatible with respect to the given left action of $A \times_{\alpha|} G/N$ on $X \times_{\zeta|} G/N$, it suffices to show that the homomorphism $\kappa \times G/N\colon A \times_{\alpha|} G/N \to M(K \times_{\mu|} G/N)$ is $\delta^{\mathrm{dec}} - \mu^{\mathrm{dec}}$ equivariant, where $\kappa\colon A \to M(K) \cong \mathcal{L}_B(X)$ is determined by the left A-action on X. Using the equations

$$(\kappa \times G/N) \circ j_A = j_K \circ \kappa, \quad (\kappa \times G/N) \circ j^A_{G/N} = j^K_{G/N}, \quad \text{and} \quad (\kappa \otimes \mathrm{id}_G) \circ \delta = \mu \circ \kappa$$

(which follow from the $\delta - \mu$ equivariance of $\kappa\colon A \to M(K)$ and the definition of $\kappa \times G/N$), we can compute

$$\begin{aligned}
((\kappa \times G/N) &\otimes \mathrm{id}_G) \circ \delta^{\mathrm{dec}}(j_A(a)j^A_{G/N}(f)) \\
&= ((\kappa \times G/N) \otimes \mathrm{id}_G)((j_A \otimes \mathrm{id}_G) \circ \delta(a)(j^A_{G/N}(f) \otimes 1)) \\
&= ((j_K \circ \kappa) \otimes \mathrm{id}_G) \circ \delta(a)(j^K_{G/N}(f) \otimes 1) \\
&= (j_K \otimes \mathrm{id}_G) \circ (\kappa \otimes \mathrm{id}_G) \circ \delta(a)(j^K_{G/N}(f) \otimes 1) \\
&= (j_K \otimes \mathrm{id}_G) \circ \mu(\kappa(a))(j^K_{G/N}(f) \otimes 1) \\
&= \mu^{\mathrm{dec}}(j_K(\kappa(a))j^K_{G/N}(f)) \\
&= \mu^{\mathrm{dec}} \circ (\kappa \times G/N)(j_A(a)j^A_{G/N}(f))
\end{aligned}$$

for all $a \in A$ and $f \in C_0(G/N)$. Thus $((\kappa \times G/N) \otimes \mathrm{id}_G) \circ \delta^{\mathrm{dec}} = \mu^{\mathrm{dec}} \circ (\kappa \times G/N)$ and the proof is complete. $\square$

THEOREM 3.26. *The object and morphism maps*

$$(A, \delta) \mapsto (A \times_{\delta|} G/N, \delta^{\mathrm{dec}}) \quad \text{and} \quad [_A X_B, \zeta] \mapsto [_{A \times G/N}(X \times_{\zeta|} G/N)_{B \times G/N}, \zeta^{\mathrm{dec}}]$$

define a functor from $\mathcal{C}(G)$ to itself.

PROOF. The proof is almost the same as that of Theorem 3.24; we merely indicate the differences. We know already that the composition $(A,\delta) \mapsto (A,\delta|) \mapsto (A \times_{\delta|} G/N)$ is a functor from $\mathcal{C}(G)$ to $\mathcal{C}$; we must trace the decomposition coactions through the various isomorphisms. First, given an isomorphism Φ of $_{(A,\delta)}(X,\zeta)_{(B,\epsilon)}$ into $_{(A,\delta)}(Y,\eta)_{(B,\epsilon)}$, the associated isomorphism $\Phi \otimes \mathrm{id}\colon X \times_{\zeta|} G/N \to Y \times_{\eta|} G/N$ (from the proof of Theorem 3.13) is $\zeta^{\mathrm{dec}} - \eta^{\mathrm{dec}}$ equivariant:

$$\begin{aligned}
((\Phi \otimes \mathrm{id}) &\otimes \mathrm{id}) \circ \zeta^{\mathrm{dec}}(j_X(x) \cdot j_{G/N}(f)) \\
&= (\Phi \otimes \mathrm{id} \otimes \mathrm{id})((j_X \otimes \mathrm{id}) \circ \zeta(x) \cdot (j_{G/N}(f) \otimes 1)) \\
&= (j_Y \otimes \mathrm{id}) \circ \eta \circ \Phi(x) \cdot (j_{G/N}(f) \otimes 1) \\
&= \eta^{\mathrm{dec}}(j_Y(\Phi(x)) \cdot j_{G/N}(f)) \\
&= \eta^{\mathrm{dec}} \circ (\Phi \otimes \mathrm{id})(j_X(x) \cdot j_{G/N}(f))
\end{aligned}$$

for $x \in X$, $f \in C_0(G/N)$.

Next, it follows straight from the definitions that the decomposition coactions coincide on a standard-bimodule crossed product $_A A_A \times_{\delta|} G/N = {}_{A \times G/N}(A \times_{\delta|} G/N)_{A \times G/N}$.

Finally, fix $_{(A,\delta)}(X,\zeta)_{(B,\epsilon)}$ and $_{(B,\epsilon)}(Y,\eta)_{(C,\vartheta)}$, and let

$$\Theta\colon (X \times_{\zeta|} G/N) \otimes_{B \times G/N} (Y \times_{\eta|} G/N) \to (X \otimes_B Y) \times_{\zeta|\sharp_B \eta|} G/N$$

be the bimodule isomorphism from Theorem 3.13 (see also Lemma 2.12). Further letting

$$\begin{aligned}
\Theta_G\colon \big((X \times_{\zeta|} G/N) &\otimes C^*(G)\big) \otimes_{(B \times G/N) \otimes C^*(G)} \big((Y \times_{\eta|} G/N) \otimes C^*(G)\big) \\
&\to \big((X \times_{\zeta|} G/N) \otimes_{B \times G/N} (Y \times_{\eta|} G/N)\big) \otimes C^*(G)
\end{aligned}$$

be the canonical isomorphism which appears in the definition of $\zeta^{\text{dec}} \sharp_{B \times G/N} \eta^{\text{dec}}$, we have

$$(\Theta \otimes \text{id}) \circ (\zeta^{\text{dec}} \sharp_{B \times G/N} \eta^{\text{dec}})(j_{G/N}(f) \cdot j_X(x) \otimes j_Y(y) \cdot j_{G/N}(g))$$
$$= (\Theta \otimes \text{id}) \circ \Theta_G\big((j_{G/N}(f) \otimes 1) \cdot (j_X \otimes \text{id}) \circ \zeta(x) \otimes (j_Y \otimes \text{id}) \circ \eta(y)$$
$$\cdot (j_{G/N}(g) \otimes 1)\big)$$
$$= (\Theta \otimes \text{id})\Big((j_{G/N}(f) \otimes 1) \cdot \Theta_G\big((j_X \otimes \text{id}) \circ \zeta(x) \otimes (j_Y \otimes \text{id}) \circ \eta(y)\big)$$
$$\cdot (j_{G/N}(g) \otimes 1)\Big)$$
$$= (j_{G/N}(f) \otimes 1) \cdot (j_{X \otimes_B Y} \otimes \text{id}) \circ \Theta_G(\zeta(x) \otimes \eta(y)) \cdot (j_{G/N}(g) \otimes 1)$$
$$= (j_{G/N}(f) \otimes 1) \cdot (j_{X \otimes_B Y} \otimes \text{id}) \circ (\zeta \sharp_B \eta)(x \otimes y) \cdot (j_{G/N}(g) \otimes 1)$$
$$= (\zeta \sharp_B \eta)^{\text{dec}}\big((j_{G/N}(f) \cdot j_{X \otimes_B Y}(x \otimes y) \cdot j_{G/N}(g)\big)$$
$$= (\zeta \sharp_B \eta)^{\text{dec}} \circ \Theta\big(j_{G/N}(f) \cdot j_X(x) \otimes j_Y(y) \cdot j_{G/N}(g)\big)$$

for all $f, g \in C_0(G/N)$, $x \in X$, and $y \in Y$. Thus Θ is $(\zeta^{\text{dec}} \sharp_{B \times G/N} \eta^{\text{dec}}) - (\zeta \sharp_B \eta)^{\text{dec}}$ equivariant, which shows that the morphism map preserves compositions. □

3.4. Induced actions

Let H be a closed subgroup of G. We will define a functor from $\mathcal{A}(H)$ to $\mathcal{A}(G)$ with object map $(A, \alpha) \mapsto (\text{Ind}_H^G A, \text{Ind}\,\alpha)$. (See (B.3) in Appendix B for the definition of the induced action $\text{Ind}\,\alpha$.) Recall from [48, Corollary 3.2] that if (A, H, α) is an action, then

$$M(\text{Ind}_H^G A) = \{x \in M(A \otimes C_0(G)) \mid x(sh) = \alpha_{h^{-1}}(x(s)) \text{ for all } s \in G, h \in H\}.$$

It follows that if $\varphi \colon A \to M(B)$ is a nondegenerate homomorphism which is equivariant for actions of H, there is an associated nondegenerate homomorphism $\text{Ind}\,\varphi \colon \text{Ind}\,A \to M(\text{Ind}\,B)$ defined by

$$\text{Ind}\,\varphi(f)(s) = \varphi(f(s)).$$

For the morphism map we need the corresponding notion of an induced right-Hilbert bimodule. Let $_{(A,\alpha)}(X, \gamma)_{(B,\beta)}$ be a right-Hilbert bimodule action of H. We define

$$\text{Ind}_H^G X = \{x \in C_b(G, X) \mid x(sh) = \gamma_{h^{-1}}(x(s)) \text{ and } (sH \mapsto \|x(s)\|) \in C_0(G/H)\},$$

and for $s \in G$ we define $\text{Ind}\,\gamma_s \colon \text{Ind}\,X \to \text{Ind}\,X$ by

$$\text{Ind}\,\gamma_s(x)(t) = x(s^{-1}t).$$

LEMMA 3.27. *With the above notation,* $\text{Ind}\,X$ *becomes a right-Hilbert* $\text{Ind}\,A - \text{Ind}\,B$ *bimodule with pointwise operations*

$$(f \cdot x)(s) = f(s) \cdot x(s), \quad (x \cdot g)(s) = x(s) \cdot g(s), \quad \text{and} \quad \langle x, y \rangle_{\text{Ind}\,B}(s) = \langle x(s), y(s) \rangle_B,$$

and $\text{Ind}\,\gamma$ *is an* $\text{Ind}\,\alpha - \text{Ind}\,\beta$ *compatible action of* G *on* $\text{Ind}\,X$. *Moreover, if* $B_X = \overline{\langle X, X \rangle}_B$ *is the range of the* B-*valued inner product on* X, *then* $\overline{\langle \text{Ind}\,X, \text{Ind}\,X \rangle}_{\text{Ind}\,B} = \text{Ind}\,B_X$.

PROOF. Routine calculations show that the above formulas satisfy the algebraic properties of a right-Hilbert bimodule. The inner product $\langle \cdot, \cdot \rangle_{\text{Ind}\,B}$ is positive-definite because $\langle \cdot, \cdot \rangle_B$ is. It is also straightforward to check that for each $s \in G$ the map $\text{Ind}\,\gamma_s$ is a right-Hilbert bimodule automorphism with coefficient maps $\text{Ind}\,\alpha_s$

and $\operatorname{Ind}\beta_s$, and that for each $x \in \operatorname{Ind} X$ we have $\operatorname{Ind}\gamma_s(x) \to x$ uniformly as $s \to e$ in G.

We next observe that there are enough elements in $\operatorname{Ind} X$. For $f \in C_c(G, X)$ define $F(s) = \int_H \gamma_h(f(sh)) \, dh$; then $F \in \operatorname{Ind} X$. Moreover, if $x \in X$ and if we choose $f = g \otimes x$ with $g \in C_c(G)^+$ supported in a small neighborhood of $e \in G$ such that $\int_H g(h)\,dh = 1$, we see that $F(e) = \int_H \gamma_h(f(h))\,dh$ is close to x. Using translation, we see that there are enough elements in $\operatorname{Ind} X$ such that all evaluations at the points $s \in G$ have dense range in X.

To see that $\overline{\langle \operatorname{Ind} X, \operatorname{Ind} X \rangle}_{\operatorname{Ind} B} = \operatorname{Ind} B_X$ (viewed canonically as a closed ideal of $\operatorname{Ind} B$), we appeal to the Lemma in [**13**]. Let C_0 be the linear span of the range of the $\operatorname{Ind} B$-valued inner product. To see that C_0 is dense in $\operatorname{Ind} B_X$ it suffices to show that for each $s \in G$ the set $\{g(s) \mid g \in C_0\}$ is dense in B_X, and that C_0 is closed under multiplication by $C_c(G/H)$. But the first follows from the density in X of the images of the evaluation maps on $\operatorname{Ind} X$ and $\overline{\langle X, X \rangle}_B = B_X$, and the second from closure of $\operatorname{Ind} X$ under the right module action of $C_c(G/H) \subseteq M(\operatorname{Ind} B)$. $\square$

REMARK 3.28. If X is a partial $A - B$ imprimitivity bimodule, we also obtain a left $\operatorname{Ind} A$-valued inner product on X which turns X into a partial $\operatorname{Ind} A - \operatorname{Ind} B$ imprimitivity bimodule and, as in the proof for the range of the $\operatorname{Ind} B$-valued inner product on $\operatorname{Ind} X$ given above, we see that $_{\operatorname{Ind} A}\overline{\langle \operatorname{Ind} X, \operatorname{Ind} X \rangle} = \operatorname{Ind} A_X$ with $A_X = {}_A\langle X, X \rangle$. This shows that $\operatorname{Ind} X$ is a right-partial imprimitivity bimodule, or a left-partial imprimitivity bimodule, whenever X is.

Moreover, if $L(X) = \begin{pmatrix} A & X \\ \tilde{X} & B \end{pmatrix}$ is the linking algebra of X, equipped with the H-action induced from the H-action on X (compare with Lemma 3.3), then it is straightforward to check that the canonical identification of $C_b(G, L(X))$ with $\begin{pmatrix} C_b(G,A) & C_b(G,X) \\ C_b(G,X)\widetilde{} & C_b(G,B) \end{pmatrix}$ restricts to a canonical identification

$$\operatorname{Ind} L(X) = \begin{pmatrix} \operatorname{Ind} A & \operatorname{Ind} X \\ \widetilde{\operatorname{Ind} X} & \operatorname{Ind} B \end{pmatrix}.$$

We use the above observation for the proof of:

LEMMA 3.29. *Let H be a closed subgroup of G, $({}_AX_B, H, \gamma)$ a right-Hilbert bimodule action, and $\mathcal{F}$ a vector subspace of $\operatorname{Ind} X$ such that for each $s \in G$ the set $\{x(s) \mid x \in \mathcal{F}\}$ is dense in X, and such that $\mathcal{F}$ is closed under multiplication by $C_c(G/H)$. Then $\mathcal{F}$ is dense in $\operatorname{Ind} X$.*

PROOF. This follows very quickly from the C^*-version, which is the Lemma in [**13**], modulo a linking-algebra connection. Forgetting about the left A-module structure, we may view X as a right-partial $K - B$ imprimitivity bimodule, where $K = \mathcal{K}_B(X)$. Let $L(X)$ be the associated linking algebra. Then $\operatorname{Ind} L(X) = L(\operatorname{Ind} X)$, so by [**13**], the set

$$\mathcal{F}_1 = \begin{pmatrix} \operatorname{Ind} K & \mathcal{F} \\ \mathcal{F}^* & \operatorname{Ind} B \end{pmatrix}$$

is dense in $\operatorname{Ind} L(X)$. But then $\mathcal{F}$ is dense in the upper right corner $\operatorname{Ind} X$ of $\operatorname{Ind} L(X)$, and we are done. $\square$

PROPOSITION 3.30. *The object and morphism maps*

$$(A, \alpha) \mapsto (\operatorname{Ind} A, \operatorname{Ind} \alpha) \quad \text{and} \quad [{}_AX_B, \gamma] \mapsto [{}_{\operatorname{Ind} A}(\operatorname{Ind} X)_{\operatorname{Ind} B}, \operatorname{Ind} \gamma]$$

define a functor from $\mathcal{A}(H)$ to $\mathcal{A}(G)$.

PROOF. We first show that the morphism map is well-defined. Suppose $\Phi\colon X \to Y$ is an isomorphism of right-Hilbert $A - B$ bimodules which is equivariant for $\alpha - \beta$ compatible actions γ and ρ of G. Then the map $\operatorname{Ind}\Phi\colon \operatorname{Ind} X \to \operatorname{Ind} Y$ defined by $\operatorname{Ind}\Phi(x)(s) = \Phi(x(s))$ is easily seen to give a right-Hilbert $\operatorname{Ind} A - \operatorname{Ind} B$ bimodule isomorphism of $\operatorname{Ind} X$ onto $\operatorname{Ind} Y$. It is trivial to see that the morphism map preserves identities.

For compositions, let $({}_A X_B, \gamma)$ be $\alpha - \beta$ compatible and let $({}_B Y_C, \rho)$ be $\beta - \epsilon$ compatible. Then $\Phi(x \otimes y)(s) = x(s) \otimes y(s)$ is easily seen to give a right-Hilbert $\operatorname{Ind} A - \operatorname{Ind} C$ bimodule homomorphism $\Phi\colon \operatorname{Ind} X \otimes_{\operatorname{Ind} B} \operatorname{Ind} Y \to \operatorname{Ind}(X \otimes_B Y)$. To see that Φ is an isomorphism it remains to show that its range is dense. But this follows from Lemma 3.29, since for each $s \in G$ the set $\{x(s) \mid x \in \Phi(\operatorname{Ind} X \odot \operatorname{Ind} Y)\}$ is dense in $X \otimes_B Y$ and $\Phi(\operatorname{Ind} X \odot \operatorname{Ind} Y)$ is closed under multiplication by $C_c(G/H)$. □

When the action of the subgroup H is restricted from an action γ of the larger group G on ${}_A X_B$, the induced action on $\operatorname{Ind} X$ is isomorphic to, but not identical to, the action $\gamma \otimes \tau$ on $X \otimes C_0(G/H)$, where τ is the canonical action of G on $C_0(G/H)$ by left translation. So to deduce from Proposition 3.30 that this latter, more common construction is also functorial, we just need to make sure that this isomorphism — which is defined for $x \in C_b(G, X)$ by $\varphi(x)(sH) = \gamma_s(x(s))$ — and the related coefficient isomorphisms transport the induced actions to the respective diagonal actions on the tensor products. Unsurprisingly, this is completely straightforward, and has nothing particularly to do with the nature of the translation action τ. The result could also be proved directly via elementary means.

COROLLARY 3.31. *For any closed subgroup H of G, the object map*

$$(A, \alpha) \mapsto (A \otimes C_0(G/H), \alpha \otimes \tau)$$

and the morphism map

$$[{}_A X_B, \gamma] \mapsto [{}_{A \otimes C_0(G/H)}(X \otimes C_0(G/H))_{B \otimes C_0(G/H)}, \gamma \otimes \tau]$$

define a functor from $\mathcal{A}(G)$ to itself.

3.5. Combined functors

In the next chapter we will need to combine several functors we have already constructed, involving the category $\mathcal{AC}(G)$:

PROPOSITION 3.32. *Let N be closed normal subgroup of G.*
  (i) *The object and morphism maps*

$$(A, \alpha) \mapsto (A \times_{\alpha|, r} N, \alpha^{\mathrm{dec}}, \operatorname{Inf} \widehat{\alpha|_N}) \quad \text{and} \quad [X, \gamma] \mapsto [X \times_{\gamma|, r} N, \gamma^{\mathrm{dec}}, \operatorname{Inf} \widehat{\gamma|_N}]$$

  define a functor from $\mathcal{A}(G)$ to $\mathcal{AC}(G)$.
  (ii) *The object and morphism maps*

$$(A, \delta) \mapsto (A \times_{\delta|} G/N, \operatorname{Inf} \widehat{\delta|_{G/N}}, \delta^{\mathrm{dec}}) \quad \text{and} \quad [X, \zeta] \mapsto [X \times_{\zeta|} G/N, \operatorname{Inf} \widehat{\zeta|_{G/N}}, \zeta^{\mathrm{dec}}]$$

  define a functor from $\mathcal{C}(G)$ to $\mathcal{AC}(G)$.

PROOF. (i) We know that $(A, \alpha) \mapsto (A \times_r N, \alpha^{\text{dec}})$ is a functor from $\mathcal{A}(G)$ to $\mathcal{A}(G)$. On the other hand, $(A, \alpha) \mapsto (A \times_r N, \operatorname{Inf} \widehat{\alpha|_N})$ is a functor from $\mathcal{A}(G)$ to $\mathcal{C}(G)$ because it is the composition of the restriction, crossed-product, and inflation functors:

$$(A, G, \alpha) \mapsto (A, N, \alpha|_N) \mapsto (A \times_r N, \widehat{\alpha|_N}) \mapsto (A \times_r N, \operatorname{Inf} \widehat{\alpha|_N}).$$

To see that the combined map $(A, \alpha) \mapsto (A \times_r N, \alpha^{\text{dec}}, \operatorname{Inf} \widehat{\alpha|_N})$ is a functor, we need (as usual) to check that the morphism map is well-defined and preserves identities and compositions. But this will be easy; for example, suppose we have an equivariant isomorphism $(X, \gamma) \cong (Y, \rho)$. Then the proof of Theorem 3.24 gives a specific isomorphism $X \times_r N \cong Y \times_r N$. On the other hand, the proofs of Corollary 3.17, Theorem 3.7, and Corollary 3.22 give corresponding isomorphisms which, when composed, produce the same isomorphism as Theorem 3.24. Thus this common isomorphism is simultaneously equivariant for both the decomposition action γ^{dec} and the inflated coaction $\operatorname{Inf} \widehat{\gamma|_N}$. Therefore, the morphism map of the present theorem is well-defined. Similar arguments show that it preserves identities and compositions, and (ii) is proved similarly. □

CHAPTER 4

The Natural Equivalences

4.1. Statement of the main results

This chapter contains our main results. The main idea is that any reasonable imprimitivity theorem can be viewed as expressing a natural equivalence of appropriate functors. We shall illustrate this for two versions of Green's Imprimitivity Theorem (one with induced algebras and another which is the basis for the usual characterization of induced representations of actions), and for Mansfield's Imprimitivity Theorem (for an extensive survey of these theorems we refer to Appendix B). More precisely, as our first main result we shall prove:

THEOREM 4.1. *Let H be a closed subgroup of a locally compact group G. The Green imprimitivity bimodules $V_H^G(A, \alpha)$ implement a natural equivalence between the functors*

$$(A, \alpha) \mapsto ((\operatorname{Ind}_H^G A) \times_{\operatorname{Ind} \alpha, r} G, \widehat{\operatorname{Ind} \alpha}) \quad \text{and} \quad (A, \alpha) \mapsto (A \times_{\alpha, r} H, \operatorname{Inf}_H^G \hat{\alpha})$$

from $\mathcal{A}(H)$ to $\mathcal{C}(G)$.

Here $V_H^G(A, \alpha)$ denotes the completion of the $C_c(G, \operatorname{Ind}_H^G A) - C_c(H, A)$ pre-imprimitivity bimodule $C_c(G, A)$ with actions and inner products described in Equation (B.4). We say "implement" here because there are some implicit assertions being made. The natural equivalence must assign to each object $(A, \alpha) \in \mathcal{A}(H)$ an equivalence in the category $\mathcal{C}(G)$ from the object $((\operatorname{Ind} A) \times_r G, \widehat{\operatorname{Ind} \alpha})$ to the object $(A \times_r H, \operatorname{Inf} \hat{\alpha})$. Green's theorem says that the isomorphism class $[V_H^G(A, \alpha)]$ is an equivalence in the category $\mathcal{C}$, so we need to construct a coaction δ_V of G on $V_H^G(A, \alpha)$ such that $[V_H^G(A, \alpha), \delta_V]$ is an equivalence in $\mathcal{C}(G)$. Then to prove that the assignment $(A, \alpha) \mapsto [V_H^G(A, \alpha), \delta_V]$ is a natural equivalence we shall need to show that any morphism $[_{(A,\alpha)}(X, \gamma)_{(B,\beta)}]$ in $\mathcal{A}(H)$ gives a commutative diagram

$$\begin{array}{ccc}
((\operatorname{Ind}_H^G A) \times_r G, \widehat{\operatorname{Ind} \alpha}) & \xrightarrow{[V_H^G(A), \delta_V]} & (A \times_r H, \operatorname{Inf} \hat{\alpha}) \\
{\scriptstyle [(\operatorname{Ind} X) \times_r G, \widehat{\operatorname{Ind} \gamma}]} \downarrow & & \downarrow {\scriptstyle [X \times_r H, \operatorname{Inf} \hat{\gamma}]} \\
((\operatorname{Ind}_H^G B) \times_r G, \widehat{\operatorname{Ind} \beta}) & \xrightarrow[{[V_H^G(B), \delta_V]}]{} & (B \times_r H, \operatorname{Inf} \hat{\beta})
\end{array}$$

in the category $\mathcal{C}(G)$ (see Theorem 4.15). For obvious reasons we shall state this more simply as "the diagram

$$(4.1) \qquad \begin{array}{ccc} (\operatorname{Ind}_H^G A) \times_r G & \xrightarrow{V_H^G(A)} & A \times_r H \\ {\scriptstyle (\operatorname{Ind} X) \times_r G} \downarrow & & \downarrow {\scriptstyle X \times_r H} \\ (\operatorname{Ind}_H^G B) \times_r G & \xrightarrow[V_H^G(B)]{} & B \times_r H \end{array}$$

commutes equivariantly for the appropriate coactions." We shall construct the coaction δ_V in Section 4.3 below and we shall also give a proof of Theorem 4.1 in that section.

If we start with an action α of G on a C^*-algebra A, then $(\operatorname{Ind}_H^G A) \times_r G$ is naturally isomorphic to $C_0(G/H, A) \times_{\alpha \otimes \tau, r} G$, and (after identifying the corresponding coaction on this algebra) Theorem 4.1 immediately gives another natural equivalence between crossed-product functors; equivalently, commutativity of the diagram

$$\begin{array}{ccc} C_0(G/H, A) \times_r G & \xrightarrow{X_H^G(A)} & A \times_r H \\ {\scriptstyle C_0(G/H,X) \times_r G} \downarrow & & \downarrow {\scriptstyle X \times_r H} \\ C_0(G/H, B) \times_r G & \xrightarrow[X_H^G(B)]{} & B \times_r H, \end{array}$$

equivariantly for the appropriate coactions, where $X_H^G(A)$ is the $C_0(G/H, A) \times_{\alpha \otimes \tau, r} G - A \times_{\alpha|, r} H$ imprimitivity bimodule of Theorem B.3.

If we require the subgroup to be *normal*, we can also get equivariance for appropriate *actions*. (In the following discussion we call the subgroup N instead of H to emphasize that it will be normal in G.) In this situation there exists a natural isomorphism between the crossed product $C_0(G/N, A) \times_{\alpha \otimes \tau, r} G$ and the iterated crossed product $A \times_{\alpha, r} G \times_{\hat{\alpha}|} G/N$, which transforms the dual coaction $\widehat{\alpha \otimes \tau}$ into the decomposition coaction $\hat{\alpha}^{\text{dec}}$ (see the proof of Lemma 4.18 below). The G-action on the iterated crossed product is the action $\operatorname{Inf} \widehat{\hat{\alpha}|}$: we first take the dual coaction $\hat{\alpha}$, then the restriction $\hat{\alpha}|$ to G/N, then the dual action $\widehat{\hat{\alpha}|}$ of G/N, and then inflate this to an action of G. The counterpart on the $A \times_r N$-side is given by the decomposition action α^{dec} as described in Section 3.3.3. Mainly as a consequence of Theorem 4.1 we then derive:

THEOREM 4.2. *Let N be a closed normal subgroup of a locally compact group G. The Green imprimitivity bimodules $X_N^G(A, \alpha)$ implement a natural equivalence between the functors*

$$(A, \alpha) \mapsto (A \times_{\alpha, r} G \times_{\hat{\alpha}|, r} G/N, \operatorname{Inf} \widehat{\hat{\alpha}|}, \hat{\alpha}^{\text{dec}}) \quad \text{and} \quad (A, \alpha) \mapsto (A \times_{\alpha, r} N, \alpha^{\text{dec}}, \operatorname{Inf} \widehat{\hat{\alpha}|})$$

from $\mathcal{AC}(G)$ to itself.

Finally, in Section 4.5 we shall derive a complete dual version of Theorem 4.2. For this let δ be a nondegenerate normal coaction of G on the C^*-algebra A, and let N be a closed normal subgroup of G. Using the generalization of Mansfield's imprimitivity theorem for coactions as presented in [**28**], we obtain an imprimitivity bimodule $Y_{G/N}^G(A, \delta)$ for the crossed products $A \times_\delta G \times_{\hat{\delta}|, r} N$ and $A \times_{\delta|} G/N$ (see Theorem B.6). Again, both crossed products carry canonical actions and coactions of G: the action on $A \times_\delta G \times_{\hat\delta|,r} N$ is the decomposition action $\hat\delta^{\text{dec}}$, and the action on $A \times_{\delta|} G/N$ is the inflation $\operatorname{Inf} \widehat{\hat\delta|}$ of the dual action $\widehat{\hat\delta|}$ of G/N on $A \times_{\delta|} G/N$. Similarly, the coaction on $A \times_\delta G \times_{\hat\delta|,r} N$ is the inflation $\operatorname{Inf} \widehat{\hat\delta|}$, and the coaction on $A \times_{\delta|} G/N$ is the decomposition coaction δ^{dec} as described in Appendix A (see Lemma A.49).

THEOREM 4.3. *Let N be a closed normal subgroup of a locally compact group G. The Mansfield imprimitivity bimodules $Y_{G/N}^G(A,\delta)$ implement a natural equivalence between the functors*

$$(A,\delta) \mapsto (A \times_{\delta,r} G \times_{\hat\delta|,r} N, \hat\delta^{\text{dec}}, \operatorname{Inf} \widehat{\hat\delta|}) \quad \text{and} \quad (A,\delta) \mapsto (A \times_{\delta,r} G/N, \operatorname{Inf} \widehat{\hat\delta|}, \delta^{\text{dec}})$$

from $\mathcal{AC}(G)$ to itself.

The construction of the appropriate actions and coactions on the bimodules, and the proof of Theorem 4.3, will be given in Section 4.5. The strategy of the proofs of all three theorems discussed above is to factor the right-Hilbert bimodule X which appears in Diagram (4.1) (or in the appropriate diagrams corresponding to Theorem 4.2 and Theorem 4.3) into the tensor product of a standard bimodule and a right-partial imprimitivity bimodule. Then we prove the theorem in these special cases, and put the results together (using functoriality) to get the desired result for general right-Hilbert bimodules.

4.2. Some further linking algebra techniques

LEMMA 4.4. *Suppose Z is a right-Hilbert $E-F$ bimodule. Let $p \in M(E)$ and $q \in M(F)$ be projections. Then, via restriction of the actions and inner product, pZq becomes a right-Hilbert $pEp-qFq$ bimodule. If p is a full projection, then pZq is a full right-Hilbert $pEp-qFq$ bimodule.*

Suppose in addition that ${}_\varphi\Phi_\psi \colon {}_EZ_F \to M({}_RW_S)$ is a nondegenerate right-Hilbert bimodule homomorphism and that $q \in M(F)$ is full. Then $\psi(q)$ is a full projection in $M(S)$, and Φ restricts to give a nondegenerate right-Hilbert bimodule homomorphism

$$_{pEp}(pZq)_{qFq} \to M\Big({}_{\varphi(p)R\varphi(p)}(\varphi(p)W\psi(q))_{\psi(q)S\psi(q)}\Big).$$

PROOF. First note that $pEpE = pE$, which follows from the fact that pE is a left pEp-Hilbert module with respect to the canonical module operations. Similarly, we have $FqFq = Fq$.

It is clear that $pEp \cdot pZq \subseteq pZq$ and $pZq \cdot qFq \subseteq pZq$, so the actions restrict. Likewise, $\langle pZq, pZq \rangle_F \subseteq qFq$, so the right inner product also restricts. Nondegeneracy of the left pEp-module action follows from the nondegeneracy of the left module action of E on Z and

$$pEp \cdot pZq = pEpE \cdot Zq = pE \cdot Zq = pZq.$$

If p is full, *i.e.*, if EpE is dense in E, it follows that

$$\langle pZq, pZq \rangle_{qFq} = q\langle pE \cdot Z, pE \cdot Z \rangle_F q = q\langle Z, EpE \cdot Z \rangle_F q$$

is dense in qFq. Hence pZq is full.

For the other part, first note that

$$R\varphi(p)R = R\varphi(E)\varphi(p)\varphi(E)R = R\varphi(EpE)R$$

is dense in R, because p is full and φ is nondegenerate.

Now clearly Φ maps pEp into $M(\varphi(p)R\varphi(p))$, pZq into $M(\varphi(p)W\psi(q))$, and qFq into $M(\psi(q)S\psi(q))$; the only remaining issue is the nondegeneracy of the restrictions of φ, Φ, and ψ. This is clear for φ and ψ: for example, using $pEpE = pE$ and the nondegeneracy of φ we get

$$\varphi(pEp)R = \varphi(pEp)(\varphi(E)R) = \overline{\varphi(pEpE)R} = \varphi(pE)R = \varphi(p)R,$$

which clearly implies $\varphi(pEp)R\varphi(p) = \varphi(p)R\varphi(p)$. To prove nondegeneracy of Φ we need fullness of q: we then get

$$\overline{\Phi(pZq)\cdot\psi(q)S\psi(q)} = \overline{\Phi(pZ\cdot Fq)\cdot\psi(F)S\psi(q)} = \overline{\varphi(p)\Phi(Z)\cdot\psi(FqF)S\psi(q)}$$
$$= \overline{\varphi(p)\Phi(Z)\cdot\psi(F)S\psi(q)} = \overline{\varphi(p)\Phi(Z)\cdot S\psi(q)} = \varphi(p)W\psi(q).$$

□

REMARK 4.5. In the above lemma, if Z is a (partial) $E-F$ imprimitivity bimodule, then pZq becomes a (partial) $pEp - qFq$ imprimitivity bimodule—just argue as above for the left inner product as well. If, in addition, $_EZ_F$ and $_RW_S$ are imprimitivity bimodules, the restriction of Φ to pZq is also a nondegenerate imprimitivity bimodule homomorphism, by Remark 1.19.

The following result generalizes [**21**, Lemma 4.6].

LEMMA 4.6. *Suppose $_AX_B$ and $_CY_D$ are partial imprimitivity bimodules, and Z is a right-Hilbert $L(X) - L(Y)$ bimodule. Let $p = \begin{pmatrix} 1 & 0 \\ 0 & 0 \end{pmatrix}$ and $q = \begin{pmatrix} 0 & 0 \\ 0 & 1 \end{pmatrix}$ denote the canonical full projections in both $M(L(X))$ and $M(L(Y))$. Then:*

(i) *Via restriction of the actions and inner products, pZp becomes a right-Hilbert $A-C$ bimodule, pZq becomes a right-Hilbert $A-D$ bimodule, and qZq becomes a right-Hilbert $B-D$ bimodule.*

(ii) *There are natural isometric right-Hilbert $A-D$ bimodule homomorphisms*

$$\Phi\colon X\otimes_B qZq \to pZq \quad \text{and} \quad \Psi\colon pZp\otimes_C Y \to pZq$$

given by $\Phi(x\otimes z) = x\cdot z$ and $\Psi(w\otimes y) = w\cdot y$.

(iii) *Let us now denote by q_X the q-projection in $M(L(X))$ and by p_Y the p-projection in $M(L(Y))$. Then, if q_X is full, the homomorphism Φ of (ii) is an isomorphism. Similarly, if p_Y is full, then Ψ is an isomorphism. In particular, both maps are isomorphisms if $_AX_B$ and $_CY_D$ are imprimitivity bimodules.*

(iv) *If Z is a partial $L(X)-L(Y)$ imprimitivity bimodule and if $_A\langle pZq,pZq\rangle \subseteq \overline{_A\langle X,X\rangle}$, then the map Φ of (ii) is an isomorphism. Similarly, Ψ is an isomorphism if $\langle pZq,pZq\rangle_D \subseteq \overline{\langle Y,Y\rangle_D}$.*

PROOF. Since $pL(X)p = A$, $qL(X)q = B$, and so on, (i) follows directly from Lemma 4.4.

In order to prove (ii), first observe that Φ clearly intertwines the left actions. To check that Φ preserves the D-valued inner products, fix $x,x' \in X$ and $z,z' \in qZq$, and compute:

$$\langle x\otimes z, x'\otimes z'\rangle_D = \langle z,\langle x,x'\rangle_B\cdot z'\rangle_D = \langle z, x^*x'\cdot z'\rangle_D$$
$$= \langle x\cdot z, x'\cdot z'\rangle_D = \langle \Phi(x\otimes z),\Phi(x'\otimes z')\rangle_D.$$

A similar argument applies to Ψ.

4.2. SOME FURTHER LINKING ALGEBRA TECHNIQUES

Suppose now that q_X is full. Then $L(X)q_X L(X)$ is dense in $L(X)$, and using $L(X) \cdot Z = Z$ by nondegeneracy of the left action, we see that

$$\Phi(X \otimes qZq) = pL(X)q_X \cdot Zq = pL(X)q_X L(X) \cdot Zq$$

is dense in pZq. Similarly, the fullness of p_Y implies that Ψ is surjective. This proves (iii).

For the proof of (iv) assume that Z is a partial $L(X) - L(Y)$ imprimitivity bimodule. Let $A_X = {}_A\overline{\langle X, X \rangle}$ and let $L^r(X) = \begin{pmatrix} A_X & X \\ \tilde{X} & B \end{pmatrix}$ be the linking algebra of the right-partial imprimitivity bimodule ${}_{A_X}X_B$. Then $L^r(X)$ is a closed ideal of $L(X)$, and we can consider the partial $L^r(X) - L(Y)$ bimodule $L^r(X) \cdot Z \subseteq Z$. We claim that $pL^r(X)Zq = pZq$. To see this we simply note that $A_X pZq = pZq$, since pZq is a left A_X-Hilbert module by assumption; this implies that $p(L^r(X) \cdot Z)q \subseteq A_X \cdot pZq = pZq$. Since ${}_{A_X}X_B$ is a right-partial imprimitivity bimodule, it follows from Proposition 1.48 that q_X is a full projection in $M(L^r(X))$, and thus it follows from the proof of (iii) that $pL(X)q_X Zq \supseteq pL^r(X)q_X Zq = pZq$. But this implies surjectivity of Φ. A similar argument shows that if $\langle pZq, pZq \rangle_D \subseteq \overline{\langle Y, Y \rangle}_D$, then Ψ is surjective. $\square$

Before continuing to develop our general techniques, let us first explain how a result like Lemma 4.6 can be used in the proofs of the main theorems. For this we go back to the situation of Theorem 4.1. As explained earlier, the proof of this theorem requires us to show the equivariant commutativity of Diagram Example 4.1, i.e., that the bimodule compositions $V_H^G(A) \otimes_{A \times_r H} (X \times_r H)$ and $((\operatorname{Ind} X) \times_r G) \otimes_{(\operatorname{Ind} B) \times_r G} V_H^G(B)$ are equivariantly isomorphic. As mentioned at the end of the previous section, our strategy is to factor the problem into two cases, and it is the partial imprimitivity bimodule case where the above lemma becomes extremely helpful. To see this, assume now that X is a partial $A - B$ imprimitivity bimodule, and let $V_H^G(L)$ be Green's imprimitivity bimodule for the associated G-action on the linking algebra $L(X)$. By Lemma 3.3 and Remark 3.28 we have the identifications

$$(\operatorname{Ind} L(X)) \times_r G = L((\operatorname{Ind} X) \times_r G) \quad \text{and} \quad L(X) \times_r H = L(X \times_r H).$$

If we now apply part (ii) of Lemma 4.6 to the right-Hilbert $L((\operatorname{Ind} X) \times_r G) - L(X \times_r H)$ bimodule $Z = V_H^G(L)$, we get $pZp = V_H^G(A)$ and $qZq = V_H^G(B)$ (which follows easily from the construction of V_H^G as given in Theorem B.2). If we can show that in this situation the maps Φ and Ψ are both surjective, we directly obtain the desired isomorphism

$$V_H^G(A) \otimes_{A \times_r H} (X \times_r H) \cong ((\operatorname{Ind} X) \times_r G) \otimes_{(\operatorname{Ind} B) \times_r G} V_H^G(B),$$

and it only remains to show the equivariance of the diagram with respect to appropriate actions and coactions. So we pause to obtain the desired surjectivity result. It follows from item (iv) of Lemma 4.6 together with:

LEMMA 4.7. *Suppose that $({}_A X_B, \gamma)$ is an $\alpha - \beta$ compatible action on the partial $A - B$ imprimitivity bimodule X, and let $L(X)$ denote the linking algebra of ${}_A X_B$ with corresponding action ν. As above, we regard $V_H^G(L)$ as a right-Hilbert $L((\operatorname{Ind} X) \times_r G) - L(X \times_r H)$ bimodule. Then, with the usual meanings of p and q, we have*

$${}_{(\operatorname{Ind} A) \times_r G}\langle pV_H^G(L)q, pV_H^G(L)q \rangle \subseteq {}_{(\operatorname{Ind} A) \times_r G}\overline{\langle (\operatorname{Ind} X) \times_r G, (\operatorname{Ind} X) \times_r G \rangle} \quad \text{and}$$

$$\langle pV_H^G(L)q, pV_H^G(L)q\rangle_{B\times_r H} \subseteq \overline{\langle X\times_r H, X\times_r H\rangle}_{B\times_r H}.$$

PROOF. By construction (see Theorem B.2), $V_H^G(L)$ is a completion of

$$C_c(G, L(X)) = \begin{pmatrix} C_c(G,A) & C_c(G,X) \\ C_c(G,\widetilde{X}) & C_c(G,B) \end{pmatrix}$$

with module actions given by the formulas (B.4). It follows that $pV_H^G(L)q$ is the closure of the upper right corner $C_c(G,X)$. Using the the multiplication rule in $L(X)$ together with the formulas for the $L(X)\times_r H$-valued inner product on $C_c(G, L(X))$ (and the fact that this inner product restricted to $C_c(G,X)$ takes values in $C_c(H,B) \subseteq B\times_r H$) we get for all $x, y \in C_c(G, X)$

$$\langle x, y\rangle_{C_c(H,B)}(h) = \Delta_H(h)^{-1/2} \int_G \langle x(t^{-1}), \gamma_h(y(t^{-1}h))\rangle_B \, dt,$$

which determines an element in $C_c(H, B_X)$, with $B_X = \overline{\langle X, X\rangle}_B$. But this implies that

$$\langle C_c(G,X), C_c(G,X)\rangle_{B\times_r H} \subseteq C_c(G, B_X) \subseteq B_X\times_r H = \overline{\langle X\times_r H, X\times_r H\rangle}_{B\times_r H},$$

where the last equation follows from Lemma 3.3. A very similar argument shows that

$$_{(\operatorname{Ind} A)\times_r G}\langle C_c(G,X), C_c(G,X)\rangle \subseteq C_c(G, \operatorname{Ind} A_X)$$
$$\subseteq {}_{(\operatorname{Ind} A)\times_r G}\overline{\langle (\operatorname{Ind} X)\times_r G, (\operatorname{Ind} X)\times_r G\rangle}.$$

□

Of course, as the statement of Theorem 4.1 indicates, we also need to show that the above isomorphism is equivariant with respect to the given coactions. For this to make sense we first have to define the appropriate coactions on Green's bimodules V_H^G, which we shall do in the next section. However, as soon as this has been done in a coherent way (i.e., in such a way that the coaction on $V_H^G(L)$ compresses to the corresponding coactions on the corners $V_H^G(A)$ and $V_H^G(B)$, respectively), then this equivariance will follow from:

LEMMA 4.8. *Suppose ${}_AX_B$ and ${}_CY_D$ are right-partial imprimitivity bimodules and Z is a right-Hilbert $L(X) - L(Y)$ bimodule, and let p and q be as in Lemma 4.6. Suppose in addition that τ is a $\mu - \nu$ compatible coaction of G on Z, where μ and ν are coactions of G on $L(X)$ and $L(Y)$ arising from coactions ζ and η on X and Y, respectively (as in Equation (3.4)). Then:*

  (i) *τ restricts to give right-Hilbert bimodule coactions of G on pZq and qZq which are compatible with the appropriate coefficient coactions.*
  (ii) *Assume further that τ restricts to a right-Hilbert bimodule coaction on pZp. Then the isometric homomorphisms $\Phi\colon X\otimes_B qZq \to pZq$ and $\Psi\colon pZp\otimes_C Y \to pZq$ of Lemma 4.6 are equivariant for the appropriate coactions.*

PROOF. By definition, if δ is the left coefficient coaction of ζ, then

$$\mu(p) = \begin{pmatrix}\delta(1) & 0 \\ 0 & 0\end{pmatrix} = \begin{pmatrix}1 & 0 \\ 0 & 0\end{pmatrix}\otimes 1 = p\otimes 1$$

in $M(L(X)\otimes C^*(G))$, and similarly $\mu(q) = q\otimes 1$, $\nu(p) = p\otimes 1$, and $\nu(q) = q\otimes 1$. Thus, since q is full (since the left inner products on ${}_AX_B$ and ${}_CY_D$ are full by assumption), it follows from Lemma 4.4 that the restriction of τ to pZq is a nondegenerate right-Hilbert bimodule homomorphism into $M((p\otimes 1)(Z\otimes C^*(G))(p\otimes 1)) = M(pZp \otimes C^*(G))$, and similarly for qZq. Since it is immediate that each of these restrictions satisfies the appropriate versions of conditions (i)–(ii) in Definition 2.10, this establishes (i) above.

To check the equivariance of Φ in (ii), first note that if

$$\Theta \colon (X \otimes C^*(G)) \otimes_{B\otimes C^*(G)} (qZq \otimes C^*(G)) \xrightarrow{\cong} (X \otimes_B qZq) \otimes C^*(G)$$

is the isomorphism from Lemma 2.12, then the extension to the G-multiplier bimodules satisfies

$$(\Phi \otimes \mathrm{id}) \circ \Theta(u \otimes v) = u \cdot v$$

for $u \in M_G(X \otimes C^*(G)) \subseteq M_G(L(X) \otimes C^*(G))$ and $v \in M_G(qZq \otimes C^*(G)) \subseteq M_G(Z \otimes C^*(G))$. This can be seen by taking elementary tensors $x \otimes c \in X \odot C^*(G)$ and $z \otimes d \in qZq \odot C^*(G)$ and computing

$$(\Phi \otimes \mathrm{id}) \circ \Theta\big((x\otimes c) \otimes (z \otimes d)\big) = (\Phi \otimes \mathrm{id})\big((x \otimes z) \otimes cd\big) = x \cdot z \otimes cd = (x \otimes c) \cdot (z \otimes d),$$

and then extending to the G-multiplier bimodules.

Now, by definition, $(\zeta \, \natural_B \, \tau)(x \otimes z) = \Theta(\zeta(x) \otimes \tau(z))$ for $x \in X$ and $z \in qZq$. Thus, we can compute:

$$(\Phi\otimes\mathrm{id})\circ(\zeta \, \natural_B \, \tau)(x\otimes z) = (\Phi\otimes\mathrm{id})\circ\Theta(\zeta(x)\otimes\tau(z)) = \zeta(x)\cdot\tau(z) = \tau(x\cdot z) = \tau\circ\Phi(x\otimes z).$$

The equivariance of Ψ is proved similarly. $\square$

For the proofs of Theorem 4.2 and Theorem 4.3 we shall also need a version of Lemma 4.8 which handles actions:

LEMMA 4.9. *Suppose $_AX_B$ and $_CY_D$ are partial imprimitivity bimodules and Z is a right-Hilbert $L(X) - L(Y)$ bimodule, and let p and q be as in Lemma 4.6. Suppose in addition that σ is a $\mu - \nu$ compatible action of G on Z, where μ and ν are actions of G on $L(X)$ and $L(Y)$ arising from actions γ and ρ on X and Y, respectively. Then:*

  (i) *σ restricts to give right-Hilbert bimodule actions of G on pZp, pZq, and qZq which are compatible with the appropriate coefficient actions.*
  (ii) *The isometric homomorphisms $\Phi \colon X \otimes_B qZq \to pZq$ and $\Psi \colon pZp \otimes_C Y \to pZq$ of Lemma 4.6 are equivariant for the appropriate actions.*

PROOF. If α is the left coefficient action of γ, then for all $s \in G$ we have

$$\mu_s(p) = \begin{pmatrix} \alpha_s(1) & 0 \\ 0 & 0 \end{pmatrix} = \begin{pmatrix} 1 & 0 \\ 0 & 0 \end{pmatrix} = p,$$

and similarly $\mu_s(q) = q$, $\nu_s(p) = p$, and $\nu_s(q) = q$. Thus

$$\sigma_s(pZp) = \mu_s(p)\sigma_s(Z)\nu_s(p) = pZp,$$

and similarly for pZq and qZq, so σ restricts appropriately to the various corners. It is immediate that these restrictions are compatible with the coefficient actions as indicated, so this establishes (i).

For (ii), fix $x \in X$, $z \in qZq$, and $s \in G$ and compute, for example, that

$$\Phi \circ (\gamma \otimes \sigma)_s(x \otimes z) = \Phi(\gamma_s(x) \otimes \sigma_s(z)) = \gamma_s(x) \cdot \sigma_s(z) = \sigma_s(x \cdot z) = \sigma_s \circ \Phi(x \otimes z).$$

Similar calculations show that Ψ is appropriately equivariant; this completes the proof. $\square$

Finally, the following lemma will help to do the nondegenerate homomorphism parts of our main theorems.

LEMMA 4.10. *Suppose* $_\varphi\Phi_\psi\colon {}_AX_B \to M({}_CY_D)$ *is a nondegenerate imprimitivity bimodule homomorphism. Then*:

(i) *The following diagram (of morphisms) commutes*:

(4.2)
$$\begin{array}{ccc} A & \xrightarrow{X} & B \\ \varphi \downarrow & & \downarrow \psi \\ C & \xrightarrow{Y} & D. \end{array}$$

(ii) *If* Φ *is equivariant for imprimitivity bimodule coactions of G on X and Y, then Diagram (4.2) commutes equivariantly for the appropriate coactions.*
(iii) *If* Φ *is equivariant for imprimitivity bimodule actions of G on X and Y, then Diagram (4.2) commutes equivariantly for the appropriate actions.*

PROOF. Part (i) is exactly [**29**, Lemma 5.3]; we re-prove it here for completeness, and because it follows easily from Lemma 4.6. Simply note that, by [**20**, Remark (2) of Appendix], Φ gives rise to a nondegenerate homomorphism

$$\begin{pmatrix} \varphi & \Phi \\ \tilde\Phi & \psi \end{pmatrix} \colon L(X) \to M(L(Y)).$$

This makes $L(Y)$ into a right-Hilbert $L(X) - L(Y)$ bimodule Z such that pZp is C with the $A - C$ bimodule structure from φ and qZq is D with the $B - D$ bimodule structure from ψ. Hence Lemma 4.6 (iii) provides an isomorphism

(4.3) $$X \otimes_B D \cong C \otimes_C Y$$

of right-Hilbert $A - D$ bimodules, which establishes the commutativity of Diagram (4.2).

For (ii), suppose the coactions on X and Y are ζ and η. The $\zeta - \eta$ equivariance of Φ implies that ν is $\mu - \nu$ compatible, where μ and ν are the associated coactions on the linking algebras, so Lemma 4.8 gives the equivariance of Diagram (4.2) for the coactions. Similarly, (iii) follows from Lemma 4.9. $\square$

4.3. Green's Theorem for induced algebras

This section is devoted to the proof of Theorem 4.1. We start with the construction of the appropriate coaction δ_V on Green's $((\operatorname{Ind}_H^G A) \times_{\operatorname{Ind}\alpha,r} G) - (A \times_{\alpha,r} H)$ imprimitivity bimodule $V_H^G(A,\alpha)$. Theorem B.2 tells us that $V_H^G(A)$ (we streamline the notation when confusion is unlikely) is a completion of the $C_c(G, \operatorname{Ind}_H^G A) - C_c(H, A)$ pre-imprimitivity bimodule $C_c(G, A)$ with actions and inner products given in Equation (B.4). We shall make extensive use of the embedding of $C_c(G, M^\beta(A))$ into $M(A \times_r G)$ as provided by Corollary C.7. Moreover, Proposition C.9 shows that $C_c(G, M^\beta(A))$ also embeds in $M(V_H^G(A))$. As we shall make more precise in Lemma 4.14 below, we may similarly view $C_c(G, M^\beta(A \otimes C^*(G)))$ as a subspace of $M(V_H^G(A) \otimes C^*(G))$, which then allows us to state:

THEOREM 4.11. *Let G be a locally compact group, and let (A, H, α) be an action of a closed subgroup H of G. There is a unique imprimitivity bimodule coaction δ_V*

of G on $V_H^G(A, \alpha)$ implementing a Morita equivalence between the coactions $\widehat{\operatorname{Ind} \alpha}$ on $(\operatorname{Ind} A) \times_{\operatorname{Ind}\alpha, r} G$ and $\operatorname{Inf} \hat{\alpha}$ on $A \times_{\alpha, r} H$, such that for $x \in C_c(G, A)$ we have

(4.4) $\qquad \delta_V(x) \in C_c(G, M^\beta(A \otimes C^*(G))) \subseteq M(V_H^G(A) \otimes C^*(G)),$

with $\delta_V(x)(s) = x(s) \otimes s$ for $s \in G$.

REMARK 4.12. We should point out that a Morita equivalence for the coactions $\operatorname{Inf} \hat{\alpha}$ and $\widehat{\operatorname{Ind} \alpha}$ has been obtained also in [**22**] as a corollary of a more general equivariance result for the dual coactions appearing in the symmetric imprimitivity theorem. However, the realization of the coaction on the bimodule $V_H^G(A)$ was quite different from that given in the above theorem; the realization we give here makes it much easier to obtain certain equivariance results for our bimodules.

Before we come to the proof of this theorem, we have to explain a bit more how to work with strictly continuous compactly-supported functions in the various multiplier algebras and bimodules. Observe first of all that if H is a closed subgroup of G and (A, H, α) is an action, then for any C^*-algebra C the canonical embedding of $C_b(G, A) \odot C$ into $C_b(G, A \otimes C)$ determines an isomorphism ψ of $(\operatorname{Ind} A) \otimes C$ onto $\operatorname{Ind}(A \otimes C)$ which is equivariant for the actions $(\operatorname{Ind} \alpha) \otimes \operatorname{id}$ and $\operatorname{Ind}(\alpha \otimes \operatorname{id})$. We use the (inverse of the) composition

$$((\operatorname{Ind} A) \times_r G) \otimes C \xrightarrow{\varphi} ((\operatorname{Ind} A) \otimes C) \times_r G \xrightarrow{\psi \times_r G} (\operatorname{Ind}(A \otimes C)) \times_r G$$

to embed $C_c(G, \operatorname{Ind}(A \otimes C))$ (regarded as functions of two variables as in the discussion following Equation (B.2) in Appendix B) into $((\operatorname{Ind} A) \times_r G) \otimes C$, and to embed $C_c(G, M^\beta(\operatorname{Ind}(A \otimes C)))$ into $M(((\operatorname{Ind} A) \times_r G) \otimes C)$[1]. This allows us to work with the dual coaction $\widehat{\operatorname{Ind} \alpha}$ in terms of two-variable functions:

LEMMA 4.13. *With notation as above, for $f \in C_c(G, \operatorname{Ind} A)$ we have*

$$\widehat{\operatorname{Ind} \alpha}(f) \in C_c(G, M^\beta(\operatorname{Ind}(A \otimes C^*(G)))) \subseteq M(((\operatorname{Ind} A) \times_r G) \otimes C^*(G)),$$

with $\widehat{\operatorname{Ind} \alpha}(f)(s, t) = f(s, t) \otimes s$ for $s, t \in G$.

PROOF. By Lemma 3.1 we have $\widehat{\operatorname{Ind} \alpha}(f) \in C_c(G, M^\beta((\operatorname{Ind} A) \otimes C^*(G)))$, so via ψ we can identify $\widehat{\operatorname{Ind} \alpha}(f)$ with an element of $C_c(G, M^\beta(\operatorname{Ind}(A \otimes C^*(G))))$. Using our conventions and the formula from Lemma 3.1, we compute:

$$\widehat{\operatorname{Ind} \alpha}(f)(s, t) = \widehat{\operatorname{Ind} \alpha}(f)(s)(t) = (f(s) \otimes s)(t) = f(s)(t) \otimes s = f(s, t) \otimes s.$$

$\square$

The following lemma justifies (4.4): for any C^*-algebra C, it allows us to embed $C_c(G, A \otimes C)$ into $V_H^G(A) \otimes C$, and to embed $C_c(G, M^\beta(A \otimes C))$ into $M(V_H^G(A) \otimes C)$.

LEMMA 4.14. *With notation as above, the canonical embedding of $C_c(G, A) \odot C$ in $C_c(G, A \otimes C)$ extends to an imprimitivity bimodule isomorphism*

$$_{(\operatorname{Ind} A \times_r G) \otimes C}(V_H^G(A, \alpha) \otimes C)_{(A \times_r H) \otimes C} \cong {}_{\operatorname{Ind}(A \otimes C) \times_r G} V_H^G(A \otimes C, \alpha \otimes \operatorname{id})_{(A \otimes C) \times_r H}.$$

---

[1]Note that for $f \in C_c(G, \operatorname{Ind} A)$ and $c \in C$ we have
$$(\psi \times_r G) \circ \varphi(f \otimes c)(s, t) = \psi(\varphi(f \otimes c)(s, \cdot))(t) = \psi(f(s, \cdot) \otimes c)(t) = f(s, t) \otimes c.$$

PROOF. We have introduced (in the above discussion and in Chapter 3) the canonical isomorphisms φ_L and φ_R of the left and right coefficient algebras of $V_H^G(A) \otimes C$ onto those of $V_H^G(A \otimes C)$. A routine calculation now shows that the embedding of $C_c(G, A) \odot C \subseteq V_H^G(A) \otimes C$ into $C_c(G, A \otimes C) \subseteq V_H^G(A \otimes C)$ respects both inner products. By Lemma 1.24, this embedding extends uniquely to an imprimitivity bimodule homomorphism of $V_H^G(A) \otimes C$ onto $V_H^G(A \otimes C)$; by Remark 1.21, since the coefficient maps are isomorphisms, we are done. □

PROOF OF THEOREM 4.11. The rule $\delta_V(x)(s) = x(s) \otimes s$ certainly defines a map from $C_c(G, A) \subseteq V_H^G(A)$ to $C_c(G, M^\beta(A \otimes C^*(G))) \subseteq M(V_H^G(A) \otimes C^*(G))$. We first show that it preserves both inner products. By Lemma 1.24, we will then know that δ_V extends uniquely to a nondegenerate imprimitivity bimodule homomorphism from $V_H^G(A)$ to $M(V_H^G(A) \otimes C^*(G))$.

Using the formulas given in Equation (B.4) (extended to the various C_c-functions in the multiplier algebras and bimodules) we compute, for $x, y \in C_c(G, A) \subseteq V_H^G(A)$ and $s, t \in G$:

$$_{M(((\operatorname{Ind} A) \times_r G) \otimes C^*(G))}\langle \delta_V(x), \delta_V(y) \rangle(s, t)$$
$$= \int_H (\alpha \otimes \operatorname{id})_h \big(\delta_V(x)(th)\delta_V(y)(s^{-1}th)^*\big) \, dh \, \Delta(s)^{-1/2}$$
$$= \int_H (\alpha_h \otimes \operatorname{id})\big((x(th) \otimes th)(y(s^{-1}th)^* \otimes h^{-1}t^{-1}s)\big) \, dh \, \Delta(s)^{-1/2}$$
$$= \int_H (\alpha_h \otimes \operatorname{id})\big(x(th)y(s^{-1}th)^* \otimes s\big) \, dh \, \Delta(s)^{-1/2}$$
$$= \int_H \alpha_h(x(th)y(s^{-1}th)^*) \, dh \, \Delta(s)^{-1/2} \otimes s$$
$$= {}_{(\operatorname{Ind} A) \times_r G}\langle x, y \rangle(s, t) \otimes s$$
$$= \widehat{\operatorname{Ind} \alpha}({}_{(\operatorname{Ind} A) \times_r G}\langle x, y \rangle)(s, t),$$

and for $h \in H$,

$$\langle \delta_V(x), \delta_V(y) \rangle_{M((A \times_r H) \otimes C^*(G))}(h)$$
$$= \int_G \delta_V(x)(s^{-1})^* (\alpha \otimes \operatorname{id})_h (\delta_V(y)(s^{-1}h)) \, ds \, \Delta(h)^{-1/2}$$
$$= \int_G (x(s^{-1})^* \otimes s)\alpha_h \otimes \operatorname{id}(y(s^{-1}h) \otimes s^{-1}h) \, ds \, \Delta(h)^{-1/2}$$
$$= \int_G x(s^{-1})^* \alpha_h(y(s^{-1}h)) \otimes h \, ds \, \Delta(h)^{-1/2}$$
$$= \int_G x(s^{-1})^* \alpha_h(y(s^{-1}h)) \, ds \, \Delta(h)^{-1/2} \otimes h$$
$$= \langle x, y \rangle_{A \times_r H}(h) \otimes h = \hat{\alpha}(\langle x, y \rangle_{A \times_r H})(h).$$

By Remark 2.11, it now suffices to verify the coaction identity. For $s \in G$:
$$(\delta_V \otimes \operatorname{id}) \circ \delta_V(x)(s) = x(s) \otimes s \otimes s = (\operatorname{id} \otimes \delta_G) \circ \delta_V(x)(s).$$
□

We are finally ready to prove Theorem 4.1. As discussed earlier, the result is equivalent to:

THEOREM 4.15. *Let G be a locally compact group, and let H be a closed subgroup of G. If $({}_A X_B, H, \gamma)$ is a right-Hilbert bimodule action, then the diagram*

(4.5)
$$\begin{array}{ccc} (\mathrm{Ind}_H^G A) \times_r G & \xrightarrow{V_H^G(A)} & A \times_r H \\ {\scriptstyle (\mathrm{Ind}_H^G X) \times_r G} \downarrow & & \downarrow {\scriptstyle X \times_r H} \\ (\mathrm{Ind}_H^G B) \times_r G & \xrightarrow[V_H^G(B)]{} & B \times_r H \end{array}$$

commutes equivariantly for the coactions $\delta_{V(A)}$, $\delta_{V(B)}$, $\widehat{\mathrm{Ind}\,\gamma}$, and $\mathrm{Inf}\,\hat{\gamma}$ of G.

PROOF. As indicated previously, our strategy is to factor the right-Hilbert bimodule. If X is a partial imprimitivity bimodule, Lemma 4.8 together with the discussion preceding that lemma show that our construction of the coaction δ_V on V_H^G completes the proof in this case, since it follows from the formula for δ_V as given in Theorem 4.11 that the associated coaction $\delta_{V(L)}$ on the linking algebra $L = L(X)$ compresses to the coactions $\delta_{V(A)}$ and $\delta_{V(B)}$ on the corners.

For the nondegenerate homomorphism version, assume that $\varphi \colon A \to M(B)$ is a nondegenerate homomorphism which is equivariant for actions α and β of H. We must show that the diagram

(4.6)
$$\begin{array}{ccc} (\mathrm{Ind}_H^G A) \times_r G & \xrightarrow{V_H^G(A)} & A \times_r H \\ {\scriptstyle \mathrm{Ind}_H^G \varphi \times_r G} \downarrow & & \downarrow {\scriptstyle \varphi \times_r H} \\ (\mathrm{Ind}_H^G B) \times_r G & \xrightarrow[V_H^G(B)]{} & B \times_r H \end{array}$$

commutes equivariantly for the appropriate coactions. For this we show that there is a nondegenerate imprimitivity bimodule homomorphism $\Psi \colon V_H^G(A) \to M(V_H^G(B))$ which has coefficient maps $\mathrm{Ind}_H^G \varphi \times_r G$ and $\varphi \times_r H$ and also is $\delta_{V(A)} - \delta_{V(B)}$ equivariant. The commutativity of the diagram will then follow from Lemma 4.10.

For $x \in C_c(G, A) \subseteq V_H^G(A)$, define $\Psi(x) \colon G \to M(B)$ by

(4.7)
$$\Psi(x)(s) = \varphi(x(s)).$$

Then Ψ is a map from $C_c(G, A)$ to $C_c(G, M^\beta(B)) \subseteq M(V_H^G(B))$. By Lemma 1.24, if we can show that Ψ preserves both inner products, then we will know that it extends uniquely to a nondegenerate imprimitivity bimodule homomorphism from $V_H^G(A)$ to $M(V_H^G(B))$. For $x, y \in C_c(G, A)$ and $s, t \in G$ we have

$$\begin{aligned} {}_{M((\mathrm{Ind}\,B) \times_r G)}\langle \Psi(x), \Psi(y) \rangle(s,t) &= \int_H \beta_h\bigl(\Psi(x)(th)\Psi(y)(s^{-1}th)^*\bigr)\,dh\,\Delta_G(s)^{-1/2} \\ &= \int_H \beta_h\bigl(\varphi(x(th))\varphi(y(s^{-1}th)^*\bigr)\,dh\,\Delta_G(s)^{-1/2} \\ &= \varphi\left(\int_H \alpha_h\bigl(x(th)y(s^{-1}th)^*\bigr)\,dh\,\Delta_G(s)^{-1/2}\right) \\ &= \varphi\bigl({}_{(\mathrm{Ind}\,A) \times_r G}\langle x, y \rangle(s,t)\bigr) \\ &= (\mathrm{Ind}_H^G \varphi \times_r G)\bigl({}_{(\mathrm{Ind}\,A) \times_r G}\langle x, y \rangle\bigr)(s,t), \end{aligned}$$

and for $h \in H$,

$$\langle \Psi(x), \Psi(y)\rangle_{M(B\times_r H)}(h) = \int_G \Psi(x)(s^{-1})^* \beta_h(\Psi(y)(s^{-1}h))\, ds\, \Delta(h)^{-1/2}$$
$$= \int_G \varphi(x(s))^* \beta_h(\varphi(y(s^{-1}h)))\, ds\, \Delta(h)^{-1/2}$$
$$= \varphi\left(\int_G x(s^{-1})^* \alpha_h(y(s^{-1}h))\, ds\, \Delta(h)^{-1/2}\right)$$
$$= \varphi(\langle x,y\rangle_{A\times_r H}(h)) = (\varphi \times_r H)(\langle x,y\rangle_{A\times_r H})(h).$$

To check equivariance of the coactions, for $x \in C_c(G,A)$ and $s \in G$ we have:
$$(\Psi \otimes \mathrm{id}) \circ \delta_{V(A)}(x)(s) = (\varphi \otimes \mathrm{id})(\delta_{V(A)}(x)(s)) = \varphi(x(s)) \otimes s$$
$$= \Psi(x)(s) \otimes s = \delta_{V(B)}(\Psi(x))(s).$$

Assume now that ${}_A X_B$ is an arbitrary right-Hilbert $A - B$ bimodule. Using Proposition 2.27, we equivariantly factor ${}_A X_B$ as the composition of a right-partial imprimitivity bimodule ${}_K X_B$ and a nondegenerate homomorphism $\varphi \colon A \to M(K)$. By our previous results we now know that for both pieces the appropriate diagrams commute. This tells us the upper and lower trapezoids of the diagram

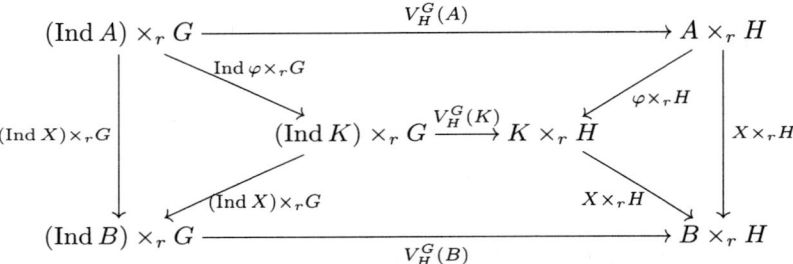

commute equivariantly for the appropriate coactions—that is, these trapezoids commute in the category $\mathcal{C}(G)$. The left and right triangles commute in the category $\mathcal{C}(G)$ by functoriality of Ind and $\times_r$. We conclude that the outer rectangle commutes equivariantly for the appropriate coactions, as desired. $\square$

4.4. Green's Theorem for induced representations

In this section we shall derive our second main theorem, Theorem 4.2, mainly as a consequence of Theorem 4.1. As mentioned in the discussion preceding Equation (B.5), if we start with an action α of G on A, then there is a natural isomorphism between $(\mathrm{Ind}_H^G A) \times_{\mathrm{Ind}\,\alpha,r} G$ and $C_0(G/H, A) \times_{\alpha\otimes\tau, r} G$, where $\mathrm{Ind}\,\alpha$ denotes the action of G induced from the restriction of α to the closed subgroup H. This isomorphism is the crossed product $\varphi \times_r G$ of the isomorphism φ of $\mathrm{Ind}_H^G A$ onto $C_0(G/H, A)$ given by $\varphi(f)(sH) = \alpha_s(f(s))$. In particular, $\varphi \times_r G$ is equivariant for the dual coactions $\widehat{\mathrm{Ind}\,\alpha}$ and $\widehat{\alpha \otimes \tau}$. Together with the map of $C_c(G,A)$ into itself given by $\Psi(x)(s) = \alpha_s(x(s))$, we obtain an isomorphism

$$\Psi \colon {}_{(\mathrm{Ind}_H^G A)\times_{\mathrm{Ind}\,\alpha,r}G} V_H^G(A)_{A\times_{\alpha,r}H} \xrightarrow{\cong} {}_{C_0(G/H,A)\times_{\alpha\otimes\tau,r}G} X_H^G(A)_{A\times_{\alpha,r}H}$$

(see Theorem B.3). This isomorphism carries the coaction δ_V on $V_H^G(A)$, as constructed in Theorem 4.11, to a unique coaction, δ_X, of G on $X_H^G(A)$, which makes the diagram

$$V_H^G(A) \xrightarrow{\delta_V} M(V_H^G(A) \otimes C^*(G))$$
$$\Psi \downarrow \qquad \qquad \downarrow \Psi \otimes \mathrm{id}$$
$$X_H^G(A) \xrightarrow{\delta_X} M(X_H^G(A) \otimes C^*(G))$$

commute. Thus δ_X implements a Morita equivalence between the G-coactions $\widehat{\alpha \otimes \tau}$ on $C_0(G/H, A) \times_r G$ and $\operatorname{Inf} \widehat{\alpha|}$ on $A \times_r H$. In this special situation, Theorem 4.15 becomes:

THEOREM 4.16. *Let G be a locally compact group, and let H be a closed subgroup of G. If $({}_A X_B, G, \gamma)$ is a right-Hilbert bimodule action, then the diagram*

(4.8)
$$\begin{array}{ccc}
C_0(G/H, A) \times_{\alpha \otimes \tau, r} G & \xrightarrow{X_H^G(A)} & A \times_{\alpha, r} H \\
C_0(G/H, X) \times_{\gamma \otimes \tau, r} G \downarrow & & \downarrow X \times_{\gamma, r} H \\
C_0(G/H, B) \times_{\beta \otimes \tau, r} G & \xrightarrow[X_H^G(B)]{} & B \times_{\beta, r} H
\end{array}$$

commutes equivariantly for the coactions $\delta_{X(A)}$, $\delta_{X(B)}$, $\widehat{\gamma \otimes \tau}$, and $\operatorname{Inf} \widehat{\gamma|}$ of G.

REMARK 4.17. Later (in the proof of Theorem 5.1) we will need a C_c-formula for the coaction δ_X of G on $X_H^G(A)$. Using the isomorphism $\Psi \colon V \to X$, it is routine to check that in fact δ_X is given by the same formula as δ_V: for $x \in C_c(G, A) \subseteq X_H^G(A)$ we have

$$\delta_X(x) \in C_c\big(G, M^\beta(A \otimes C^*(G))\big) \subseteq M\big(X_H^G(A) \otimes C^*(G)\big),$$

and $\delta_X(x)(s) = x(s) \otimes s$ for $s \in G$.

We now want to specialize to the case where H is *normal* in G, and in order to make this more evident we shall from now on write N instead of H for our subgroup. In this situation the above results may be restated in terms of duality theory and dual coactions, as follows from:

LEMMA 4.18. *If (A, G, α) is an action and N is a closed normal subgroup of G, then*

$$C_0(G/N, A) \times_{\alpha \otimes \tau, r} G \cong A \times_{\alpha, r} G \times_{\widehat{\alpha|}} G/N,$$

equivariantly for the coactions $\widehat{\alpha \otimes \tau}$ and $\hat{\alpha}^{\mathrm{dec}}$ of G.

PROOF. Theorem A.65 gives an isomorphism ψ of $C_0(G/N, A) \times_{\alpha \otimes \tau, r} G$ onto $A \times_{\alpha, r} G \times_{\widehat{\alpha|}} G/N$ such that

$$\psi\big(i^r_{C_0(G/N, A)}(a \otimes f) i^r_G(g)\big) = j_{G/N}(f) j_{A \times_r G}(i^r_A(a) i^r_G(g))$$

for $a \in A$, $f \in C_0(G/N)$, and $g \in C_c(G)$. We only have to check that this isomorphism is equivariant for the coactions $\widehat{\alpha \otimes \tau}$ and $\hat{\alpha}^{\mathrm{dec}}$:

$$\hat{\alpha}^{\mathrm{dec}} \circ \psi\bigl(i^r_{C_0(G/N,A)}(a \otimes f) i^r_G(g)\bigr) = \hat{\alpha}^{\mathrm{dec}}\bigl(j_{G/N}(f) j_{A \times_r G}(i^r_A(a) i^r_G(g))\bigr)$$
$$= (j_{G/N}(f) \otimes 1)(j_{A \times_r G} \otimes \mathrm{id}) \circ \hat{\alpha}(i^r_A(a) i^r_G(g))$$
$$= (j_{G/N}(f) \otimes 1)(j_{A \times_r G} \otimes \mathrm{id})\Bigl(\int_G g(s) i^r_A(a) i^r_G(s) \otimes s\, ds\Bigr)$$
$$= (j_{G/N}(f) \otimes 1) \int_G g(s) j_{A \times_r G}(i^r_A(a) i^r_G(s)) \otimes s\, ds$$
$$= \int_G g(s) j_{G/N}(f) j_{A \times_r G}(i^r_A(a) i^r_G(s)) \otimes s\, ds$$
$$= \int_G g(s) \psi\bigl(i^r_{C_0(G/N,A)}(a \otimes f) i^r_G(s)\bigr) \otimes s\, ds$$
$$= \int_G g(s)(\psi \otimes \mathrm{id})\bigl(i^r_{C_0(G/N,A)}(a \otimes f) i^r_G(s) \otimes s\bigr) ds$$
$$= (\psi \otimes \mathrm{id})\Bigl(\int_G g(s) \widehat{\alpha \otimes \tau}\bigl(i^r_{C_0(G/N,A)}(a \otimes f) i^r_G(s)\bigr) ds\Bigr)$$
$$= (\psi \otimes \mathrm{id}) \circ \widehat{\alpha \otimes \tau}\bigl(i^r_{C_0(G/N,A)}(a \otimes f) i^r_G(g)\bigr).$$

$\square$

Of course it follows from the above lemma that, in case of a normal subgroup N of G, we may replace Diagram (4.8) by the diagram

(4.9)
$$\begin{array}{ccc} A \times_r G \times G/N & \xrightarrow{X^G_N(A)} & A \times_r N \\ {\scriptstyle X \times_r G \times G/N}\Big\downarrow & & \Big\downarrow{\scriptstyle X \times_r N} \\ B \times_r G \times G/N & \xrightarrow[X^G_N(B)]{} & B \times_r N, \end{array}$$

which commutes equivariantly for the coactions $\hat{\gamma}^{\mathrm{dec}}$ and $\mathrm{Inf}\,\widehat{\gamma|}$ (where γ denotes the given action on ${}_A X_B$). However, in order to get a complete proof of Theorem 4.2 we need to show that the above diagram is also equivariant with respect to the G-actions $\mathrm{Inf}\,\widehat{\hat{\gamma}|}$ on $X \times_r G \times G/N$ and γ^{dec} on $X \times_r N$. For this we have to construct a G-action on the bimodule $X^G_N(A)$ which is compatible with the actions $\mathrm{Inf}\,\widehat{\hat{\alpha}|}$ and α^{dec}. Such an action was constructed in [**15**, Theorem 1], for actions twisted over N, full (twisted) crossed products, and with slightly different modular functions in the formulas for the module actions and inner products. The following lemma and its proof are an adaptation of that construction to the present situation.

LEMMA 4.19. *If (A, G, α) is an action and N is a closed normal subgroup of G, there is a unique imprimitivity bimodule action α^X of G on $X^G_N(A)$, implementing a Morita equivalence between the actions $\mathrm{Inf}\,\widehat{\hat{\alpha}|}$ on $A \times_r G \times G/N$ and α^{dec} on $A \times_r N$, such that for $x \in C_c(G, A)$ and $t \in G$ we have*

(4.10)
$$\alpha^X_s(x)(t) = \Delta_{G,N}(s)^{1/2} x(ts).$$

Recall that the modular function $\Delta_{G,N}$ for conjugation of G on N is given by

$$\Delta_{G,N}(s) = \Delta_G(s) \Delta_{G/N}(sN)^{-1}.$$

PROOF. Formula (4.10) certainly defines a homomorphism of G into the group of linear automorphisms of the vector space $C_c(G, A)$. We need to show that for $x, y \in C_c(G, A) \subseteq X_N^G(A)$, $s \in G$, and $b \in C_c(G \times G/N, A) \subseteq A \times_r G \times G/N$ we have

(i) $\alpha_s^X(b \cdot x) = \operatorname{Inf} \widehat{\hat{\alpha}}|_s(b) \cdot \alpha_s^X(x)$,
(ii) $\langle \alpha_s^X(x), \alpha_s^X(y) \rangle_{A \times_r N} = \alpha_s^{\mathrm{dec}}(\langle x, y \rangle_{A \times_r N})$, and
(iii) $s \mapsto \alpha_s^X(x)$ is continuous for the norm of $X_N^G(A)$.

Note that Theorem A.65 not only allows us to identify $C_c(G \times G/N, A)$ as a dense subalgebra of $A \times_r G \times G/N$, but also implies that the action $\operatorname{Inf} \widehat{\hat{\alpha}}|$ is given on C_c-functions by

(4.11) $$\operatorname{Inf} \widehat{\hat{\alpha}}|_s(f)(t, rN) = f(t, rsN).$$

For (i) and (ii), we compute:

$$\alpha_s^X(b \cdot x)(t) = \Delta_{G,N}(s)^{1/2}(b \cdot x)(ts)$$
$$= \Delta_{G,N}(s)^{1/2} \int_G b(r, tsN) \alpha_r(x(r^{-1}ts)) \Delta_G(r)^{1/2} \, dr$$
$$= \int_G b(r, tsN) \alpha_r\big(\Delta_{G,N}(s)^{1/2} x(r^{-1}ts)\big) \Delta_G(r)^{1/2} \, dr$$
$$= \int_G \operatorname{Inf} \widehat{\hat{\alpha}}|_s(b)(r, tN) \alpha_r\big(\alpha_s^X(x)(r^{-1}t)\big) \Delta_G(r)^{1/2} \, dr$$
$$= \big(\operatorname{Inf} \widehat{\hat{\alpha}}|_s(b) \cdot \alpha_s^X(x)\big)(t),$$

and for $n \in N$,

$$\langle \alpha_s^X(x), \alpha_s^X(y) \rangle_{A \times_r N}(n) = \Delta_N(n)^{-1/2} \int_G \alpha_t\big(\alpha_s^X(x)(t^{-1})^* \alpha_s^X(y)(t^{-1}n)\big) \, dt$$
$$= \Delta_N(n)^{-1/2} \Delta_{G,N}(s) \int_G \alpha_t\big(x(t^{-1}s)^* y(t^{-1}ns)\big) \, dt$$
$$\stackrel{t \mapsto st}{=} \Delta_N(n)^{-1/2} \Delta_{G,N}(s) \int_G \alpha_{st}\big(x(t^{-1})^* y(t^{-1}s^{-1}ns)\big) \, dt$$
$$= \Delta_{G,N}(s) \alpha_s\left(\Delta_N(n)^{-1/2} \int_G \alpha_t\big(x(t^{-1})^* y(t^{-1}s^{-1}ns)\big) \, dt\right)$$
$$= \Delta_{G,N}(s) \alpha_s\big(\langle x, y \rangle_{A \times_r N}(s^{-1}ns)\big)$$
$$= \alpha_s^{\mathrm{dec}}\big(\langle x, y \rangle_{A \times_r N}\big)(n).$$

For (iii), it suffices to observe that if $s \to e$ in G then $\alpha_s^X(x) \to x$ in the inductive limit topology of $C_c(G, A)$, which is stronger than the norm topology from $X_N^G(A)$. □

We are now almost done with the proof of Theorem 4.2, which is equivalent to:

THEOREM 4.20. *Let G be a locally compact group, and let N be a closed normal subgroup of G. If $({}_A X_B, G, \gamma)$ is a right-Hilbert bimodule action, then Diagram (4.9) commutes equivariantly for the actions $\alpha^{X(A)}$, $\alpha^{X(B)}$, $\operatorname{Inf} \widehat{\hat{\gamma}}|$, and γ^{dec} of G, and also for the coactions $\delta_{X(A)}$, $\delta_{X(B)}$, $\widehat{\gamma}^{\mathrm{dec}}$, and $\operatorname{Inf} \widehat{\gamma}|$ of G.*

PROOF. We already observed that Diagram (4.9) commutes equivariantly for the coactions. We need to know that it is also equivariant for the actions, and for this we must trace through the steps of the argument of Theorem 4.15, where we factored ${}_A X_B$ into a right-partial imprimitivity bimodule ${}_K X_B$ and a standard bimodule ${}_A K_K$. In the proof of the partial imprimitivity bimodule case we appealed to Lemmas 4.6, 4.7, and 4.8 to get an isomorphism which is equivariant for the coactions (see the discussion preceding Lemma 4.7). Lemma 4.9 does the same job for the actions.

Next, for the standard bimodule case, *i.e.*, for a nondegenerate equivariant homomorphism $\varphi \colon A \to M(B)$, we appealed to Lemma 4.10 to see that coaction-equivariance of the bimodule homomorphism $V(A) \to M(V(B))$ implied that the diagram commuted equivariantly for the coactions. The action part of that lemma does the same job for our actions as soon as we can show that the bimodule homomorphism $X_N^G(A) \to M(X_N^G(B))$ is equivariant for the appropriate actions. For this first note that the isomorphism $V_N^G \cong X_N^G$ transforms the bimodule homomorphism $V(A) \to M(V(B))$ of Equation (4.7) to a bimodule homomorphism $\Psi \colon X(A) \to M(X(B))$ given by exactly the same formula. Thus, for $x \in C_c(G, A) \subseteq X_H^G(A)$ and $t \in G$ we can compute:

$$\Psi(\alpha_s^{X(A)}(x))(t) = \varphi(\alpha_s^{X(A)}(x)(t)) = \varphi(x(ts)) = \Psi(x)(ts) = \alpha_s^{X(B)}(\Psi(x))(t).$$

This establishes the action-equivariance in the standard bimodule case.

Exactly as in the proof of Theorem 4.15 we combine both special cases in order to get the desired result. □

4.5. Mansfield's Theorem

In this section we are going to prove Theorem 4.3. The basic idea is the same as for the corresponding results concerning actions as given in the previous sections: using Proposition 2.30, we equivariantly factor a given equivariant right-Hilbert $A - B$ bimodule X into a right-partial imprimitivity bimodule and a nondegenerate homomorphism and show that the appropriate diagrams commute equivariantly in both special cases.

THEOREM 4.21. *Suppose that $({}_A X_B, G, \zeta)$ is a right-Hilbert bimodule coaction whose coefficient coactions are nondegenerate and normal, and that N is a closed normal subgroup of G. Then there is an action α^Y and a coaction δ_Y of G on Mansfield's bimodule $Y_{G/N}^G$ (given by the formulas (4.13) and (4.14) below) such that the diagram*

(4.12)
$$\begin{array}{ccc} A \times G \times_r N & \xrightarrow{Y_{G/N}^G(A)} & A \times G/N \\ {\scriptstyle X \times G \times_r N} \downarrow & & \downarrow {\scriptstyle X \times G/N} \\ B \times G \times_r N & \xrightarrow[Y_{G/N}^G(B)]{} & B \times G/N \end{array}$$

commutes equivariantly for the actions $\alpha^{Y(A)}$, $\alpha^{Y(B)}$, $\hat{\zeta}^{\mathrm{dec}}$, and $\operatorname{Inf} \widehat{\zeta|}$ of G, and also for the coactions $\delta_{Y(A)}$, $\delta_{Y(B)}$, $\operatorname{Inf} \widehat{\hat{\zeta}|}$, and ζ^{dec} of G.

As already discussed, this theorem is equivalent to Theorem 4.3. We start with the construction of the action α^Y. Recall from Appendix B that for a nondegenerate normal coaction δ of G on a C^*-algebra A and closed normal subgroup N of G, the $(A \times_\delta G \times_{\hat{\delta}|, r} N) - (A \times_{\delta|} G/N)$ imprimitivity bimodule $Y_{G/N}^G(A)$ is obtained as the

4.5. MANSFIELD'S THEOREM

completion of the $C_c(N, \mathcal{D}) - \mathcal{D}_N$ pre-imprimitivity bimodule $\mathcal{D}$ with operations given by Equation (B.7).

PROPOSITION 4.22. *For any nondegenerate normal coaction (A, G, δ), there is a unique imprimitivity bimodule action α^Y of G on $Y^G_{G/N}(A)$, implementing a Morita equivalence between the actions $\hat{\delta}^{\mathrm{dec}}$ on $A \times_\delta G \times_{\hat{\delta}|, r} N$ and $\operatorname{Inf} \hat{\delta}|$ on $A \times_{\delta|} G/N$, such that for $x \in \mathcal{D}$ we have*

$$(4.13) \qquad \alpha^Y_s(x) = \Delta_{G,N}(s)^{1/2} \hat{\delta}_s(x).$$

PROOF. Formula (4.13) certainly defines a homomorphism of G into the group of linear automorphisms of the vector space $\mathcal{D}$. By Lemma 1.24 and Remark 2.6 it suffices to show that for $x, y \in \mathcal{D}$, $s \in G$, and $g \in C_c(N, \mathcal{D})$ we have

(i) ${}_{A \times G \times_r N}\langle \alpha^Y_s(x), \alpha^Y_s(y) \rangle = \hat{\delta}^{\mathrm{dec}}_s({}_{A \times G \times_r N}\langle x, y \rangle)$,
(ii) $\langle \alpha^Y_s(x), \alpha^Y_s(y) \rangle_{A \times G/N} = \operatorname{Inf} \hat{\delta}|_s(\langle x, y \rangle_{A \times G/N})$, and
(iii) $s \mapsto \alpha^Y_s(x)$ is continuous for the norm of $Y^G_{G/N}$.

For (i) we compute, with $n \in N$:

$$\begin{aligned}
{}_{A \times G \times_r N}\langle \alpha^Y_s(x), \alpha^Y_s(y) \rangle(n) &= \alpha^Y_s(x) \hat{\delta}_n(\alpha^Y_s(y)^*) \Delta(n)^{-1/2} \\
&= \Delta_{G,N}(s) \hat{\delta}_s(x) \hat{\delta}_{ns}(y^*) \Delta(n)^{-1/2} \\
&= \Delta_{G,N}(s) \hat{\delta}_s\big(x \hat{\delta}_{s^{-1}ns}(y^*) \Delta(s^{-1}ns)^{-1/2}\big) \\
&= \Delta_{G,N}(s) \hat{\delta}_s\big({}_{A \times G \times_r N}\langle x, y \rangle (s^{-1}ns)\big) \\
&= \hat{\delta}^{\mathrm{dec}}_s\big({}_{A \times G \times_r N}\langle x, y \rangle\big)(n).
\end{aligned}$$

In order to prove (ii), we mention first that the embedding $j_A \times j_G| : A \times_{\delta|} G/N \to M(A \times_\delta G)$ is $\operatorname{Inf} \hat{\delta}| - \hat{\delta}$ equivariant. Thus, since we identify $A \times_{\delta|} G/N$ with its image in $M(A \times_\delta G)$ when working with the dense subalgebra $\mathcal{D}_N$, we may compute

$$\langle \alpha^Y_s(x), \alpha^Y_s(y) \rangle_{A \times G/N} = \int_N \hat{\delta}_n(\alpha^Y_s(x)^* \alpha^Y_s(y)) \, dn = \int_N \hat{\delta}_n\big(\Delta_{G,N}(s) \hat{\delta}_s(x^*y)\big) \, dn$$

$$\stackrel{n \mapsto sns^{-1}}{=} \int_N \hat{\delta}_{sn}(x^*y) \, dn = \hat{\delta}_s \left(\int_N \hat{\delta}_n(x^*y) \, dn \right) = \operatorname{Inf} \hat{\delta}|_s(\langle x, y \rangle_{A \times G/N}).$$

For (iii) it suffices to observe that if $s \to e$ in G then

$${}_{A \times G \times_r N}\langle \hat{\delta}_s(x) - x, \hat{\delta}_s(x) - x \rangle \to 0$$

in the inductive limit topology of $C_c(N, A \times_\delta G)$. □

The $\operatorname{Inf} \widehat{\hat{\delta}|} - \delta^{\mathrm{dec}}$ equivariant coaction δ_Y on the bimodule $Y^G_{G/N}$ was actually constructed in [21, Proposition 3.2]. Translating the construction given there into a more abstract setting, we get the formula

$$(4.14) \qquad \delta_Y(x) = (x \otimes 1)(j_G \otimes \operatorname{id})(w_G^*) \qquad \text{for } x \in \mathcal{D}.$$

PROOF OF THEOREM 4.21. We first address Diagram (4.12) in the case where ${}_A X_B$ is a right-partial imprimitivity bimodule. As in the action case, commutativity follows almost automatically from our general linking algebra techniques. We want to apply Lemmas 4.6, 4.9, and 4.8 to the Mansfield bimodule $Y^G_{G/N}(L)$ of the linking algebra $L(X)$, using the identifications

$$L(X) \times G \times_r N = L(X \times G \times_r N) \quad \text{and} \quad L(X) \times G/N = L(X \times G/N).$$

Since $qY^G_{G/N}(L)q = Y^G_{G/N}(B)$ and $pY^G_{G/N}(L)p = Y^G_{G/N}(A)$, item (ii) of Lemma 4.6 provides isometric $A \times G \times_r N - B \times G/N$ bimodule homomorphisms

$$\Phi \colon (X \times G \times_r N) \otimes_{B \times G \times_r N} Y^G_{G/N}(B) \to pY^G_{G/N}(L)q$$

and

$$\Psi \colon Y^G_{G/N}(A) \otimes_{A \times G/N} (X \times G/N) \to pY^G_{G/N}(L)q,$$

which, by Lemmas 4.9 and 4.8, are equivariant for the appropriate coactions. We have to show that these maps are surjective in order to obtain an action- and coaction-equivariant isomorphism

$$(X \times G \times_r N) \otimes_{B \times G \times_r N} Y^G_{G/N}(B) \cong Y^G_{G/N}(A) \otimes_{A \times G/N} (X \times G/N).$$

To see this we appeal to item (iv) of Lemma 4.6 and the following lemma, which is an analogue of Lemma 4.7.

LEMMA 4.23. *The ranges of the $A \times G \times_r N$- and $B \times G/N$- valued inner products on $pY^G_{G/N}(L)q$ lie in range of the $A \times G \times_r N$-valued inner product on $X \times G \times_r N$ and the range of the $B \times G/N$-valued inner product of $X \times G/N$, respectively.*

PROOF. Since X is a right-partial $A - B$ imprimitivity bimodule, it follows from Lemma 3.10 and Lemma 3.3 that $X \times G \times_r N$ is a right-partial $A \times G \times_r N - B \times G \times_r N$ imprimitivity bimodule. Thus the $A \times G \times_r N$-valued inner product on $X \times G \times_r N$ is full, which clearly implies that it contains the $A \times G \times_r N$-valued inner product on $pY^G_{G/N}(L)q$. To see the other inclusion, let B_X be the range of the B-valued inner product on X. It then follows from Lemma 3.10 that the $B \times G/N$-valued inner product on $X \times G/N$ has range $B_X \times G/N$.

On the other hand, if we follow the construction of $Y^G_{G/N}(L)$ as described in Section B.2 of Appendix B, we see that it is the closed linear span (with respect to the Hilbert-module norm) of certain elements of the form

$$j_L\left(\begin{pmatrix} a & x \\ \tilde{y} & b \end{pmatrix}\right) j_G(f)$$

with $f \in C_c(G)$, from which it follows that $pY^G_{G/N}(L)q$ is generated by certain elements of the form $j_X(x) j_G(f)$. It follows then from the formula for the right inner product on $Y^G_{G/N}$ as given in Equation (B.7), that the $B \times G/N$-valued inner product of two such elements $j_X(x)j_G(f), j_X(y)j_G(g) \in pY^G_{G/N}(L)q$ is given by

$$\langle j_X(x)j_{G/N}(f), j_X(y)j_{G/N}(g) \rangle_{B \times G/N} = j_G(\varphi(\bar{f})) j_B(\langle x, y \rangle_B) j_G(\varphi(g)),$$

with $\varphi(h)(sN) = \int_N h(sn)\,dn$ for $h \in C_c(G)$. Since $\langle x, y \rangle_B \in B_X$, it follows that

$$\langle j_X(x)j_{G/N}(f), j_X(y)j_{G/N}(g) \rangle_{B \times G/N} \in B_X \times G/N.$$

□

BACK TO THE PROOF OF THEOREM 4.21. It's a bit harder to prove the commutativity of Diagram (4.12) in the nondegenerate homomorphism case. Given a nondegenerate homomorphism $\varphi \colon A \to M(B)$ which is equivariant for nondegenerate normal coactions δ and ϵ of G on A and B, respectively, we shall show that the restriction of $\varphi \times G \colon A \times_\delta G \to M(B \times_\epsilon G)$ to $\mathcal{D}(A) \subseteq A \times_\delta G$ extends to a nondegenerate imprimitivity bimodule homomorphism

$$\Psi \colon Y^G_{G/N}(A) \to M(Y^G_{G/N}(B))$$

which has coefficient maps $\varphi \times G \times_r N$ and $\varphi \times G/N$, and is also both $\delta_{Y(A)} - \delta_{Y(B)}$ and $\alpha^{Y(A)} - \alpha^{Y(B)}$ equivariant. The result will then follow from Lemma 4.10.

We first claim that if $x \in \mathcal{D}(A)$ and $c \in \mathcal{D}_N(B)$, then

$$(\varphi \times G)(x)c \in \mathcal{D}(B).$$

By the definition of $\mathcal{D}(A)$, $\mathcal{D}(B)$, and $\mathcal{D}_N(B)$ (see Definition B.8), it suffices to show that for each pair $u, v \in A_c(G)$ and each pair of compact subsets $E, E' \subseteq G$, there exist $w \in A_c(G)$ and a compact set $F \subseteq G$ such that $(\varphi \times G)(x)c \in \mathcal{D}_{(w,F)}(B)$ for each $x \in \mathcal{D}_{(u,E)}(A)$ and $c \in \mathcal{D}_{(v,E',N)}(B)$. Since the pairing $(x, c) \mapsto (\varphi \times G)(x)c$ is certainly norm-continuous in both variables, it suffices to take

$$x = j_A(\delta_u(a))j_G^A(f) \in \mathcal{D}_{(u,E)}(A) \quad \text{and} \quad c = j_B(\epsilon_v(b))j_G^B(g) \in \mathcal{D}_{(v,E',N)}(B).$$

To verify the claim note first that by [38, Lemma 9] there exists a compact subset F' (only depending on E) such that x can be approximated in norm by elements of the form $\sum_{i=1}^n j_G^A(f_i)j_A(\delta_u(a_i))$, with $\operatorname{supp} f_i \subseteq F'$ for all i. By the norm continuity of the pairing $(x, c) \to (\varphi \times G)(x)c$ it follows from this that we may as well assume that $x = j_G^A(f)j_A(\delta_u(a))$ with $\operatorname{supp} f \subseteq F'$. Then

$$(\varphi \times G)(x)c = j_G^B(f)j_B(\varphi(\delta_u(a)))j_B(\epsilon_v(b))j_G^B(g) = j_G^B(f)j_B\bigl(\epsilon_u(\varphi(a))\epsilon_v(b)\bigr)j_G^B(g).$$

Choose $w \in A_c(G)$ which is identically 1 on $(\operatorname{supp} u)(\operatorname{supp} v)$. If $\{e_i\}$ is a bounded approximate identity for B, then

$$\epsilon_w\bigl(\epsilon_u(e_i\varphi(a))\epsilon_v(b)\bigr) = \epsilon_u(e_i\varphi(a))\epsilon_v(b) \quad \text{for all } i,$$

by [38, Lemma 1 (iii)]. Since $\epsilon_u(e_i\varphi(a)) \to \epsilon_u(\varphi(a))$ by strict continuity of ϵ_u on $M(B)$, we get $\epsilon_w(\epsilon_u(\varphi(a))\epsilon_v(b)) = \epsilon_u(\varphi(a))\epsilon_v(b)$ by norm continuity of ϵ_w on $B \times_\epsilon G$. Thus

$$(\varphi \times G)(x)c = j_G^B(f)j_B(\epsilon_w(d))j_G^B(g),$$

with $d = \epsilon_u(\varphi(a))\epsilon_v(b) \in B$. Another use of [38, Lemma 9] reveals that there exists a compact subset $F \subseteq G$, only depending on F' (and hence on E) such that $j_G^B(f)j_B(\epsilon_w(d))j_G^B(g) \in \mathcal{D}(B)_{(w,F)}$. This proves the claim. For later use note that when $N = \{e\}$ the above computations give

(4.15) $\qquad (\varphi \times G)(x)y \in \mathcal{D}(B) \qquad$ for $x \in \mathcal{D}(A)$, $y \in \mathcal{D}(B)$.

In the next step we show that the pairing $(x, c) \mapsto (\varphi \times G)(x)c$ is actually continuous with respect to the norms on $\mathcal{D}(A)$ and $\mathcal{D}(B)$ inherited from the bimodules $Y^G_{G/N}(A)$ and $Y^G_{G/N}(B)$, respectively. For this recall from Equation (B.7) that the $A \times_{\delta|} G/N$-valued inner product is given by

$$\langle x, y\rangle_{A \times G/N} = \int_N \hat{\delta}_n(x^* y) \, dn,$$

where for x in $\mathcal{D}(A)$, the integral $\int_N \hat{\delta}_n(x) \, dn$ is determined by the equations

$$\omega\left(\int_N \hat{\delta}_n(x) \, dn\right) = \int_N \omega(\hat{\delta}_n(x)) \, dn, \qquad \omega \in (A \times_\delta G)^*.$$

Using this, and the identity $\hat{\epsilon}_n(c) = c$ for all $c \in \mathcal{D}_N(B)$ and $n \in N$, we now compute for any $\omega \in (B \times_\epsilon G)^*$

$$\omega(\langle (\varphi \times G)(x)c, (\varphi \times G)(x)c \rangle_{B \times G/N}) = \omega\left(\int_N \hat{\epsilon}_n(c^*(\varphi \times G)(x^*x)c)\, dn\right)$$

$$= \int_N \omega(c^* \hat{\epsilon}_n((\varphi \times G)(x^*x))c)\, dn = \int_N c \cdot \omega \cdot c^*((\varphi \times G)(\hat{\delta}_n(x^*x)))\, dn$$

$$= \int_N (\varphi \times G)^*(c \cdot \omega \cdot c^*)(\hat{\delta}_n(x^*x))\, dn = (\varphi \times G)^*(c \cdot \omega \cdot c^*)\left(\int_N \hat{\delta}_n(x^*x)\, dn\right)$$

$$= \omega(c^*(\varphi \times G/N)(\langle x, x \rangle_{A \times G/N})c),$$

so that

(4.16)
$$\|(\varphi \times G)(x)c\|^2_{Y^G_{G/N}(B)} = \|\langle (\varphi \times G)(x)c, (\varphi \times G)(x)c \rangle_{B \times G/N}\|$$
$$= \|c^*(\varphi \times G)(\langle x, x \rangle_{A \times G/N})c\| \leq \|\langle x, x \rangle_{A \times G/N}\| \|c^*c\| = \|x\|^2_{Y^G_{G/N}(A)} \|c\|^2.$$

It follows that for every $x \in \mathcal{D}(A)$ the formula $T(c) = (\varphi \times G)(x)c$ defines a bounded linear map from $\mathcal{D}_N(B)$ to $\mathcal{D}(B)$, where the latter is given the Hilbert module norm of $Y^G_{G/N}(B)$. We show that T is adjointable. For $y \in \mathcal{D}(B)$ and $c \in \mathcal{D}_N(B)$ we have

$$\langle T(c), y \rangle_{B \times G/N} = \int_N \hat{\epsilon}_n(c^*(\varphi \times G)(x^*)y)\, dn = \int_N c^* \hat{\epsilon}_n((\varphi \times G)(x^*)y)\, dn$$

$$= c^* \int_N \hat{\epsilon}_n((\varphi \times G)(x^*)y)\, dn$$

$$= \left\langle c, \int_N \hat{\epsilon}_n((\varphi \times G)(x^*)y)\, dn \right\rangle_{B \times G/N},$$

where we use (4.15) to see that all integrals above are well-defined. Thus

$$T^*(y) = \int_N \hat{\epsilon}_n((\varphi \times G)(x^*)y)\, dn$$

defines an adjoint T^* for T. It follows that T extends uniquely to an adjointable linear map from $B \times_{\epsilon|} G/N$ to $Y^G_{G/N}(B)$, i.e., to a multiplier of the $(B \times_\epsilon G \times_{\hat{\epsilon}|,r} N) - (B \times_{\epsilon|} G/N)$ imprimitivity bimodule $Y^G_{G/N}(B)$, by [20, Proposition 1.3]. We have shown that for all $x \in \mathcal{D}(A)$ there exists $\Psi(x) \in M(Y^G_{G/N}(B))$ such that $\Psi(x)c = (\varphi \times G)(x)c$ for $c \in \mathcal{D}_N(B)$.

A computation similar to the derivation of the inequality (4.16) shows that for $x, y \in \mathcal{D}(A)$ and $c, d \in \mathcal{D}_N(B)$ we have

$$\langle \Psi(x)c, \Psi(y)d \rangle_{B \times G/N} = c^*(\varphi \times G/N)(\langle x, y \rangle_{A \times G/N})d,$$

so

$$\langle \Psi(x), \Psi(y) \rangle_{M(B \times G/N)} = (\varphi \times G/N)(\langle x, y \rangle_{A \times G/N}).$$

By Lemma 1.24, the following computation allows us to extend Ψ uniquely to a nondegenerate imprimitivity bimodule homomorphism, which we still denote by Ψ, from $Y^G_{G/N}(A)$ to $M(Y^G_{G/N}(B))$, with coefficient maps $\varphi \times G \times_r N$ and $\varphi \times G/N$:

for $x, y \in \mathcal{D}(A)$ and $z \in \mathcal{D}(B)$ we have

$$_{M(B \times G \times_r N)}\langle \Psi(x), \Psi(y) \rangle z = \Psi(x) \langle \Psi(y), z \rangle_{B \times G/N} = \Psi(x) \Psi(y)^* z$$

$$= (\varphi \times G)(x) \int_N \hat{\epsilon}_n \big((\varphi \times G)(y^*) z \big) \, dn$$

$$= \int_N (\varphi \times G)(x) \hat{\epsilon}_n \big((\varphi \times G)(y^*) \big) \hat{\epsilon}_n(z) \, dn$$

$$= \int_N (\varphi \times G)(x)(\varphi \times G)\big(\hat{\delta}_n(y^*) \big) \hat{\epsilon}_n(z) \, dn$$

$$= \int_N (\varphi \times G)\big(x \hat{\delta}_n(y^*) \Delta(n)^{-1/2} \big) \hat{\epsilon}_n(z) \Delta(n)^{1/2} \, dn$$

$$= \int_N (\varphi \times G) \big(_{A \times G \times_r N}\langle x, y \rangle(n) \big) \hat{\epsilon}_n(z) \Delta(n)^{1/2} \, dn$$

$$= \int_N (\varphi \times G \times_r N) \big(_{A \times G \times_r N}\langle x, y \rangle \big)(n) \hat{\epsilon}_n(z) \Delta(n)^{1/2} \, dn$$

$$= (\varphi \times G \times_r N) \big(_{A \times G \times_r N}\langle x, y \rangle \big) z,$$

so

$$_{M(B \times G \times_r N)}\langle \Psi(x), \Psi(y) \rangle = (\varphi \times G \times_r N) \big(_{A \times G \times_r N}\langle x, y \rangle \big).$$

We now check the coaction equivariance. For $x \in \mathcal{D}(A)$ we have

$$(\Psi \otimes \mathrm{id}) \circ \delta_{Y(A)}(x) = ((\varphi \times G) \otimes \mathrm{id})\big((x \otimes 1)(j_G^A \otimes \mathrm{id})(w_G^*)\big)$$
$$= ((\varphi \times G)(x) \otimes 1)\big((\varphi \times G) \circ j_G^A \otimes \mathrm{id}\big)(w_G^*) = ((\varphi \times G)(x) \otimes 1)(j_G^B \otimes \mathrm{id})(w_G^*).$$

Now, for any $c \in \mathcal{D}_N(B)$,

$$\delta_{Y(B)}((\varphi \times G)(x)) \epsilon^{\mathrm{dec}}(c)$$
$$= \delta_{Y(B)}((\varphi \times G)(x)c) = ((\varphi \times G)(x)c \otimes 1)(j_G^B \otimes \mathrm{id})(w_G^*)$$
$$= ((\varphi \times G)(x) \otimes 1)(c \otimes 1)(j_G^B \otimes \mathrm{id})(w_G^*)$$
$$= ((\varphi \times G)(x) \otimes 1)(j_G^B \otimes \mathrm{id})(w_G^*) \epsilon^{\mathrm{dec}}(c),$$

so

$$(\Psi \otimes \mathrm{id}) \circ \delta_{Y(A)}(x) = \delta_{Y(B)}((\varphi \times G)(x)) = \delta_{Y(B)} \circ \Psi(x).$$

This gives the coaction equivariance. For the action equivariance, if $x \in \mathcal{D}(A)$ we have

$$\Psi \circ \alpha_s^{Y(A)}(x) = (\varphi \times G)(\Delta_{G,N}(s)^{1/2} \hat{\delta}_s(x))$$
$$= \Delta_{G,N}(s)^{1/2} \hat{\epsilon}_s((\varphi \times G)(x)) = \alpha_s^{Y(B)} \circ \Psi(x).$$

As with Theorems 4.15 and 4.20, this suffices to complete the proof. $\square$

CHAPTER 5

Applications

In this chapter we give some applications of the naturality theorems from the preceding chapter to our motivating problem of understanding the relationships between induction and duality. First we uncover some new relationships between Green and Mansfield induction. Important special cases of these results say that the Green bimodules $X^G_{\{e\}}(A)$ and Mansfield bimodules $Y^G_{G/G}(A)$ are in duality:

$$X^G_{\{e\}}(A) \times G \cong Y^G_{G/G}(A \times G) \quad \text{and} \quad X^G_{\{e\}}(A \times G) \cong Y^G_{G/G}(A) \times G.$$

Results of this type require several applications of our main theorems, and it is vital that we know everything is appropriately equivariant. The same is true of our other applications to the restriction-induction duality program of [**14, 29, 18**]. We close with a new application of linking algebra techniques to the symmetric imprimitivity theorem of [**51**].

5.1. Equivariant triangles

Our first application concerns a curious relationship between the Green and Mansfield bimodules. We prove two results (Theorems 5.1 and 5.9) which say, very roughly, that Green and Mansfield imprimitivity are "inverse to each other", at least up to crossed product duality.

5.1.1. Dual Mansfield equivariant triangle.
We begin with an action (A, G, α) and a closed normal subgroup N of G. We consider various imprimitivity bimodules arising from this data: first of all, there is the Green bimodule $_{A \times_r G \times G/N} X^G_N(A)_{A \times_r N}$. Temporarily forgetting about N, we also have the Green bimodule $_{A \times_r G \times G} X^G_e(A)_A$. Recall from Lemma 4.19 that every Green bimodule carries an action α^X; in the particular case of X^G_e, the action α^X is $\hat{\hat{\alpha}} - \alpha$ compatible. Restricting these actions to N and taking crossed products gives an $(A \times_r G \times G \times_r N) - (A \times_r N)$ imprimitivity bimodule $X^G_e(A) \times_r N$. On the other hand, the dual coaction $\hat{\alpha}$ of G on $A \times_r G$ gives rise to the Mansfield bimodule $_{A \times_r G \times G \times_r N} Y^G_{G/N}(A \times_r G)_{A \times_r G \times G/N}$. The following theorem ties these bimodules together:

THEOREM 5.1. *For any action (A, G, α) and any closed normal subgroup N of G, the diagram*

(5.1)
$$A \times_{\alpha,r} G \times_{\hat{\alpha}} G \times_{\hat{\hat{\alpha}},r} N \xrightarrow{Y^G_{G/N}(A \times_r G)} A \times_{\alpha,r} G \times_{\hat{\alpha}|} G/N$$

$$X^G_e(A) \times_r N \Big\downarrow \qquad \nearrow X^G_N(A)$$

$$A \times_{\alpha|,r} N$$

commutes equivariantly for the appropriate actions and coactions of G.

PROOF. First of all, recall that the dual coaction $\hat{\alpha}$ is automatically normal and nondegenerate, so the Mansfield imprimitivity bimodule $Y_{G/N}^G(A \times_r G)$ exists. Commutativity as a diagram of imprimitivity bimodules, but at the level of full crossed products (and without the equivariance), is in [**18**, Theorem 3.1]. We will adapt the isomorphism of [**18**] to the present context of reduced crossed products.

Before we get into the details of this isomorphism, we should make sure we know what the "appropriate" actions and coactions are: on $X_N^G(A)$ we take the action α^X from Lemma 4.19, given on x in the dense subspace $C_c(G, A)$ by
$$\alpha_s^X(x)(t) = \Delta_{G,N}(s)^{1/2} x(ts).$$
(Recall that $\Delta_{G,N}(s) = \Delta_G(s)\Delta_{G/N}(sN)^{-1}$.) The coaction on $X_N^G(A)$ will be the δ_X from Remark 4.17, given on x in $C_c(G, A)$ by
$$\delta_X(x)(s) = x(s) \otimes s,$$
where we regard $\delta_X(x)$ as an element of $C_c(G, M^\beta(A \otimes C^*(G))) \subseteq M(X_N^G(A) \otimes C^*(G))$. The crossed product $X_e^G(A) \times_r N$ carries a decomposition action, which for ease of writing in this proof we denote simply by $\alpha^{X \times N}$. Note that $C_c(N \times G, A)$ embeds in $X_e^G(A) \times_r N$ via the chain of inclusions
$$C_c(N \times G, A) \hookrightarrow C_c(N, C_c(G, A)) \hookrightarrow C_c(N, X) \hookrightarrow X \times_r N,$$
and $\alpha^{X \times N}$ is given on z in the dense subspace $C_c(N \times G, A)$ by
$$\alpha_s^{X \times N}(z)(n, t) = \Delta_{G,N}(s) z(s^{-1} n s, t s).$$
The coaction on $X_e^G(A) \times_r N$, which we shall simply denote by $\delta_{X \times N}$, will be the inflation to G of the dual coaction of N. However, before we can give a formula for this coaction which will suit our purposes in this proof we must prepare some more tools involving C_c-functions.

LEMMA 5.2. *The* $(A \times_r G \times G \times_r N) - (A \times_r N)$ *imprimitivity bimodule* $X_e^G \times_r N$ *is the completion of the* $C_c(N \times G \times G, A) - C_c(N, A)$ *pre-imprimitivity bimodule* $C_c(N \times G, A)$, *with operations*
$$f \cdot x(n, t) = \int_N \int_G f(k, s, t) \alpha_s(x(k^{-1}n, s^{-1}tk)) \Delta(s)^{1/2} \, ds \, dk$$
$$x \cdot g(n, t) = \int_N x(k, t) \alpha_{tk}(g(k^{-1}n)) \, dk$$
$$_{A \times_r G \times G \times_r N}\langle x, y \rangle(n, s, t) = \Delta(s)^{-1/2} \Delta_{G,N}(n)^{1/2}$$
$$\int_N x(nk, t) \alpha_s(y(k, s^{-1}tn)^*) \Delta(k) \, dk$$
$$\langle x, y \rangle_{A \times_r N}(n) = \int_N \int_G \alpha_{k^{-1}s}(x(k, s^{-1})^* y(kn, s^{-1})) \, ds \, dk,$$
for $x, y \in C_c(N \times G, A)$, $f \in C_c(N \times G \times G, A)$, *and* $g \in C_c(N, A)$.

Just to be clear about how $C_c(N \times G \times G, A)$ is sitting inside $A \times_r G \times G \times_r N$, note that $C_c(N \times G \times G, A)$ embeds continuously in $C_c(N, C_c(G \times G, A))$ (for the respective inductive limit topologies), hence in $A \times_r G \times G \times_r N$ since $C_c(G \times G, A)$ embeds continuously in $A \times_r G \times G$ via the isomorphism $A \times_r G \times G \cong C_0(G, A) \times_r G$ from (the special case $N = \{e\}$ of) Theorem A.65. Also recall that in this latter special case we must remember that the second variable comes from $C_0(G, A)$.

5.1. EQUIVARIANT TRIANGLES

PROOF. We just compute:
$$f \cdot x(n,t) = f \cdot x(n, \cdot)(t)$$
$$= \left(\int_N f(k, \cdot, \cdot) \cdot \alpha_k^X \left(x(k^{-1}n, \cdot) \right) dk \right)(t)$$
$$= \int_N \left(f(k, \cdot, \cdot) \cdot \alpha_k^X (x(k^{-1}n, \cdot)) \right)(t) \, dk$$
$$= \int_N \int_G f(k,s,t) \alpha_s \left(\alpha_k^X (x(k^{-1}n, \cdot))(s^{-1}t) \right) \Delta(s)^{1/2} \, ds \, dk$$
$$= \int_N \int_G f(k,s,t) \alpha_s \left(x(k^{-1}n, s^{-1}tk) \right) \Delta(s)^{1/2} \, ds \, dk,$$

$$x \cdot g(n,t) = x \cdot g(n)(t) = \left(\int_N x(k, \cdot) \cdot \alpha_k(g(k^{-1}n)) \, dk \right)(t)$$
$$= \int_N \left(x(k, \cdot) \cdot \alpha_k(g(k^{-1}n)) \right)(t) \, dk$$
$$= \int_N x(k,t) \alpha_t \left(\alpha_k(g(k^{-1}n)) \right) dk$$
$$= \int_N x(k,t) \alpha_{tk}(g(k^{-1}n)) \, dk,$$

$${}_{A \times_r G \times G \times_r N} \langle x, y \rangle (n, s, t) = {}_{A \times_r G \times G \times_r N} \langle x, y \rangle (n, \cdot, \cdot)(s, t)$$
$$= \left(\int_N {}_{A \times_r G \times G} \langle x(nk, \cdot), \alpha_n^X(y(k, \cdot)) \rangle \Delta(k) \, dk \right)(s, t)$$
$$= \int_N {}_{A \times_r G \times G} \langle x(nk, \cdot), \alpha_n^X(y(k, \cdot)) \rangle (s, t) \Delta(k) \, dk$$
$$= \int_N \Delta(s)^{-1/2} x(nk, t) \alpha_s \left(\alpha_n^X(y(k, \cdot))(s^{-1}t)^* \right) \Delta(k) \, dk$$
$$= \int_N \Delta(s)^{-1/2} x(nk, t) \alpha_s \left(\Delta_{G,N}(n)^{1/2} y(k, s^{-1}tn)^* \right) \Delta(k) \, dk$$
$$= \Delta(s)^{-1/2} \Delta_{G,N}(n)^{1/2} \int_N x(nk, t) \alpha_s (y(k, s^{-1}tn)^*) \Delta(k) \, dk,$$

and

$$\langle x, y \rangle_{A \times_r N}(n) = \int_N \alpha_{k^{-1}} \left(\langle x(k, \cdot), y(kn, \cdot) \rangle_A \right) dk$$
$$= \int_N \alpha_{k^{-1}} \left(\int_G \alpha_s \left(x(k, s^{-1})^* y(kn, s^{-1}) \right) ds \right) dk$$
$$= \int_N \int_G \alpha_{k^{-1}s} \left(x(k, s^{-1})^* y(kn, s^{-1}) \right) ds \, dk.$$
□

LEMMA 5.3. *With notation as above, for* $x \in C_c(N \times G, A)$ *we have*
$$\delta_{X \times N}(x) \in C_c \big(N \times G, M^\beta(A \otimes C^*(G)) \big) \subseteq M \big((X_e^G \times_r N) \otimes C^*(G) \big),$$
with $\delta_{X \times N}(x)(n, s) = x(n, s) \otimes n$ *for* $(n, s) \in N \times G$.

PROOF. Since $x \in C_c(N, X_e^G)$, by Proposition 3.5 the dual coaction of N takes x to the element of $C_c(N, M^\beta(X_e^G \otimes C^*(N)))$ given by $n \mapsto x(n, \cdot) \otimes n$. After inflating to a coaction of G, we get

$$\delta_{X \times N}(x) \in C_c(N, M^\beta(X_e^G \otimes C^*(G))) \quad \text{and} \quad \delta_{X \times N}(x)(n) = x(n, \cdot) \otimes n.$$

Further, since $x(n, \cdot) \in C_c(G, A)$, our usual canonical embedding gives

$$x(n, \cdot) \otimes n \in C_c(G, M^\beta(A \otimes C^*(N))) \quad \text{and} \quad (x(n, \cdot) \otimes n)(s) = x(n, s) \otimes n.$$

Now, the map $(n, s) \mapsto x(n, s) \otimes n$ is in $C_c(N \times G, M^\beta(A \otimes C^*(G)))$, and we have just seen that $\delta_{X \times N}(x)$ agrees with this map, so we are done. □

We must now do similar (and a little harder) work to obtain formulas for the action α^Y from Proposition 4.22 and the coaction δ_Y from (4.14) on the Mansfield bimodule $Y_{G/N}^G(A \times_r G)$ at the level of C_c-functions. When N is amenable, [**18**] shows that $C_c(G \times G, A)$ embeds in Mansfield's bimodule $Y_{G/N}^G(A \times_r G)$. The techniques there involve the symmetric imprimitivity theorem of [**51**], but this can be avoided:

LEMMA 5.4. *The $\hat{\hat\alpha}$-invariant $*$-subalgebra $C_c(G \times G, A)$ of $A \times_r G \times G$ is contained in $\mathcal{D}(A \times_r G)$ (see Appendix B), and has algebraic operations*

$$xy(s, t) = \int_G x(r, t) \alpha_r(y(r^{-1}s, r^{-1}t)) \, dr$$
$$x^*(s, t) = \alpha_s(x(s^{-1}, s^{-1}t)^*) \Delta(s)^{-1}$$
$$\hat{\hat\alpha}_r(x)(s, t) = x(s, tr).$$

PROOF. Let $x \in C_c(G \times G, A)$. Then $\operatorname{supp} x$ is contained in $F \times F$ for some compact subset F of G. Choose a compact subset E of G whose interior contains F, and then choose $u \in A_c(G)$ which is identically 1 on E. We claim that x is (u, E). Approximate x in the norm of $A \times_r G \times G$ by a finite sum

$$\sum_k j_G(f_k) j_{A \times_r G}(i_A^r(a_k) i_G^r(c_k)) \quad \text{with } a_k \in A, f_k, c_k \in C_E(G).$$

This sum is (u, E) since $\operatorname{supp} f_k \subseteq E$ and

$$i_A^r(a_k) i_G^r(c_k) = i_A^r(a_k) i_G^r(uc_k) = \hat\alpha_u(i_A^r(a_k) i_G^r(c_k)).$$

This shows that $C_c(G \times G, A) \subseteq \mathcal{D}(A \times_r G)$.

The next two assertions follow quite easily from the isomorphism $A \times_{\alpha, r} G \times_{\hat\alpha} G \cong C_0(G, A) \times_{\alpha \otimes \tau, r} G$, which we freely abuse:

$$xy(s, t) = xy(s)(t) = \left(\int_G x(r)(\alpha \otimes \tau)_r(y(r^{-1}s)) \, dr \right)(t)$$
$$= \int_G x(r)(t)(\alpha_r \otimes \tau_r)(y(r^{-1}s))(t) \, dr$$
$$= \int_G x(r, t) \alpha_r(y(r^{-1}s)(r^{-1}t)) \, dr$$
$$= \int_G x(r, t) \alpha_r(y(r^{-1}s, r^{-1}t)) \, dr,$$

and

$$x^*(s,t) = x^*(s)(t) = (\alpha \otimes \tau)_s\big(x(s^{-1})^*\big)(t)\Delta(s)^{-1}$$
$$= \alpha_s\big(x(s^{-1})^*(s^{-1}t)\big)\Delta(s)^{-1}$$
$$= \alpha_s\big(x(s^{-1})(s^{-1}t)^*\big)\Delta(s)^{-1}$$
$$= \alpha_s(x(s^{-1}, s^{-1}t)^*)\Delta(s)^{-1}.$$

The formula for $\hat{\alpha}$ follows from Equation (4.11). $\square$

The above lemma and Proposition 4.22 immediately give us the following C_c-formula for the bimodule action α^Y: for $x \in C_c(G \times G, A) \subseteq \mathcal{D}(A \times_r G)$ we have

$$\alpha_s^Y(x)(t,r) = \Delta_{G,N}(s)^{1/2} x(t, rs).$$

The following C_c-description of the Mansfield bimodule is essentially [**18**, Proposition 1.1], but we do it for reduced crossed products, and we give a different argument.

LEMMA 5.5. *For any action (A, G, α) and any closed normal subgroup N of G, the $(A \times_r G \times G \times_r N) - (A \times_r G \times G/N)$ imprimitivity bimodule $Y_{G/N}^G(A \times_r G)$ is the completion of the $C_c(N \times G \times G, A) - C_c(G \times G/N, A)$ pre-imprimitivity bimodule $C_c(G \times G, A)$, with operations*

$$f \cdot x(s,t) = \int_N \int_G f(n,r,t)\alpha_r(x(r^{-1}s, r^{-1}tn))\Delta(n)^{1/2}\, dr\, dn$$
$$x \cdot g(s,t) = \int_G x(r,t)\alpha_r(g(r^{-1}s, r^{-1}tN))\, dr$$
$$_{A \times_r G \times G \times_r N}\langle x, y\rangle(n,s,t) = \Delta(n)^{-1/2}\int_G x(r,t)\alpha_{s^{-1}}(y(sr, stn)^*)\Delta(r)\, dr$$
$$\langle x, y\rangle_{A \times_r G \times G/N}(s, tN) = \int_N \int_G \alpha_{r^{-1}}\big(x(r, rtn)^* y(rs, rtn)\big)\, dr\, dn,$$

for $f \in C_c(N \times G \times G, A)$, $x, y \in C_c(G \times G, A)$, and $g \in C_c(G \times G/N, A)$.

PROOF. Note first of all that $C_c(N \times G \times G, A)$, $C_c(G \times G, A)$, and $C_c(G \times G/N, A)$ are dense in the respective normed spaces $A \times_r G \times G \times_r N$, $Y_{G/N}^G(A \times_r G)$, and $A \times_r G \times G/N$. We have

$$f \cdot x(s,t) = \int_N \big(f(n,\cdot,\cdot) \cdot n \cdot x\big)(s,t)\, dn = \int_N \big(f(n,\cdot,\cdot)\hat{\alpha}_n(x)\Delta(n)^{1/2}\big)(s,t)\, dn$$
$$= \int_N \int_G f(n,r,t)\alpha_r\big(\hat{\alpha}_n(x)(r^{-1}s, r^{-1}t)\big)\Delta(n)^{1/2}\, dr\, dn$$
$$= \int_N \int_G f(n,r,t)\alpha_r(x(r^{-1}s, r^{-1}tn))\Delta(n)^{1/2}\, dr\, dn,$$

$$_{A\times_r G\times G\times_r N}\langle x,y\rangle(n,s,t) = {}_{A\times_r G\times G\times_r N}\langle x,y\rangle(n,\cdot,\cdot)(s,t)$$

$$= \Delta(n)^{-1/2}(x\hat{\tilde{\alpha}}_n(y^*))(s,t) = \Delta(n)^{-1/2}\int_G x(r,t)\alpha_r(\hat{\tilde{\alpha}}_n(y^*)(r^{-1}s,r^{-1}t))\,dr$$

$$= \Delta(n)^{-1/2}\int_G x(r,t)\alpha_r(y^*(r^{-1}s, r^{-1}tn))\,dr$$

$$= \Delta(n)^{-1/2}\int_G x(r,t)\alpha_r\bigl(\alpha_{r^{-1}s}(y(s^{-1}r,s^{-1}rr^{-1}tn)^*)\bigr)\Delta(r^{-1}s)^{-1}\,dr$$

$$= \Delta(n)^{-1/2}\int_G x(r,t)\alpha_{s^{-1}}(y(sr,stn)^*)\Delta(r)\,dr,$$

and

$$\langle x,y\rangle_{A\times_r G\times G/N}(s,tN) = \int_N \hat{\tilde{\alpha}}_n(x^*y)(s,t)\,dn = \int_N x^*y(s,tn)\,dn$$

$$= \int_N\int_G x^*(r,tn)\alpha_r(y(r^{-1}s, r^{-1}tn))\,dr\,dn$$

$$= \int_N\int_G \alpha_r\bigl(x(r^{-1},r^{-1}tn)^*y(r^{-1}s,r^{-1}tn)\bigr)\Delta(r)^{-1}\,dr\,dn$$

$$= \int_N\int_G \alpha_{r^{-1}}\bigl(x(r,rtn)^*y(rs,rtn)\bigr)\,dr\,dn.$$

The formula for $x\cdot g$ follows from membership of g in $M(A\times_r G\times G)$, since $C_c(G\times G/N,A)$ embeds continuously in $A\times_r G\times G/N$, hence also continuously in $M^\beta(A\times_r G\times G)$. □

The following lemma prepares us to deal with the coaction δ_Y in terms of C_c-functions; as usual, the trick is to allow the extra copy of $C^*(G)$ to be a freely moving object.

LEMMA 5.6. *For any action (A,G,α), any closed normal subgroup N of G, and any C^*-algebra C, the canonical embedding of $C_c(G\times G, A)\odot C$ in $C_c(G\times G, A\otimes C)$ extends to an imprimitivity bimodule isomorphism*

$$\Psi\colon {}_{(A\times_r G\times G\times_r N)\otimes C}\bigl(Y^G_{G/N}(A\times_{\alpha,r} G)\otimes C\bigr)_{(A\times_r G\times G/N)\otimes C}$$
$$\stackrel{\cong}{\longrightarrow} {}_{(A\otimes C)\times_r G\times G\times_r N}\bigl(Y^G_{G/N}((A\otimes C)\times_{\alpha\otimes\mathrm{id},r} G)\bigr)_{(A\otimes C)\times_r G\times G/N}.$$

PROOF. We first must exhibit isomorphisms

$$\varphi_L\colon (A\times_r G\times G\times_r N)\otimes C \xrightarrow{\cong} (A\otimes C)\times_r G\times G\times_r N \quad\text{and}$$
$$\varphi_R\colon (A\times_r G\times G/N)\otimes C \xrightarrow{\cong} (A\otimes C)\times_r G\times G/N$$

between the left and right coefficient algebras. This mainly involves a few applications of Lemma A.20 and Theorem A.65: for the right coefficients, we have

$$(A\times_r G\times G/N)\otimes C \cong \bigl((A\otimes C_0(G/N))\times_r G\bigr)\otimes C$$
$$\cong \bigl((A\otimes C_0(G/N)\otimes C)\times_r G\bigr)$$
$$\cong \bigl((A\otimes C\otimes C_0(G/N))\times_r G\bigr)$$
$$\cong \bigl((A\otimes C)\times_r G\bigr)\times G/N,$$

giving the isomorphism φ_R. Note that φ_R is equivariant for the actions $(\operatorname{Inf}\widehat{\hat\alpha|})\otimes\mathrm{id}$ and $\operatorname{Inf}(\widehat{\alpha\otimes\mathrm{id}|})\hat{\ }$. For the left coefficients, we use Lemma A.20 once more, together

with what we have already shown about φ_R:
$$(A \times_r G \times G \times_r N) \otimes C \cong \big((A \times_r G \times G) \otimes C\big) \times_r N \cong \big((A \otimes C) \times_r G \times G\big) \times_r N,$$
which gives φ_L.

We are now ready to prove the required bimodule isomorphism. By Lemma 1.24 and Remark 1.21, it suffices to show that Ψ respects both inner products. For $y, z \in C_c(G \times G, A)$ and $c, d \in C$ we have

$$\begin{aligned}
&{}_L\langle \Psi(y \otimes c), \Psi(z \otimes d)\rangle(n,s,t) \\
&= \Delta(n)^{-1/2} \int_G \Psi(y \otimes c)(r,t)(\alpha \otimes \mathrm{id})_{s^{-1}}\big(\Psi(z \otimes d)(sr, stn)^*\big)\Delta(r)\, dr \\
&= \Delta(n)^{-1/2} \int_G \big(y(r,t) \otimes c\big)(\alpha_{s^{-1}} \otimes \mathrm{id})\big(z(sr,stn)^* \otimes d^*\big)\Delta(r)\, dr \\
&= \Delta(n)^{-1/2} \int_G y(r,t)\alpha_{s^{-1}}(z(sr,stn)^*)\Delta(r)\, dr \otimes cd^* \\
&= {}_L\langle y,z\rangle(n,s,t) \otimes cd^* = \varphi_L\big({}_L\langle y,z\rangle \otimes cd^*\big)(n,s,t) = \varphi_L\big({}_L\langle y \otimes c, z \otimes d\rangle\big)(n,s,t)
\end{aligned}$$

and

$$\begin{aligned}
&\langle \Psi(y \otimes c), \Psi(z \otimes d)\rangle_R(s, tN) \\
&= \int_N \int_G (\alpha \otimes \mathrm{id})_{r^{-1}}\big(\Psi(y \otimes c)(r, rtn)^* \Psi(z \otimes d)(rs, rtn)\big)\, dr\, dn \\
&= \int_N \int_G (\alpha_{r^{-1}} \otimes \mathrm{id})\big((y(r,rtn)^* \otimes c^*)(z(rs,rtn) \otimes d)\big)\, dr\, dn \\
&= \int_N \int_G \alpha_{r^{-1}}\big(y(r,rtn)^* z(rs,rtn)\big)\, dr\, dn \otimes c^* d \\
&= \langle y,z\rangle_R(s,tN) \otimes c^* d = \varphi_R\big(\langle y,z\rangle_R \otimes c^* d\big)(s,tN) = \varphi_R\big(\langle y \otimes c, z \otimes d\rangle_R\big)(s,tN).
\end{aligned}$$
$\square$

We use Lemma 5.6 to embed $C_c(G \times G, A \otimes C)$ in $Y^G_{G/N}((A \times_r G) \otimes C)$ and $C_c(G \times G, M^\beta(A \otimes C))$ in $M(Y^G_{G/N}((A \times_r G) \otimes C))$. Recall from (4.14) that the decomposition coaction δ_Y of G on $Y^G_{G/N}$ from [**21**] is given by

$$\delta_Y(y) = (y \otimes 1)(j_G \otimes \mathrm{id})(w_G^*)$$

for $y \in \mathcal{D}$, and in particular for $y \in C_c(G \times G, A)$, and the multiplication takes place in $M((A \times_r G \times G) \otimes C^*(G))$.

LEMMA 5.7. *With notation as above, for $y \in C_c(G \times G, A)$ we have*
$$\delta_Y(y) \in C_c(G \times G, M^\beta(A \otimes C^*(G))),$$
with $\delta_Y(y)(s,t) = y(s,t) \otimes t^{-1}s$ for $s, t \in G$.

PROOF. We first claim that for $f \in C_0(G)$ we have $y j_G(f) \in C_c(G \times G, A)$ and
$$(y j_G(f))(s,t) = y(s,t) f(s^{-1}t).$$
Because of the isomorphism from Theorem A.65, we can work in $C_0(G, A) \times_{\alpha \otimes \tau, r} G$ instead of $A \times_{\alpha, r} G \times_{\hat{\alpha}} G$. First, for $s \in G$ we have
$$\big(y i^r_{C_0(G,A)}(1 \otimes f)\big)(s) = y(s, \cdot)(\alpha \otimes \tau)_s(1 \otimes f) = y(s, \cdot)(1 \otimes \tau_s(f)),$$

and evaluating this at $t \in G$ we get
$$\left(yi^\tau_{C_0(G,A)}(1 \otimes f)\right)(s)(t) = y(s,t)\tau_s(f)(t) = y(s,t)f(s^{-1}t).$$
This is jointly continuous in (s,t) and has compact support (because y does), and this verifies the claim.

Hence, if we further have $c \in C^*(G)$ then
$$(y \otimes 1)(j_G \otimes \mathrm{id})(f \otimes c) = yj_G(f) \otimes c \in C_c(G \times G, A \otimes C^*(G))$$
and
$$\begin{aligned}((y \otimes 1)(j_G \otimes \mathrm{id})(f \otimes c))(s,t) &= (yj_G(f) \otimes c)(s,t) = (yj_G(f))(s,t) \otimes c \\ &= y(s,t)f(s^{-1}t) \otimes c = y(s,t) \otimes f(s^{-1}t)c = y(s,t) \otimes (f \otimes c)(s^{-1}t).\end{aligned}$$
The lemma follows by density and continuity, since $w_G \in M(C_0(G) \otimes C^*(G))$. $\square$

BACK TO THE PROOF OF THEOREM 5.1. The isomorphism Φ of $Y \otimes_{A \times_r G \times G/N} X$ onto $X \times_r N$ from [18] is given on elementary tensors $y \otimes x \in C_c(G \times G, A) \odot C_c(G, A)$ by
$$\Phi(y \otimes x)(n,t) = \Delta(n)^{-1/2} \int_G y(s,t)\alpha_s(x(s^{-1}tn))\Delta(s)^{1/2}\,ds.$$
We show that this is equivariant for the appropriate actions: for $y \in C_c(G \times G, A)$ and $x \in X^G_N(A)$ we have
$$\begin{aligned}\Phi &\circ \left(\alpha^Y \otimes \alpha^X\right)_r(y \otimes x)(n,t) \\ &= \Delta(n)^{-1/2} \int_G \alpha^Y(y)(s,t)\alpha_s(\alpha^X_r(x)(s^{-1}tn))\Delta(s)^{1/2}\,ds \\ &= \Delta(n)^{-1/2} \int_G \Delta_{G,N}(r)^{1/2}y(s,tr)\alpha_s\bigl(\Delta_{G,N}(r)^{1/2}x(s^{-1}tnr)\bigr)\Delta(s)^{1/2}\,ds \\ &= \Delta(n)^{-1/2}\Delta_{G,N}(r) \int_G y(s,tr)\alpha_s(x(s^{-1}tnr))\Delta(s)^{1/2}\,ds \\ &= \Delta_{G,N}(r)\Phi(y \otimes x)(r^{-1}nr, tr) \\ &= \alpha^{X \times N}_r \circ \Phi(y \otimes x)(n,t).\end{aligned}$$

We turn to the coaction equivariance, and for this we must show that the diagram
$$\begin{array}{ccc} Y \otimes_{A \times_r G \times G/N} X & \xrightarrow{\delta_Y \sharp_{A \times_r G \times G/N} \delta_X} & M\bigl((Y \otimes_{A \otimes_r G \times G/N} X) \otimes C^*(G)\bigr) \\ \Phi \downarrow & & \downarrow \Phi \otimes \mathrm{id} \\ X \times_r N & \xrightarrow{\delta_{X \times N}} & M\bigl((X \times_r N) \otimes C^*(G)\bigr) \end{array}$$
commutes. Recall from Proposition 2.13 that
$$\delta_Y \sharp_{A \times_r G \times G/N} \delta_X = \Theta \circ (\delta_Y \otimes_{A \times_r G \times G/N} \delta_X),$$
where
$$\Theta \colon \bigl(Y \otimes C^*(G)\bigr) \otimes_{(A \times_r G \times G/N) \otimes C^*(G)} \bigl(X \otimes C^*(G)\bigr) \xrightarrow{\cong} (Y \otimes_{A \times_r G \times G/N} X) \otimes C^*(G)$$
is the isomorphism defined by
$$\Theta\bigl((y \otimes c) \otimes (x \otimes d)\bigr) = (y \otimes x) \otimes cd \qquad \text{for } y \in Y,\ x \in X,\ c,d \in C^*(G).$$

For $y \in C_c(G \times G, A)$ and $x \in C_c(G, A)$ we have

$$(\Phi \otimes \mathrm{id}) \circ (\delta_Y \,\natural_{A \times_r G \times G/N}\, \delta_X)(y \otimes x)(n,t) = (\Phi \otimes \mathrm{id}) \circ \Theta(\delta_Y(y) \otimes \delta_X(x))(n,t)$$
$$\stackrel{(*)}{=} \Delta(n)^{-1/2} \int_G \delta_Y(y)(s,t)(\alpha_s \otimes \mathrm{id})\big(\delta_X(x)(s^{-1}tn)\big)\Delta(s)^{1/2}\,ds$$
$$= \Delta(n)^{-1/2} \int_G \big(y(s,t) \otimes t^{-1}s\big)(\alpha_s \otimes \mathrm{id})\big(x(s^{-1}tn) \otimes s^{-1}tn\big)\Delta(s)^{1/2}\,ds$$
$$= \Delta(n)^{-1/2} \int_G y(s,t)\alpha_s(x(s^{-1}tn))\Delta(s)^{1/2}\,ds \otimes n$$
$$= \Phi(y \otimes x)(n,t) \otimes n = \delta_{X \times N} \circ \Phi(y \otimes x)(n,t),$$

where the equality at $(*)$ is justified by replacing $\delta_Y(y)$ and $\delta_X(x)$ by elementary tensors $u \otimes c \in C_c(G \times G, A) \odot C^*(G)$ and $v \otimes d \in C_c(G, A) \odot C^*(G)$, respectively, and appealing to density and continuity:

$$(\Phi \otimes \mathrm{id}) \circ \Theta\big((u \otimes c) \otimes (v \otimes d)\big)(n,t) = (\Phi \otimes \mathrm{id})\big((u \otimes v) \otimes cd\big)(n,t)$$
$$= \big(\Phi(u \otimes v) \otimes cd\big)(n,t) = \Phi(u \otimes v)(n,t) \otimes cd$$
$$= \Delta(n)^{-1/2} \int_G u(s,t)\alpha_s(v(s^{-1}tn))\Delta(s)^{1/2}\,ds \otimes cd$$
$$= \Delta(n)^{-1/2} \int_G (u(s,t) \otimes c)(\alpha_s \otimes \mathrm{id})\big(v(s^{-1}tn) \otimes d\big)\Delta(s)^{1/2}\,ds$$
$$= \Delta(n)^{-1/2} \int_G (u \otimes c)(s,t)(\alpha_s \otimes \mathrm{id})\big((v \otimes d)(s^{-1}tn)\big)\Delta(s)^{1/2}\,ds.$$

□

As an immediate and interesting corollary, taking $N = G$ we have:

COROLLARY 5.8. *Let (A, G, α) be an action. Then*
$$X_e^G(A) \times_{\alpha^X, r} G \cong Y_{G/G}^G(A \times_{\alpha, r} G)$$

as $(A \times_{\alpha,r} G \times_{\hat\alpha} G \times_{\hat{\hat\alpha},r} G) - (A \times_{\alpha,r} G)$ imprimitivity bimodules, equivariantly for the appropriate actions and coactions.

Note that in this case the appropriate coactions are $(\alpha^X)\check{\,}$ and $(\hat\alpha)_Y$, and the appropriate actions are the trivial ones.

5.1.2. Dual Green equivariant triangle. The following result is dual to Theorem 5.1, starting with a coaction rather than an action.

THEOREM 5.9. *For any nondegenerate normal coaction (A, G, δ) and any closed normal subgroup N of G, the diagram*

(5.2)
$$\begin{array}{c}
A \times_\delta G \times_{\hat\delta, r} G \times_{\hat{\hat\delta}|} G/N \\
\swarrow^{X_N^G(A \times G)} \qquad \downarrow^{Y_{G/G}^G(A) \times G/N} \\
A \times_\delta G \times_{\hat\delta|,r} N \xrightarrow[Y_{G/N}^G(A)]{} A \times_{\delta|} G/N
\end{array}$$

commutes equivariantly for the appropriate actions and coactions.

PROOF. First apply Theorem 5.1 to the dual action $(A \times G, G, \hat{\delta})$ with $N = G$. This gives an $(A \times G \times_r G \times G \times_r G) - (A \times G \times_r G)$ imprimitivity bimodule isomorphism
$$X_e^G(A \times G) \times_r G \cong Y_{G/G}^G(A \times G \times_r G)$$
which is equivariant for the coactions $\widehat{\hat{\delta}^X}$ and $(\hat{\delta})_Y$ (and in this case the appropriate actions are the trivial ones). Also, Theorem 4.21 applied to the $A \times G \times_r G - A$ imprimitivity bimodule $Y_{G/G}^G(A)$ with subgroup G gives an imprimitivity bimodule isomorphism
$$Y_{G/G}^G(A \times G \times_r G) \cong Y_{G/G}^G(A) \times G \times_r G$$
which is equivariant for the dual coactions $(\hat{\hat{\delta}})_Y$ and $(\delta_Y)\hat{\,}$. Combining these, we have
(5.3) $$X_e^G(A \times G) \times_r G \cong Y_{G/G}^G(A) \times G \times_r G,$$
equivariantly for $\widehat{\hat{\delta}^X}$ and $(\delta_Y)\hat{\,}$. We wish to conclude that
(5.4) $$X_e^G(A \times G) \cong Y_{G/G}^G(A) \times G,$$
equivariantly for the actions $\hat{\delta}^X$ and $(\delta_Y)\hat{\,}$. For this, we need the following "duality" result for coactions on right-Hilbert bimodules, which is an easy corollary of functoriality of the crossed products and Theorem 4.21.

LEMMA 5.10. *Suppose Z and W are right-Hilbert $A - B$ bimodules, δ and ϵ are nondegenerate normal coactions of G on A and B, and ζ and η are $\delta - \epsilon$ compatible coactions of G on Z and W, respectively. Then*
$$(Z, \zeta) \cong (W, \eta) \quad \text{if and only if} \quad (Z \times_\zeta G, \hat{\zeta}) \cong (W \times_\eta G, \hat{\eta}).$$

PROOF. The forward implication follows from Theorem 3.13, so assume $(Z \times G, \hat{\zeta}) \cong (W \times G, \hat{\eta})$. Then Theorem 3.7 gives
(5.5) $$\left(Z \times G \times_r G, \widehat{\hat{\zeta}}\right) \cong \left(W \times G \times_r G, \widehat{\hat{\eta}}\right).$$
Now Theorem 4.21 with subgroup $N = G$ gives diagrams

$$\begin{array}{ccc} A \times G \times_r G & \xrightarrow{Y_{G/G}^G(A)} & A \\ {\scriptstyle Z \times G \times_r G}\downarrow & & \downarrow{\scriptstyle Z} \\ B \times G \times_r G & \xrightarrow[Y_{G/G}^G(B)]{} & B \end{array} \quad \text{and} \quad \begin{array}{ccc} A \times G \times_r G & \xrightarrow{Y_{G/G}^G(A)} & A \\ {\scriptstyle W \times G \times_r G}\downarrow & & \downarrow{\scriptstyle W} \\ B \times G \times_r G & \xrightarrow[Y_{G/G}^G(B)]{} & B \end{array}$$

which commute equivariantly for the various coactions. Piecing these together using the equivariant isomorphism (5.5) and canceling the equivalences $Y_{G/G}^G(A)$ and $Y_{G/G}^G(B)$ gives the proposition. □

REMARK 5.11. Although we won't need it, we remark that the result dual to Proposition 5.10, starting with actions rather than coactions, can be proved analogously by using Theorem 4.20 in place of Theorem 4.21.

BACK TO THE PROOF OF THEOREM 5.9. As indicated immediately before the statement of Lemma 5.10, the crossed product in the isomorphism (5.3) can be removed to give the isomorphism (5.4). Taking crossed products by N, we have

(5.6) $$X_e^G(A \times G) \times_r N \cong Y_{G/G}^G(A) \times G \times_r N,$$

and by functoriality (Theorem 3.32) this isomorphism is equivariant for the appropriate actions and coactions.

Next, apply Theorem 5.1 to the dual action $(A \times G, G, \hat{\delta})$ with subgroup N to get a commutative diagram

$$\begin{array}{c}
A \times G \times_r G \times G \times_r N \xrightarrow{Y_{G/N}^G(A \times G \times_r G)} A \times G \times_r G \times G/N \\
\downarrow X_e^G(A \times G) \times_r N \qquad \swarrow X_N^G(A \times G) \\
A \times G \times_r N
\end{array}$$

which on using (5.6) gives equivariant commutativity for actions and coactions of the upper left triangle of the diagram

(5.7) $$\begin{array}{c}
A \times G \times_r G \times G \times_r N \xrightarrow{Y_{G/N}^G(A \times G \times_r G)} A \times G \times_r G \times G/N \\
\downarrow Y_{G/G}^G(A) \times G \times_r N \quad X_N^G(A \times G) \quad \downarrow Y_{G/G}^G(A) \times G/N \\
A \times G \times_r N \xrightarrow{Y_{G/N}^G(A)} A \times G/N.
\end{array}$$

Now note that the outer square in (5.7) commutes equivariantly for actions and coactions by applying Theorem 4.21 to the imprimitivity bimodule $Y_{G/G}^G(A)$; we conclude that the lower right triangle of (5.7) commutes equivariantly for the appropriate actions and coactions, which proves the theorem. □

An interesting corollary of the proof is dual to Corollary 5.8:

COROLLARY 5.12. *Let* (A, G, δ) *be a nondegenerate normal coaction. Then*

$$X_e^G(A \times_\delta G) \cong Y_{G/G}^G(A) \times_{\delta_Y} G$$

as $(A \times_\delta G \times_{\hat{\delta},r} G \times_{\hat{\hat{\delta}}} G) - (A \times_\delta G)$ *imprimitivity bimodules, equivariantly for the appropriate actions and coactions.*

In this case the appropriate actions are $(\hat{\delta})^X$ and $(\delta_Y)^{\hat{}}$, and the appropriate coactions are the trivial ones.

5.2. Restriction and induction

Various combinations of the authors [14], [18], [29] have established a duality between restriction and induction for actions and coactions. A little more precisely, given an action or coaction, restricting representations in the given system is dual to inducing representations in the dual system, and similarly with restricting and inducing reversed. These "dualities" are actually expressed as commutative diagrams in the category $\mathcal{C}$, and for convenience we refer to them as "Ind-Res" and "Res-Ind" diagrams, respectively. Here we will apply the results of the current paper to give simplified proofs of the Ind-Res duality. The strategy is to deduce the Res-Ind diagram for the given system from the Ind-Res diagram for the dual system.

The Ind-Res diagram for actions, which essentially amounts to induction in stages, is largely due to Green:

PROPOSITION 5.13 ([**25**]). *For any action* (A, G, α) *and any closed normal subgroups* $N \subseteq M$ *of* G, *the diagram*

$$\begin{array}{ccc} A \times_{\alpha,r} G \times_{\widehat{\alpha}|} G/M & \xrightarrow{X_M^G} & A \times_{\alpha|,r} M \\ {\scriptstyle A\times_{\alpha,r}G\times_{\widehat{\alpha}|}G/N}\downarrow & & \downarrow{\scriptstyle X_N^M} \\ A \times_{\alpha,r} G \times_{\widehat{\alpha}|} G/N & \xrightarrow[X_N^G]{} & A \times_{\alpha|,r} N \end{array}$$

commutes in the category $\mathcal{C}$.

PROOF. We must show
$$X_M^G \otimes_{A \times M} X_N^M \cong (A \times_r G \times G/N) \otimes_{A\times_r G \times G/N} X_N^G$$
as right-Hilbert $(A \times_r G \times G/M) - (A \times_r N)$ bimodules. Momentarily leaving off the second crossed products by G/M and G/N, this becomes
(5.8) $$X_M^G \otimes_{A \times M} X_N^M \cong (A \times_r G) \otimes_{A \times_r G} X_N^G$$
as right-Hilbert $(A \times_r G) - (A \times_r N)$ bimodules. Since
$$(A \times_r G) \otimes_{A \times_r G} X_N^G \cong X_N^G$$
by cancellation, the isomorphism (5.8) is just Green's induction in stages for actions [**25**, Proposition 8], and is given on the generators by
$$\Phi(x \otimes y) = x \cdot y \qquad \text{for } x \in C_c(G, A), y \in C_c(M, A),$$
where y is viewed as an element of $A \times_r M$, acting on the right of X_M^G. We just need to check that Φ preserves the left action of $C_0(G/M)$: for $f \in C_0(G/M)$ we have
$$\Phi(f \cdot (x \otimes y)) = \Phi(f \cdot x \otimes y) = (f \cdot x) \cdot y = f \cdot (x \cdot y) = f \cdot \Phi(x \otimes y).$$
□

We now recover the Res-Ind diagram for coactions [**29**, Theorem 3.1][1]:

COROLLARY 5.14 ([**29**]). *For any nondegenerate normal coaction* (B, G, δ) *and any closed normal subgroups* $N \subseteq M$ *of* G, *the diagram*

$$\begin{array}{ccc} B \times_\delta G \times_{\widehat{\delta}|,r} M & \xrightarrow{Y_{G/M}^G(B)} & B \times_{\delta|} G/M \\ {\scriptstyle X_N^M(B\times_\delta G)}\downarrow & & \downarrow{\scriptstyle B\times_{\delta|}G/N} \\ B \times_\delta G \times_{\widehat{\delta}|,r} N & \xrightarrow[Y_{G/N}^G(B)]{} & B \times_{\delta|} G/N \end{array}$$

commutes in the category $\mathcal{C}$.

---

[1]Actually, the theorem we state here is slightly weaker than in [**29**], since there we get to assume only as much "normality" as necessary for the bits of the diagram, namely "Mansfield imprimitivity works for M". Here we need to require δ itself to be normal, because of our method of proof. Similarly for Proposition 5.15 below.

5.2. RESTRICTION AND INDUCTION

PROOF. The desired diagram is the outer rectangle of the diagram

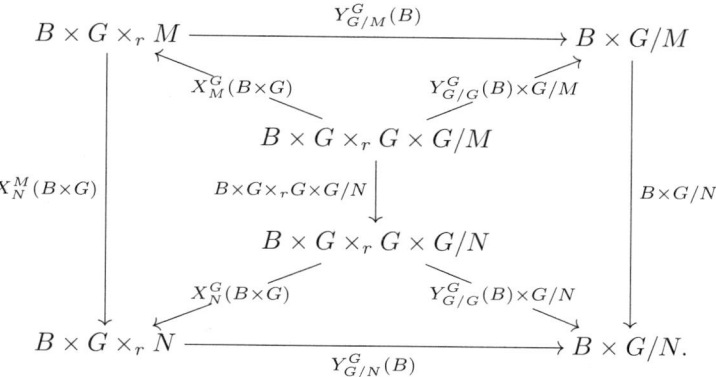

Since the slanted arrows are isomorphisms, it suffices to show the inner polygons commute.

The left quadrilateral commutes by the preceding proposition. The right quadrilateral commutes by [**29**, left face of Diagram (5.1)]. Finally, the top and bottom triangles commute by Theorem 5.9. □

Dually, suppose (B, G, δ) is a coaction. The following Ind-Res diagram is [**29**, Theorem 4.1]:

PROPOSITION 5.15 ([**29**]). *For any nondegenerate normal coaction (B, G, δ) and any closed normal subgroups $N \subseteq M$ of G, the diagram*

$$\begin{array}{ccc} B \times_\delta G \times_{\widehat{\delta}|,r} N & \xrightarrow{Y^G_{G/N}} & B \times_{\delta|} G/N \\ {\scriptstyle B \times_\delta G \times_{\widehat{\delta}|,r} M} \downarrow & & \downarrow {\scriptstyle Y^{G/M}_{G/N}} \\ B \times_\delta G \times_{\widehat{\delta}|,r} M & \xrightarrow{Y^G_{G/M}} & B \times_{\delta|} G/M \end{array}$$

commutes in the category $\mathcal{C}$.

The next theorem substantially improves Corollary 3.3 of [**18**], which covers the case $N = \{e\}$.

THEOREM 5.16. *For any action (A, G, α) and any closed normal subgroups $N \subseteq M$ of G, the diagram*

$$\begin{array}{ccc} A \times_{\alpha,r} G \times_{\widehat{\alpha}|} G/N & \xrightarrow{X^G_N(A)} & A \times_{\alpha|,r} N \\ {\scriptstyle Y^{G/N}_{G/M}(A \times_{\alpha,r} G)} \downarrow & & \downarrow {\scriptstyle A \times_{\alpha|,r} M} \\ A \times_{\alpha,r} G \times_{\widehat{\alpha}|} G/M & \xrightarrow{X^G_M(A)} & A \times_{\alpha|,r} M \end{array}$$

commutes in the category $\mathcal{C}$.

PROOF. The proof is patterned after that of Corollary 5.14. The desired diagram is the outer rectangle of the diagram

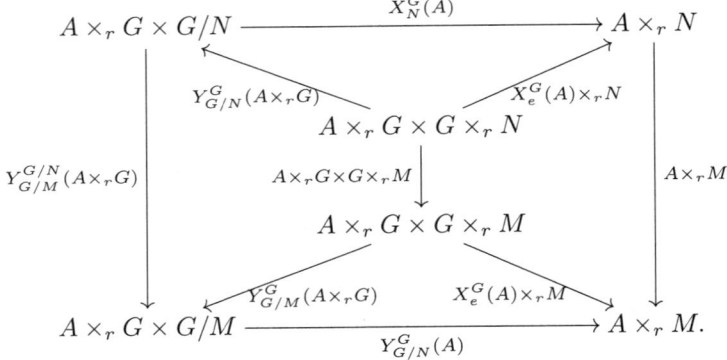

The left quadrilateral commutes by the preceding proposition. The right quadrilateral commutes by [**29**, Lemma 5.7] (which gives the analogous result for full crossed products), together with the argument at the end of the proof of [**29**, Theorem 5.6], which shows that the kernels of the regular representations match up. Finally, the top and bottom triangles commute by Theorem 5.1. □

5.3. Symmetric imprimitivity

In this section we show how to deduce the induced algebra results of [**51**, Section 4] from the main symmetric imprimitivity theorem of that paper, thus avoiding the need to repeat the arguments of [**51**, Section 1]. Suppose, therefore, that we are in the setting of [**51**, Section 4] (see also Section 1 of Appendix B): we have a locally compact space $_H P_K$ with commuting free and proper actions of locally compact groups H and K, a Morita equivalence $_D Y_E$, commuting actions (δ, σ) of (H, K) on D and (τ, γ) of (H, K) on E, and compatible actions (ν, μ) of (H, K) on $_D Y_E$. We then have commuting actions

$$L(\tau) = \begin{pmatrix} \delta & \nu \\ \bar{\nu} & \tau \end{pmatrix} : \mathcal{H} \to \operatorname{Aut} L(Y) \quad \text{and} \quad L(\sigma) = \begin{pmatrix} \sigma & \mu \\ \bar{\mu} & \gamma \end{pmatrix} : K \to \operatorname{Aut} L(Y)$$

on the linking algebra [**11**, Section 4]. The induced algebras $\operatorname{Ind}_H^P L(Y)$ and $\operatorname{Ind}_K^P L(Y)$ carry actions $L(\alpha)$ of K and $L(\beta)$ of H, respectively, and it follows from [**51**, Theorem 1.1] that $C_c(P, L(Y))$ can be made into an $\bigl((\operatorname{Ind}_H^P L(Y)) \times_{L(\alpha)} K\bigr) - \bigl((\operatorname{Ind}_K^P L(Y)) \times_{L(\beta)} H\bigr)$ imprimitivity bimodule.

If we set

$$\operatorname{Ind}_H^P Y = \left\{ f : P \to Y \;\middle|\; \begin{array}{l} f(hp) = L(\tau)_h(f(p)) \text{ for } h \in H, \text{ and} \\ Hp \mapsto \|\langle f(p), f(p)\rangle_D\| \text{ vanishes at } \infty \end{array} \right\},$$

then $\operatorname{Ind}_H^P L(Y)$ is naturally isomorphic to

$$\begin{pmatrix} \operatorname{Ind}_H^P D & \operatorname{Ind}_H^P Y \\ \operatorname{Ind}_H^P \widetilde{Y} & \operatorname{Ind}_H^P E \end{pmatrix} = L(\operatorname{Ind}_H^P Y).$$

Under this identification, the action $L(\alpha)$ restricts in the corners to the tensor product actions of K on $\operatorname{Ind}_H^P D$ and $\operatorname{Ind}_H^P E$, and hence the top left corner in the

decomposition
$$\left(\operatorname{Ind}_H^P L(Y)\right) \times_{L(\alpha)} K = \begin{pmatrix} (\operatorname{Ind}_H^P D) \times K & (\operatorname{Ind}_H^P Y) \times K \\ (\operatorname{Ind}_H^P \widetilde{Y}) \times K & (\operatorname{Ind}_H^P E) \times K \end{pmatrix}$$
is the crossed product of $\operatorname{Ind}_H^P D$ by the action α of [**51**, Section 4]: with our conventions, $\alpha_k(f)(p) = \sigma_k(f(pk))$. Since the bottom right corner in the analogous decomposition of $(\operatorname{Ind}_K^P L(Y)) \times_{L(\beta)} H$ is $(\operatorname{Ind}_K^P E) \times_\beta H$, we deduce immediately from the second part of Lemma 4.6 that the upper right corner $C_c(P, Y)$ in $C_c(P, L(Y))$ completes to give an $((\operatorname{Ind}_H^P D) \times_\alpha K) - ((\operatorname{Ind}_K^P E) \times_\beta H)$ imprimitivity bimodule. It is a straightforward matter to check that the module actions and inner products are, modulo our change in conventions, the ones described in [**51**, page 384]. For example, for $f, g \in C_c(P, Y)$, $\langle f, g \rangle_{\operatorname{Ind} E \times H}$ is the bottom right corner in
$$\left\langle \begin{pmatrix} 0 & f \\ 0 & 0 \end{pmatrix}, \begin{pmatrix} 0 & g \\ 0 & 0 \end{pmatrix} \right\rangle_{C_c(H, \operatorname{Ind} L(Y))} = \begin{pmatrix} 0 & 0 \\ 0 & r \end{pmatrix},$$
where r is the function
$$r(h, p) = \Delta(h)^{-1/2} \int_K \gamma_k \bigl(\langle f(h^{-1}p), \tau_h(g(h^{-1}pk)) \rangle_E \bigr) \, dk.$$
Thus we have shown how to deduce [**51**, Theorem 4.1] from [**51**, Theorem 1.1].

In retrospect, it was always relatively easy to deduce the existence of the Morita equivalence in [**51**, Theorem 4.1] from [**51**, Theorem 1.1], by composing the Morita equivalence of $\operatorname{Ind}_H^P D \times K$ and $\operatorname{Ind}_K^P D \times H$ given by [**51**, Theorem 1.1] with the equivalence of $\operatorname{Ind}_K^P D \times H$ and $\operatorname{Ind}_K^P E \times H$ induced by ${}_D Y_E$. The third part of Lemma 4.6 says that the tensor product bimodule thus obtained is isomorphic to the one we have just constructed. Thus the main point of [**51**, Section 4] is the specific nature of the bimodule.

REMARK 5.17. Lemma 4.9 and Lemma 4.8 would, in the case $P = G$ studied in [**22**], give an equivariant version of [**51**, Section 4].

APPENDIX A

Crossed Products by Actions and Coactions

In this appendix we give an introduction to the theory of crossed products by actions and coactions of groups on C^*-algebras. Since the theory of coactions is much newer, and since there are at least three different definitions of coactions of groups on C^*-algebras in the literature (all somehow mixed together) we decided to present an almost self-contained exposition of coactions and their crossed products, including all proofs for the basic constructions. None of the results in this appendix are new and the main sources are [26, 31, 35, 48, 45, 47, 50, 53].

Some general notation: If X is a locally compact space and E is a normed vector space, then $C_c(X,E)$, $C_0(X,E)$, and $C_b(X,E)$ denote the spaces of continuous E-valued compactly supported functions, functions which vanish at infinity, and bounded functions, respectively. If $E = \mathbb{C}$ we simply write $C_c(X), C_0(X)$, and $C_b(X)$, respectively. $M(B)$ denotes the multiplier algebra of a C^*-algebra B and $UM(B)$ denotes the group of unitary elements of $M(B)$. The *strict topology* on $M(B)$ is the locally convex topology generated by the seminorms $m \mapsto \|ma\|, \|am\|$, $a \in A$. Note that $M(A)$ is the strict completion of A, and we write $M^\beta(A)$ if we consider $M(A)$ equipped with the strict topology. For example, $C_c(X, M^\beta(A))$ will denote the strictly continuous functions of X into $M(A)$ with compact support. A homomorphism $\varphi \colon A \to M(B)$ of a C^*-algebra A into $M(B)$ is called *nondegenerate* if $\varphi(A)B = B$. We use the same letter for a nondegenerate homomorphism $\varphi \colon A \to M(B)$ and its unique (strictly continuous) extension $M(A) \to M(B)$ [35, Lemma 1.1]. Finally, if $\mathcal{H}$ is a Hilbert space, we shall always assume that the inner product on $\mathcal{H}$ is conjugate linear in the first and linear in the second variable.

A.1. Tensor products

Tensor products of C^*-algebras play a basic rôle in the theory of crossed products by actions and coactions: in a certain sense, a crossed product by a group action of G on a C^*-algebra A is just a skew tensor product of A with $C^*(G)$, and a crossed product by a coaction of G on A can be viewed as a skew tensor product of A with $C_0(G)$.

If E and F are complex vector spaces, then we denote by $E \odot F$ the *algebraic tensor product* of E and F. If A and B are C^*-algebras, then $A \odot B$ becomes a $*$-algebra in the canonical way and a C^*-cross norm on $A \odot B$ is a norm $\|\cdot\|_\nu$ which satisfies $\|a \otimes b\|_\nu = \|a\|\|b\|$ for all elementary tensors $a \otimes b \in A \odot B$ and such that the completion of $A \otimes_\nu B = \overline{A \odot B}^{\|\cdot\|_\nu}$ is a C^*-algebra. $A \otimes_\nu B$ is then called the *ν-tensor product* of A and B. We denote by $k_A^\nu \colon A \to M(A \otimes_\nu B)$ and $k_B^\nu \colon B \to M(A \otimes_\nu B)$ the canonical maps $k_A^\nu(a) = a \otimes 1$, $k_B^\nu(b) = 1 \otimes b$.

Among the (possibly) many C^*-cross norms on $A \odot B$ there is a maximal one and a minimal one. The maximal norm $\|\cdot\|_{\max}$ is characterized by the universal property that, whenever we have two nondegenerate homomorphisms $\varphi\colon A \to M(D)$ and $\psi\colon B \to M(D)$ with commuting ranges (i.e., $\varphi(a)\psi(b) = \psi(b)\varphi(a)$ for all $a \in A$, $b \in B$), then there exists a nondegenerate homomorphism $\varphi \otimes \psi\colon A \otimes_{\max} B \to M(D)$ satisfying $(\varphi \otimes \psi) \circ k_A^{\max} = \varphi$ and $(\varphi \otimes \psi) \circ k_B^{\max} = \psi$. Thus

$$\left\| \sum_{i=1}^n a_i \otimes b_i \right\|_{\max} = \sup\left\{ \left\| \sum_{i=1}^n \varphi(a_i)\psi(b_i) \right\| \right\},$$

where the supremum is taken over all commuting pairs of nondegenerate homomorphisms of A and B. In particular, if $\|\cdot\|_\nu$ is any other C^*-cross norm, then $k_A^\nu \otimes k_B^\nu\colon A \otimes_{\max} B \to A \otimes_\nu B$ is a surjection (since it is the identity on $A \odot B$), and hence $\|\cdot\|_\nu$ is dominated by $\|\cdot\|_{\max}$.

If $\pi\colon A \to \mathcal{B}(\mathcal{H})$ and $\rho\colon B \to \mathcal{B}(\mathcal{K})$ are representations of A and B on the Hilbert spaces $\mathcal{H}$ and $\mathcal{K}$, respectively, then there exists a representation $\pi \otimes \rho\colon A \odot B \to \mathcal{B}(\mathcal{H} \otimes \mathcal{K})$ satisfying $(\pi \otimes \rho)(a \otimes b) = \pi(a) \otimes \rho(b)$, and $\pi \otimes \rho$ is faithful on $A \odot B$ if π and ρ are faithful. The *minimal norm* on $A \odot B$ is defined by

$$\left\| \sum_{i=1}^n a_i \otimes b_i \right\|_{\min} = \sup\left\{ \left\| \sum_{i=1}^n \pi(a_i) \otimes \rho(b_i) \right\| \right\},$$

where the supremum is taken over all representations π and ρ of A and B, respectively. It is a nontrivial fact that $\|\cdot\|_{\min}$ is indeed smaller than any other C^*-cross norm on $A \odot B$ (however, it is clear from the definition that $\|\cdot\|_{\min} \leq \|\cdot\|_{\max}$). We shall simply write $\|\cdot\|$ for $\|\cdot\|_{\min}$ and $A \otimes B$ for the *minimal tensor product* $A \otimes_{\min} B$. Note that it follows from the minimality of $\|\cdot\|_{\min}$ that, whenever $\pi\colon A \to \mathcal{B}(\mathcal{H})$ and $\rho\colon B \to \mathcal{B}(\mathcal{K})$ are faithful representations of A and B, respectively, then $\pi \otimes \rho\colon A \otimes B \to \mathcal{B}(\mathcal{H} \otimes \mathcal{K})$ is a faithful representation of $A \otimes B$. Note also that $\pi \otimes \rho$ is nondegenerate if and only if π and ρ are nondegenerate. The following result will be used frequently.

LEMMA A.1. *Let $\varphi\colon A \to M(C)$ and $\psi\colon B \to M(D)$ be homomorphisms. Then there is a homomorphism $\varphi \otimes \psi\colon A \otimes B \to M(C \otimes D)$ satisfying $(\varphi \otimes \psi)(a \otimes b) = \varphi(a) \otimes \psi(b)$. If φ and ψ are nondegenerate (resp. faithful), then so is $\varphi \otimes \psi$.*

PROOF. Representing C and D faithfully on Hilbert spaces turns φ and ψ into $*$-representations, and the result follows from the above-mentioned properties of the minimal tensor product. $\square$

REMARK A.2. Recall that a C^*-algebra A is *nuclear* if $\|\cdot\|_{\max} = \|\cdot\|_{\min}$ on $A \odot B$ for any C^*-algebra B, i.e., $A \otimes_{\max} B = A \otimes B$ for all B. Basic examples of nuclear C^*-algebras are the commutative C^*-algebras and the algebras $\mathcal{K}(\mathcal{H})$ of compact operators on a Hilbert space $\mathcal{H}$, but there are many others.

In this paper, we often need to work with a certain subalgebra $M_C(A \otimes C)$ of the multiplier algebra $M(A \otimes C)$ of the minimal tensor product $A \otimes C$.

DEFINITION A.3. Suppose that A and C are C^*-algebras. Then we define the *C-multiplier algebra* $M_C(A \otimes C)$ of $A \otimes C$ as the set

$$M_C(A \otimes C) = \{m \in M(A \otimes C) \mid m(1 \otimes C) \cup (1 \otimes C)m \subseteq A \otimes C\}.$$

The *C-strict topology* on $M_C(A \otimes C)$ is the locally convex topology generated by the seminorms $m \mapsto \|m(1 \otimes c)\|, \|(1 \otimes c)m\|, c \in C$.

REMARK A.4. (1) It is straightforward to check that $M_C(A \otimes C)$ is a closed $*$-subalgebra of $M(A \otimes C)$. We decided to call it the C-multiplier algebra, since it somehow consists of elements which are "real" multipliers only in the C-factor. Of course, there is a similar definition of the A-multiplier algebra of $A \otimes C$.

(2) If $C = C_0(V)$ for some locally compact space V, and if we identify $A \otimes C_0(V)$ with $C_0(V, A)$ in the canonical way, then one can check, using the identification of $M(A \otimes C_0(V))$ with $C_b(V, M^\beta(A))$ (see [1, Corollary 3.4]), that $M_{C_0(V)}(X \otimes C_0(V))$ can be identified with $C_b(V, A)$.

(3) It is clear that a similar object can also be defined for the maximal tensor product $A \otimes_{\max} C$. However, since we only use minimal tensor products in the main body of this work, we stick to this case. Note that the C-multiplier algebra has appeared in several places in the literature in connection with coactions of groups and Hopf algebras (see, e.g., [**35, 2**], where it is denoted by $\widetilde{M}(A \otimes C)$).

In what follows we gather some important properties of the C-multiplier algebra.

PROPOSITION A.5. *Let A and C be C^*-algebras.*
  (i) *The C-strict topology on $M_C(A \otimes C)$ is stronger than the strict topology induced from $M(A \otimes C)$, and multiplication is separately C-strictly continuous on $M_C(A \otimes C)$. Also, the involution on $M_C(A \otimes C)$ is C-strictly continuous.*
  (ii) $M_C(A \otimes C)$ *is the C-strict completion of $A \otimes C$.*
  (iii) *We have $(1 \otimes M(C))M_C(A \otimes C) \cup M_C(A \otimes C)(1 \otimes M(C)) \subseteq M_C(A \otimes C)$.*

PROOF. Let $(m_i)_{i \in I}$ be a net in $M_C(A \otimes C)$ which converges C-strictly to m. If $z \in A \otimes C$ we can factor $z = (1 \otimes c)y$ for some $c \in C$, $y \in A \otimes C$, to conclude that
$$m_i z = m_i(1 \otimes c)y \to m(1 \otimes c)y = mz,$$
where convergence is in norm. A similar argument shows that $zm_i \to zm$. Separate C-strict continuity of multiplication and C-strict continuity of involution follows then from continuity with respect to the strict topology. Hence (i).

For the proof of (ii), let $m \in M_C(A \otimes C)$ and let $(c_i)_{i \in I}$ be a bounded approximate unit of C. We claim that $(m(1 \otimes c_i))_{i \in I} \subseteq A \otimes C$ converges C-strictly to m. In fact, if $c \in C$, then $c_i c \to c$ in norm, which implies that $m(1 \otimes c_i)(1 \otimes c) \to m(1 \otimes c)$ in norm. On the other hand, one easily checks that $z(1 \otimes c_i) \to z$ in norm for all $z \in A \otimes C$, from which it follows that $(1 \otimes c)m(1 \otimes c_i) \to (1 \otimes c)m$ in norm.

Suppose now that $(m_i)_{i \in I}$ is a C-strict Cauchy net in $M_C(A \otimes C)$. By (i) it follows that $(m_i)_{i \in I}$ is also a strict Cauchy net in $M(A \otimes C)$. Since $M(A \otimes C)$ is the strict completion of $A \otimes C$ we can find an $m \in M(A \otimes C)$ such that $m_i \to m$ strictly. We claim that $m \in M_C(A \otimes C)$ and $m_i \to m$ C-strictly. For this we let $\epsilon > 0$ and $c \in C$, and choose $i_0 \in I$ such that $\|(m_i - m_j)(1 \otimes c)\| \leq \epsilon$ for all $i, j \geq i_0$. Since $m_j(1 \otimes c)z \to m(1 \otimes c)z$ in norm, it follows that
$$\|(m_i - m)(1 \otimes c)z\| = \lim_j \|(m_i - m_j)(1 \otimes c)z\| \leq \epsilon$$
for all $i \geq i_0$. Thus $m_i(1 \otimes c) \to m(1 \otimes c)$ in norm, and a similar argument shows that $(1 \otimes c)m_i \to (1 \otimes c)m$ in norm for all $c \in C$. Thus $m_i \to m$ C-strictly. Finally, since $m_i(1 \otimes c), (1 \otimes c)m_i \in A \otimes C$, it follows from the fact that $A \otimes C$ is norm closed in $M(A \otimes C)$ that $(1 \otimes c)m, m(1 \otimes c) \in A \otimes C$ for all $c \in C$. Thus $m \in M_C(X \otimes C)$. This proves (ii).

We omit the straightforward proof of (iii). □

PROPOSITION A.6. *Suppose that A, B, C, and D are C^*-algebras. Let $\varphi\colon A \to M(B)$ be a possibly degenerate $*$-homomorphism and let $\psi\colon C \to M(D)$ be a nondegenerate $*$-homomorphism.*

(i) *There exists a unique $*$-homomorphism*
$$\overline{\varphi \otimes \psi}\colon M_C(A \otimes C) \to M(B \otimes D)$$
which extends the homomorphism $\varphi \otimes \psi \colon A \otimes C \to M(B \otimes D)$ of Lemma A.1

(ii) *The homomorphism $\overline{\varphi \otimes \psi}$ of (i) is C-strict to strict continuous.*

(iii) *If φ and ψ are faithful, then so is $\overline{\varphi \otimes \psi}$.*

(iv) *If $\varphi(A) \subseteq B$, then $\overline{\varphi \otimes \psi}(M_C(A \otimes C)) \subseteq M_D(B \otimes D)$ and $\overline{\varphi \otimes \psi}$ is C-strict to D-strict continuous.*

PROOF. We show that $\varphi \otimes \psi\colon A \otimes C \to M(B \otimes D)$ is C-strict to strict continuous. If $(a_i)_{i \in I}$ is a net in $A \otimes C$ which converges C-strictly to $a \in A \otimes C$, and if $z \in B \otimes D$, we can factor $z = (1 \otimes \psi(c))y$ for some $c \in C$ and $y \in B \otimes D$ (since ψ is nondegenerate) to conclude that
$$\varphi \otimes \psi(a_i)z = \varphi \otimes \psi(a_i(1 \otimes c))y \to \varphi \otimes \psi(a(1 \otimes c))y = \varphi \otimes \psi(a)z,$$
where convergence is in norm. A similar argument shows that $z(\varphi \otimes \psi(a_i)) \to z(\varphi \otimes \psi(a))$ in norm for all $z \in B \otimes D$.

It follows that there exists a unique C-strict to strict continuous linear extension $\overline{\varphi \otimes \psi}\colon M_C(A \otimes C) \to M(B \otimes D)$ which is automatically a $*$-homomorphism by the C-strict (resp. strict) continuity of involution and the separate C-strict (resp. strict) continuity of multiplication in both algebras.

Assume now that $\eta\colon M_C(A \otimes C) \to M(B \otimes D)$ is another extension of $\varphi \otimes \psi$ and let $m \in M_C(A \otimes C)$. If $z \in B \otimes D$, we can factor $z = (1 \otimes \psi(c))y$ for some $c \in C$ and $y \in B \otimes D$ to compute
$$\eta(m)z = \varphi \otimes \psi(m(1 \otimes c))y = \overline{\varphi \otimes \psi}(m)z,$$
which implies $\eta = \overline{\varphi \otimes \psi}$. This finishes the proof of (i) and (ii).

If φ and ψ are faithful, then so is $\varphi \otimes \psi\colon A \otimes C \to M(B \otimes D)$ by Lemma A.1. Thus, if $\overline{\varphi \otimes \psi}(m) = 0$ for $m \in M_C(A \otimes C)$, it follows that $\varphi \otimes \psi(m(A \otimes C)) = \{0\}$, which implies that $m(A \otimes C) = \{0\}$. Hence $m = 0$. This proves (iii).

Finally, if $\varphi(A) \subseteq B$, it is first clear that $\varphi \otimes \psi(A \otimes C) \subseteq M_D(B \otimes D)$ since $\varphi \otimes \psi(a \otimes c)(1 \otimes d) = \varphi(a) \otimes \psi(c)d \in B \otimes D$ for all elementary tensors $a \otimes c \in A \odot C$ and $d \in D$. But this implies that $\varphi \otimes \psi(A \otimes C)(1 \otimes D) \subseteq B \otimes D$, and applying the $*$-operation then gives $(1 \otimes D)(\varphi \otimes \psi(A \otimes C)) \subseteq B \otimes D$. Factoring $D = \psi(C)D$, a similar argument as in the proof of (i) then shows that $\varphi \otimes \psi\colon A \otimes C \to M_D(B \otimes D)$ is C-strict to D-strict continuous, which implies that there exists a C-strict to D-strict continuous extension
$$\eta\colon M_C(A \otimes C) \to M_D(B \otimes D),$$
which is a $*$-homomorphism by C-strict (resp. D-strict) continuity of the algebra operations. But then the uniqueness clause of (i) implies that $\eta = \overline{\varphi \otimes \psi}$. □

Some further information on C-multiplier algebras will be given in Section 1.4 of Chapter 1.

A.2. Actions and their crossed products

If G is a locally compact group, then ds denotes left Haar measure on G and Δ its modular function, *i.e.*, we have $\int_G f(s)\,ds = \Delta(t)\int_G f(st)\,ds$ for all $f \in C_c(G), t \in G$. An *action* of G on a C^*-algebra A is a strongly continuous homomorphism α of G into the $*$-automorphism group $\operatorname{Aut}(A)$ of A. We also call the triple (A, G, α) an action. If (A, G, α) is an action, then $C_c(G, A)$ becomes a $*$-algebra with respect to the convolution and involution defined by

$$(A.1) \qquad f * g(s) = \int_G f(t)\alpha_t(g(t^{-1}s))\,dt \quad \text{and} \quad f^*(s) = \Delta(s^{-1})\alpha_s(f(s^{-1}))^*.$$

A *covariant homomorphism* of (A, G, α) into the multiplier algebra $M(D)$ of a C^*-algebra D is a pair (π, U), where $\pi\colon A \to M(D)$ is a nondegenerate homomorphism and $U\colon G \to UM(D)$ is a strictly continuous homomorphism satisfying

$$\pi \circ \alpha_s = \operatorname{Ad} U_s \circ \pi \qquad \text{for all } s \in G,$$

where for any unitary v we let $\operatorname{Ad} v$ denote the usual conjugation automorphism $\operatorname{Ad} v(b) = vbv^*$. A *covariant representation* of (A, G, α) on a Hilbert space $\mathcal{H}$ is a covariant homomorphism into $M(\mathcal{K}(\mathcal{H})) = \mathcal{B}(\mathcal{H})$. If (π, U) is a covariant homomorphism into $M(D)$, then the *integrated form* $\pi \times U\colon C_c(G, A) \to M(D)$ is defined by

$$(A.2) \qquad (\pi \times U)(f) = \int_G \pi(f(s))U_s\,ds.$$

With the help of [**53**, Lemma 7] it is not hard to see that $\pi \times U$ is well-defined and continuous with respect to the inductive limit topology on $C_c(G, A)$. Also, for each covariant representation (π, U) and each $f \in C_c(G, A)$ we have

$$\|(\pi \times U)(f)\| \leq \int_G \|f(s)\|\,ds,$$

from which it follows that for each $f \in C_c(G, A)$ we can form

$$\|f\| = \sup\{\|(\pi \times U)(f)\| \mid (\pi, U) \text{ is a covariant representation of } (A, G, \alpha)\}.$$

One can check that $\|\cdot\|$ is a norm on $C_c(G, A)$. This leads to

DEFINITION A.7. Let (A, G, α) be an action. Then the completion $A \times_\alpha G$ of $C_c(G, A)$ with respect to $\|\cdot\|$ is called the *(full) crossed product* of (A, G, α).

REMARK A.8. (1) There is a canonical covariant homomorphism (i_A, i_G) of (A, G, α) into $M(A \times_\alpha G)$ given by the formulas

$$(i_A(a)f)(s) = af(s) \qquad (i_G(t)f)(s) = \alpha_t(f(t^{-1}s))$$
$$(fi_A(a))(s) = f(s)\alpha_s(a) \qquad (fi_G(t))(s) = \Delta(t^{-1})f(st^{-1}),$$

$f \in C_c(G, A)$. It follows that if $f \in C_c(G, A)$, then $(i_A \times i_G)(f) = f$ as elements of $A \times_\alpha G$.

(2) The triple $(A \times_\alpha G, i_A, i_G)$ enjoys the following universal property: For any covariant homomorphism (π, U) of (A, G, α) into $M(D)$ there exists a unique nondegenerate homomorphism $\pi \times U\colon A \times_\alpha G \to M(D)$ such that $(\pi \times U) \circ i_A = \pi$ and $(\pi \times U) \circ i_G = U$.

To see this observe that it follows from the definition of the greatest C^*-norm on $C_c(G, A)$ that the integrated form $\pi \times U\colon C_c(G, A) \to M(D)$ is continuous with respect to this norm and therefore extends uniquely to a homomorphism of $A \times_\alpha G$.

Using an approximate identity of $A\times_\alpha G$ which lies inside $C_c(G,A)$, it is not too hard to check that the extension $\pi\times U\colon A\times_\alpha G \to M(D)$ satisfies all properties mentioned above.[1] Conversely, if $\rho\colon A\times_\alpha G \to M(D)$ is a nondegenerate homomorphism, then $(\rho\circ i_A, \rho\circ i_G)$ is a covariant homomorphism of (A,G,α) such that $\rho=(\rho\circ i_A)\times(\rho\circ i_G)$. Thus, $(\pi,U)\mapsto \pi\times U$ gives a one-to-one correspondence between the covariant homomorphisms of (A,G,α) and the nondegenerate homomorphisms of $A\times_\alpha G$.

(3) In [**50**] the universal properties were used to *define* the full crossed product of (A,G,α) as a triple (C,k_A,k_G) satisfying

   (i) (k_A,k_G) is a covariant homomorphism of (A,G,α) into $M(C)$;
   (ii) $k_A(A)k_G(C^*(G))$ is dense in C;
   (iii) if (π,U) is any covariant homomorphism of (A,G,α) into $M(D)$, then there exists a unique nondegenerate homomorphism $\varphi_{\pi,U}\colon C\to M(D)$ such that $\varphi_{\pi,U}\circ k_A = \pi$ and $\varphi_{\pi,U}\circ k_G = U$.

Clearly, $(A\times_\alpha G, i_A, i_G)$ is a crossed product in this sense. Moreover, if (C,k_A,k_G) is any other triple satisfying (i)–(iii), then $k_A\times k_G\colon A\times_\alpha G \to C$ is an isomorphism with inverse $\varphi_{i_A,i_G}\colon C\to A\times_\alpha G$, where φ_{i_A,i_G} is the homomorphism associated to (i_A,i_G) by (iii).

(4) It is sometimes useful to be able to consider integrated forms of pairs (π,U) with $\pi\colon A\to M(D)$ a $*$-homomorphism and $U\colon G\to UM(D)$ a strictly continuous homomorphism which satisfy the covariance condition $\pi(\alpha_s(a))=U_s\pi(a)U_s^*$ for all $a\in A$, $s\in G$, but where $\pi\colon A\to M(D)$ is degenerate. We shall call such a pair a *degenerate covariant homomorphism*. As for nondegenerate homomorphisms we get a $*$-homomorphism $\pi\times U\colon C_c(G,A)\to M(D)$ by integration. So the only problem is to see that $\pi\times U$ is norm-decreasing, in order to obtain a unique extension to $A\times_\alpha G$. For this represent $M(D)$ faithfully into $B(\mathcal{H})$ for some Hilbert space $\mathcal{H}$ and write $\mathcal{H}_1=\pi(A)\mathcal{H}$. Then we get $U_s\mathcal{H}_1=U_s\pi(A)\mathcal{H}=\pi(A)U_s\mathcal{H}=\pi(A)\mathcal{H}=\mathcal{H}_1$, from which we obtain a nondegenerate representation (π_1,U_1) of (A,G,α) into $\mathcal{B}(\mathcal{H}_1)$. The desired result then follows from $\|\pi\times U(f)\|=\|\pi_1\times U_1(f)\|$ for all $f\in C_c(G,A)$.

The following (well-known) lemma serves as a first example of the usefulness of the universal properties of the full crossed product. It also indicates the conceptual similarity of full crossed products with maximal tensor products of C^*-algebras.

LEMMA A.9. *Let (A,G,α) be an action and let B be a C^*-algebra. Let $\mathrm{id}\otimes\alpha$ be the diagonal action of G on $B\otimes_{\max}A$ and let k_B and $k_{A\times_\alpha G}$ denote the canonical maps of B and $A\times_\alpha G$ into $M(B\otimes_{\max}(A\times_\alpha G))$. Further write $k_A=k_{A\times_\alpha G}\circ i_A$ and $k_G=k_{A\times_\alpha G}\circ i_G$. Then*

$$(k_B\otimes k_A)\times k_G\colon (B\otimes_{\max}A)\times_{\mathrm{id}\otimes\alpha} G \to B\otimes_{\max}(A\times_\alpha G)$$

is an isomorphism.

PROOF. Just check (using the universal properties of the crossed product and the maximal tensor product) that $(B\otimes_{\max}(A\times_\alpha G), k_B\otimes k_A, k_G)$ satisfies conditions (i)–(iii) above. $\square$

---

[1]One can construct such an approximate identity as follows: for each compact neighborhood V of the identity $e\in G$ choose $g_V\in C_c(G)^+$ such that $\operatorname{supp} g_V\subseteq V$ and $\int_G g_V(s)\,ds=1$. Then, if $\{a_i\}_i$ is an approximate identity of A, $\{i_A(a_i)i_G(g_V)\}_{(V,i)}$ with ordering $(V,i)\leq (W,j) \Leftrightarrow V\supseteq W$ and $i\leq j$ becomes an approximate identity of $A\times_\alpha G$.

If $A = \mathbb{C}$ the crossed product $\mathbb{C} \times_{\mathrm{id}} G$ is just the (full) *group C^*-algebra* $C^*(G)$ of G. We write $u \colon G \to UM(C^*(G))$ for the canonical map (whenever a name is needed). Thus $C^*(G)$ enjoys the universal property that for any strictly continuous homomorphism $V \colon G \to UM(D)$ there exists a unique nondegenerate homomorphism $\bar{V} \colon C^*(G) \to M(D)$ such that $\bar{V} \circ u = V$. In what follows we shall make no notational distinction between V and its integrated form (denoted $\bar{V}$ above), i.e., we simply write $V(z)$ for $z \in C^*(G)$. The *reduced group C^*-algebra* $C_r^*(G)$ is the image of $C^*(G)$ under the *left regular representation* $\lambda \colon G \to U(L^2(G))$ defined by $(\lambda_s \xi)(t) = \xi(s^{-1}t)$. Notice that $\lambda \colon C^*(G) \to C_r^*(G)$ is an isomorphism if and only if G is amenable [**43**, Theorem 7.3.9].

EXAMPLE A.10. One of the most interesting actions from the point of view of duality theory is the action $(C_0(G), G, \tau)$, where $\tau \colon G \to \mathrm{Aut}(C_0(G))$ is given by left translation: $\tau_s(f)(t) = f(s^{-1}t)$. If $M \colon C_0(G) \to \mathcal{B}(L^2(G))$ denotes the representation of $C_0(G)$ as multiplication operators on $L^2(G)$, then it is easily seen that (M, λ) is a covariant representation of $(C_0(G), G, \tau)$. If $f \in C_c(G \times G) \subseteq C_c(G, C_0(G))$, the integrated form of f can be written as

$$((M \times \lambda)(f)\xi)(t) = \int_G f(s, t)\xi(s^{-1}t)\,ds = \int_G \Delta(s^{-1})f(ts^{-1}, t)\xi(s)\,ds.$$

Thus, $(M \times \lambda)(f)$ is an integral operator with kernel $k(s, t) = \Delta(s^{-1})f(ts^{-1}, t)$ in $C_c(G \times G)$. Since the set of integral operators with kernels in $C_c(G \times G)$ is dense in $\mathcal{K}(L^2(G))$ it follows that $(M \times \lambda)(C_0(G) \times_\tau G) = \mathcal{K}(L^2(G))$.

It is a nontrivial result that any representation of $(C_0(G), G, \tau)$ is unitarily equivalent to a representation of the form $(M \otimes 1, \lambda \otimes 1)$ on $L^2(G) \otimes \mathcal{H}$ for some Hilbert space $\mathcal{H}$. For $G = \mathbb{R}^n$ this is a reformulation of the uniqueness of the Heisenberg commutation relations, and for arbitrary G this is a special case of Mackey's imprimitivity theorem (for a good treatment of this special case see [**55**], but we shall give an independent proof in Corollary A.66 below). It follows that $M \times \lambda \colon C_0(G) \times_\tau G \to \mathcal{K}(L^2(G))$ is faithful and $C_0(G) \times_\tau G \cong \mathcal{K}(L^2(G))$.

REMARK A.11. The first part of the above example shows in particular that

$$\overline{M(C_0(G))\lambda(C^*(G))} = \overline{\lambda(C^*(G))M(C_0(G))} = \mathcal{K}(L^2(G)).$$

A similar result is true for the *right regular representation* $\rho \colon G \to U(L^2(G))$ given by $(\rho_s \xi)(t) = \Delta(s)^{1/2}\xi(ts)$. To see this, consider the self-adjoint unitary operator U on $L^2(G)$ defined by $(U\xi)(s) = \Delta(s)^{-1/2}\xi(s^{-1})$ and observe that $\mathrm{Ad}\,U \circ M = M$ and $\mathrm{Ad}\,U \circ \lambda = \rho$.

If (A, G, α) and (B, G, β) are two actions, then a homomorphism $\varphi \colon A \to M(B)$ is called *G-equivariant* if $\varphi \circ \alpha_s = \beta_s \circ \varphi$ (where we implicitly extend each β_s to an automorphism of $M(B)$). If (i_B, i_G) denotes the canonical maps of (B, G) into $M(B \times_\beta G)$, then we obtain a (possibly degenerate) homomorphism

$$\varphi \times G = (i_B \circ \varphi) \times i_G \colon A \times_\alpha G \to M(B \times_\beta G).$$

If $\varphi \colon A \to B$ is a G-equivariant isomorphism, then $\varphi \times G$ is an isomorphism between $A \times_\alpha G$ and $B \times_\beta G$ with inverse given by $\varphi^{-1} \times G$.

EXAMPLE A.12. Let (A, G, α) be an action and let $(C_0(G), G, \tau)$ be as in Example A.10. Let $\alpha \otimes \tau$ denote the diagonal action of G on $A \otimes C_0(G) = A \otimes_{\max} C_0(G)$. We want to show that $(A \otimes C_0(G)) \times_{\alpha \otimes \tau} G$ is canonically isomorphic to $A \otimes \mathcal{K}(L^2(G))$. To see this define $\varphi \colon A \otimes C_0(G) \to A \otimes C_0(G)$ by

$\varphi(f)(s) = \alpha_{s^{-1}}(f(s))$ for $f \in C_0(G, A) \cong A \otimes C_0(G)$. Then it is straightforward to check that φ is an $(\alpha \otimes \tau) - (\mathrm{id} \otimes \tau)$ equivariant isomorphism. Thus, $(A \otimes C_0(G)) \times_{\alpha \otimes \tau} G$ is isomorphic to $(A \otimes C_0(G)) \times_{\mathrm{id} \otimes \tau} G$, which in turn is isomorphic to $A \otimes_{\max} (C_0(G) \times_\tau G) \cong A \otimes \mathcal{K}(L^2(G))$ by Lemma A.9, Example A.10 and the nuclearity of $\mathcal{K}(L^2(G))$.

We are now going to define the reduced crossed product of an action (A, G, α). Writing $\widetilde{\alpha}(a)(s) = \alpha_{s^{-1}}(a)$, we may view the action α as a homomorphism of A into $C_b(G, A) \subseteq M(A \otimes C_0(G))$. It is straightforward to check that $(i_A^r, i_G^r) = ((\mathrm{id}_A \otimes M) \circ \widetilde{\alpha}, 1 \otimes \lambda)$ is a covariant homomorphism of (A, G, α) into $M(A \otimes \mathcal{K}(L^2(G)))$.

DEFINITION A.13. Let (A, G, α) be an action and let $(i_A^r, i_G^r) = ((\mathrm{id}_A \otimes M) \circ \widetilde{\alpha}, 1 \otimes \lambda)$ be as above. The *reduced crossed product* $A \times_{\alpha, r} G$ is the image $(i_A^r \times i_G^r)(A \times_\alpha G) \subseteq M(A \otimes \mathcal{K}(L^2(G)))$. We usually regard (i_A^r, i_G^r) as maps from (A, G) into $M(A \times_{\alpha, r} G)$ and call them the *canonical maps* of (A, G) into $M(A \times_{\alpha, r} G)$.

REMARK A.14. The reduced crossed product $A \times_{\alpha, r} G$ actually lies in the $\mathcal{K}(L^2(G))$-multiplier algebra $M_{\mathcal{K}(L^2(G))}(A \otimes \mathcal{K}(L^2(G)))$. To see this it is enough to show that $i_A^r(a) i_G^r(z) \in M_{\mathcal{K}(L^2(G))}(A \otimes \mathcal{K}(L^2(G)))$ for all $a \in A$, $z \in C^*(G)$, and since $i_G^r(z) \in 1 \otimes M(\mathcal{K}(L^2(G)))$ we can use part (iii) of Proposition A.5 to see that it actually suffices to show that $i_A^r(A) \subseteq M_{\mathcal{K}(L^2(G))}(A \otimes \mathcal{K}(L^2(G)))$. For this we first observe that $\widetilde{\alpha}(A) \subseteq C_b(G, A) \cong M_{C_0(G)}(A \otimes C_0(G))$ (see part (2) of Remark A.4). Then we use part (iv) of Proposition A.6 to see that

$$i_A^r(A) = (\mathrm{id}_A \otimes M) \circ \widetilde{\alpha}(A) \subseteq (\mathrm{id}_A \otimes M)\big((M_{C_0(G)}(A \otimes C_0(G))\big)$$
$$\subseteq M_{\mathcal{K}(L^2(G))}(A \otimes \mathcal{K}(L^2(G))).$$

Let $\pi \colon A \to \mathcal{B}(\mathcal{H})$ be a (possibly degenerate) representation of A on some Hilbert space $\mathcal{H}$. The *induced representation* $\mathrm{Ind}\,\pi \colon A \times_\alpha G \to \mathcal{B}(\mathcal{H} \otimes L^2(G))$ is defined as the integrated form of the covariant representation $((\pi \otimes M) \circ \widetilde{\alpha}, 1 \otimes \lambda)$ on the Hilbert space $\mathcal{H} \otimes L^2(G) \cong L^2(G, \mathcal{H})$. Thus we have

$$\mathrm{Ind}\,\pi = (\pi \otimes \mathrm{id}) \circ (i_A^r \times i_G^r),$$

which implies that $\mathrm{Ind}\,\pi$ factors through a representation of $A \times_{\alpha, r} G$ on $\mathcal{H} \otimes L^2(G)$. Note that by Proposition A.6 the above composition makes perfect sense even if π is degenerate, since by the above remark $A \times_{\alpha, r} G = i_A^r \times i_G^r(A \times_\alpha G)$ lies in the $\mathcal{K}(L^2(G))$-multiplier algebra of $A \otimes \mathcal{K}(L^2(G))$. Since $\pi \otimes \mathrm{id} \colon A \otimes \mathcal{K}(L^2(G)) \to \mathcal{B}(\mathcal{H} \otimes L^2(G))$ is faithful if π is, it follows that $\mathrm{Ind}\,\pi$ factors through a faithful representation of $A \times_{\alpha, r} G$ whenever π is faithful. In this case we call $\mathrm{Ind}\,\pi$ the *regular representation of (A, G, α) induced from π*.

REMARK A.15. In the literature, the reduced crossed product is often defined as the image $\mathrm{Ind}\,\pi(A \times_\alpha G)$ of any regular representation $\mathrm{Ind}\,\pi$ of (A, G, α). Of course, this would determine $A \times_{\alpha, r} G$ only up to isomorphism. However, it is often convenient to work with regular representations and to identify $A \times_{\alpha, r} G$ with its image $\mathrm{Ind}\,\pi(A \times_{\alpha, r} G) \subseteq \mathcal{B}(\mathcal{H} \otimes L^2(G))$. Whenever we do this, we also identify i_A^r with $(\pi \otimes M) \circ \widetilde{\alpha}$ and i_G^r with $1 \otimes \lambda$ (although we will not do this outside of this appendix).

The following lemma is quite useful:

LEMMA A.16. *Assume that α and β are actions of G on C^*-algebras A and B, respectively. Assume further that $\varphi \colon A \to M(B)$ is a (possibly degenerate) G-equivariant $*$-homomorphism. Then there exists a $*$-homomorphism*
$$\varphi \times_r G = (i_B^r \circ \varphi) \times i_G^r \colon A \times_{\alpha,r} G \to M(B \times_{\beta,r} G).$$

PROOF. We have to check that the homomorphism $(i_B^r \circ \varphi) \times i_G^r \colon A \times_\alpha G \to M(B \times_{\beta,r} G)$ factors through $A \times_{\alpha,r} G$. For this consider the composition
$$A \times_\alpha G \xrightarrow{i_A^r \times i_G^r} M_{\mathcal{K}(L^2(G))}(A \otimes \mathcal{K}(L^2(G))) \xrightarrow{\varphi \otimes \mathrm{id}} M(B \otimes \mathcal{K}(L^2(G))).$$
Since φ is G-equivariant, we have
$$(\varphi \otimes \mathrm{id}) \circ i_A^r = (\varphi \otimes \mathrm{id}) \circ (\mathrm{id}_A \otimes M) \circ \alpha$$
$$= (\mathrm{id}_B \otimes M) \circ (\varphi \otimes \mathrm{id}) \circ \alpha = (\mathrm{id}_B \otimes M) \circ \beta \circ \varphi.$$
Since it is clear that $(\varphi \otimes \mathrm{id}) \circ i_G^r = i_G^r$ (in the appropriate sense), it follows that
$$(\varphi \otimes \mathrm{id}) \circ (i_A^r \times i_G^r) = (i_B^r \circ \varphi) \times i_G^r$$
and the result follows. If φ is faithful, then $\varphi \otimes \mathrm{id}$ is faithful on $M_{\mathcal{K}(L^2(G))}(A \otimes \mathcal{K}(L^2(G))) \supseteq A \times_{\alpha,r} G$ by Proposition A.6. Hence $\varphi \times_r G$ is faithful, too. □

REMARK A.17. It is helpful to observe that the homomorphisms
$$\varphi \times G \colon A \times_\alpha G \to M(B \times_\beta G) \quad \text{and} \quad \varphi \times_r G \colon A \times_{\alpha,r} G \to M(B \times_{\beta,r} G)$$
are given on elements of $C_c(G, A)$ by $\varphi \times G(f) = \varphi \circ f \in C_c(G, M(B))$, acting on $C_c(G, B)$ via convolution.

If $\pi \times V$ is a representation of $A \times_\alpha G$ on $\mathcal{H}$ and U is a representation of G on $\mathcal{K}$, then we write $(\pi \times U) \otimes V \colon A \times_\alpha G \to \mathcal{B}(\mathcal{H} \otimes \mathcal{K})$ for the integrated form of the covariant representation $(\pi \otimes 1_\mathcal{K}, V \otimes U)$. We shall frequently use:

LEMMA A.18. *Let $\pi \colon A \to \mathcal{B}(\mathcal{H})$ be a representation of A, and let U be a representation of G on some Hilbert space $\mathcal{K}$. Then*
  (i) *$(\mathrm{Ind}\,\pi) \otimes U$ is unitarily equivalent to $\mathrm{Ind}(\pi \otimes 1_\mathcal{K})$.*
  (ii) *If $\mathcal{K} = \mathcal{H}$ and (π, U) is covariant for (A, G, α), then $\mathrm{Ind}\,\pi$ is unitarily equivalent to $(\pi \times U) \otimes \lambda$.*

PROOF. For the proof of (i), identify $\mathcal{H} \otimes \mathcal{K} \otimes L^2(G)$ and $\mathcal{H} \otimes L^2(G) \otimes \mathcal{K}$ with $L^2(G, \mathcal{H} \otimes \mathcal{K})$ in the canonical way and check that the unitary W on $L^2(G, \mathcal{H} \otimes \mathcal{K})$ defined by $(W\xi)(s) = (1 \otimes U_s)\xi(s)$ intertwines the representations. Similarly, for the proof of (ii) check that $W \in U(L^2(G, \mathcal{H}))$ defined by $(W\xi)(s) = U_s\xi(s)$ does the job. □

As a first application we get:

PROPOSITION A.19. *If G is amenable, then $i_A^r \times i_G^r \colon A \times_\alpha G \to A \times_{\alpha,r} G$ is an isomorphism for any action $\alpha \colon G \to \mathrm{Aut}(A)$.*

PROOF. Choose any faithful representation $\pi \times U$ of $A \times_\alpha G$ on some Hilbert space $\mathcal{H}$, and let $1_G \colon C_r^*(G) \cong C^*(G) \to \mathbb{C}$ denote the integrated form of the trivial representation of G. We then have the identity
$$(\mathrm{id} \otimes 1_G) \circ \big((\pi \times U) \otimes \lambda\big) = \pi \times U$$
under the canonical identification $\mathcal{H} \otimes \mathbb{C} \cong \mathcal{H}$. In particular, it follows that $(\pi \times U) \otimes \lambda$ is injective. But Lemma A.18 implies that $(\pi \times U) \otimes \lambda$ is unitarily equivalent to a regular representation of (A, G, α). Hence $\ker(i_A^r \times i_G^r) = \ker(\pi \times U) \otimes \lambda = \{0\}$. □

Note that $\mathbb{C} \times_{\mathrm{id}} G = C^*(G) \cong C_r^*(G) = \mathbb{C} \times_{\mathrm{id},r} G$ via $i_\mathbb{C}^r \times i_G^r = \lambda$ if and only if G is amenable, so that the reduced crossed product does not coincide with the full crossed product in general. We should point out that it follows from Example A.10 that every representation of $C_0(G) \times_\tau G$ is faithful, so $C_0(G) \times_\tau G \cong C_0(G) \times_{\tau,r} G$ even if G is not amenable.[2] Anyway, reduced crossed products can be viewed in a certain sense as skew minimal tensor products of A with the reduced group algebra $C_r^*(G)$. For instance, parallel to Lemma A.9 we have:

LEMMA A.20. *Let (A, G, α) be an action and let B be a C^*-algebra. Then there exists a canonical isomorphism $(B \otimes A) \times_{\mathrm{id} \otimes \alpha, r} G \cong B \otimes (A \times_{\alpha, r} G)$.*

PROOF. Let $(k_B \otimes k_A) \times k_G : (B \otimes_{\max} A) \times_{\mathrm{id} \otimes \alpha} G \to B \otimes_{\max} (A \times_\alpha G)$ be the isomorphism of Lemma A.9. Then it is easy to check that $(k_B \otimes k_A) \times k_G$ transforms the homomorphism

$$(\mathrm{id}_B \otimes \mathrm{id}_A \otimes M) \circ (\mathrm{id}_B \otimes \alpha) \times (1 \otimes 1 \otimes \lambda) : (B \otimes_{\max} A) \times_{\mathrm{id} \otimes \alpha} G \to M(B \otimes A \otimes \mathcal{K}(L^2(G)))$$

to the homomorphism

$$\mathrm{id}_B \otimes \big((\mathrm{id}_A \otimes M) \circ \alpha \times (1 \otimes \lambda)\big) : B \otimes_{\max} (A \times_\alpha G) \to M(B \otimes A \otimes \mathcal{K}(L^2(G))).$$

But the image of the first map is $(B \otimes A) \times_{\mathrm{id} \otimes \alpha, r} G$ and the image of the second is $B \otimes (A \times_{\alpha,r} G)$. □

The combination of Lemma A.9, Lemma A.20, and Proposition A.19 implies the important fact that $A \times_\alpha G$ is nuclear whenever A is nuclear and G is amenable: if B is any C^*-algebra, then

$$B \otimes_{\max} (A \times_\alpha G) \cong (B \otimes_{\max} A) \times_{\mathrm{id} \otimes \alpha} G \cong (B \otimes A) \times_{\mathrm{id} \otimes \alpha} G$$
$$\cong (B \otimes A) \times_{\mathrm{id} \otimes \alpha, r} G \cong B \otimes (A \times_{\alpha, r} G) \cong B \otimes (A \times_\alpha G).$$

A.3. Coactions

If (A, G, α) is an action such that G is abelian, then there exists a natural action $\widehat{\alpha}$ of the dual group $\widehat{G}$ of G on $A \times_\alpha G$ given for $f \in C_c(G, A)$ by

(A.3) $$\widehat{\alpha}_\chi(f)(s) = \chi(s) f(s).$$

The famous Takesaki-Takai duality theorem asserts that the double crossed product $A \times_\alpha G \times_{\widehat{\alpha}} \widehat{G}$ is isomorphic to $A \otimes \mathcal{K}(L^2(G))$, *i.e.*, the double crossed product is stably isomorphic to A. If G is nonabelian, then $\widehat{G}$ is not a group and there is no dual action to talk about.

Coactions of groups on C^*-algebras were mainly introduced in order to overcome this problem and to obtain a reasonable duality theory for actions of non-abelian groups. In order to define coactions of groups on C^*-algebras let us first note that the group C^*-algebra $C^*(G)$ of G carries a natural *comultiplication* given by the integrated form $\delta_G : C^*(G) \to M(C^*(G) \otimes C^*(G))$ of the strictly continuous homomorphism $s \mapsto s \otimes s \in UM(C^*(G) \otimes C^*(G))$. It follows directly from the definition that δ_G satisfies the *comultiplication identity*

$$(\delta_G \otimes \mathrm{id}_G) \circ \delta_G = (\mathrm{id}_G \otimes \delta_G) \circ \delta_G.$$

---

[2] If the regular homomorphism $i_A^r \times i_G^r : A \times_\alpha G \to A \times_{\alpha, r} G$ is not faithful, can the C^*-algebras $A \times_\alpha G$ and $A \times_{\alpha, r} G$ still be isomorphic? Presumably so, but we do not know an example.

A.3. COACTIONS

DEFINITION A.21. A *(full) coaction* of a locally compact group G on a C^*-algebra A is an injective and nondegenerate homomorphism $\delta \colon A \to M(A \otimes C^*(G))$ satisfying

(i) $\delta(A)(1 \otimes C^*(G)) \subseteq A \otimes C^*(G)$, and
(ii) $(\delta \otimes \mathrm{id}_G) \circ \delta = (\mathrm{id}_A \otimes \delta_G) \circ \delta$ as maps from A into $M(A \otimes C^*(G) \otimes C^*(G))$ (the *coaction identity*).

We also call the triple (A, G, δ) a coaction. A coaction δ is called *nondegenerate* if

(iii) $\overline{\delta(A)(1 \otimes C^*(G))} = A \otimes C^*(G)$.

It might be helpful to realize that Condition (ii) above simply means that the diagram

$$\begin{array}{ccc} A & \xrightarrow{\delta} & M(A \otimes C^*(G)) \\ {\scriptstyle \delta} \downarrow & & \downarrow {\scriptstyle \mathrm{id}_A \otimes \delta_G} \\ M(A \otimes C^*(G)) & \xrightarrow[\delta \otimes \mathrm{id}_G]{} & M(A \otimes C^*(G) \otimes C^*(G)) \end{array}$$

commutes.

REMARK A.22. (1) If we apply the $*$-operation to the inclusion in item (i) of the definition, we obtain the inclusion $(1 \otimes C^*(G))\delta(A) \subseteq A \otimes C^*(G)$. This shows that condition (i) of the definition is equivalent to the requirement that $\delta(A)$ lies in the $C^*(G)$-multiplier algebra $M_{C^*(G)}(A \otimes C^*(G))$ of $A \otimes C^*(G)$ (see Definition A.3). In order to simplify notation we shall write $M_G(A \otimes C^*(G))$ for $M_{C^*(G)}(A \otimes C^*(G))$ and will call it the *G-multiplier algebra* of $A \otimes C^*(G)$.

(2) The easiest example of a coaction is δ_G itself, which is a coaction of G on $C^*(G)$. The notion of a coaction we use here is that of a *full* coaction of G on A in the sense of [**47**]. The term *full* refers to the fact that we use $C^*(G)$ instead of the reduced group algebra $C_r^*(G)$, which had for quite some time been the standard setting in the literature (*e.g.*, see [**26, 34, 31, 35**]). Full coactions were first introduced in [**53**], using maximal tensor products instead of the minimal tensor products we use here. We shall discuss the (relatively small) differences between the different approaches later (see Section A.9 below).

(3) Somehow irritatingly, the word *nondegenerate* has two meanings in connection with coactions. By definition, every coaction is a nondegenerate homomorphism of A into $M(A \otimes C^*(G))$ (*i.e.*, $\overline{\delta(A)(A \otimes C^*(G))} = A \otimes C^*(G)$), but being nondegenerate *as a coaction* is the apparently stronger condition that $\overline{\delta(A)(1 \otimes C^*(G))} = A \otimes C^*(G)$. The reason for this terminology is that a nondegenerate coaction determines a nondegenerate module action of the Fourier algebra $A(G)$ on A (see Proposition A.31 below). It is actually an open question whether every coaction is automatically nondegenerate, although it has been settled affirmatively for G amenable [**34**, Lemma 3.8], [**31**, Proposition 6] and G discrete [**2**].

EXAMPLE A.23. If G is abelian, then there is a one-to-one correspondence between coactions of G and strongly continuous actions of the dual group $\widehat{G}$. To see this let us identify $C^*(G)$ with $C_0(\widehat{G})$ via the Fourier transform $\mathcal{F} \colon C^*(G) \to C_0(\widehat{G})$ given by $\mathcal{F}(x)(\chi) = \chi(x)$. A brief calculation shows that the comultiplication δ_G is then translated to the formula

$$\delta_G(f)(\chi, \mu) = f(\chi\mu) \in C_b(\widehat{G} \times \widehat{G})$$

for $f \in C_0(\widehat{G}), \chi, \mu \in \widehat{G}$. If $\alpha\colon \widehat{G} \to \mathrm{Aut}(A)$ is an action, then α determines an injective and nondegenerate homomorphism $\delta^\alpha\colon A \to C^b(\widehat{G}, A) \subseteq M(A \otimes C_0(\widehat{G}))$ by the formula
$$\delta^\alpha(a)(\chi) = \alpha_\chi(a) \qquad \text{for } a \in A, \chi \in \widehat{G}.$$
Condition (i) of the definition is, in this setting, equivalent to δ^α taking values in the subalgebra $C^b(\widehat{G}, A)$ of $M(A \otimes C_0(\widehat{G}))$. The coaction identity (ii) follows from a straightforward computation, using multiplicativity of α. Thus δ^α is a coaction of G on A. Conversely, if $\delta\colon A \to C_b(\widehat{G}, A)$ is any injective nondegenerate homomorphism which satisfies the coaction identity, then we obtain an action of $\widehat{G}$ on A by putting $\beta_\chi(b) = \delta(b)(\chi)$.

Before we present some other important examples of coactions we need

LEMMA A.24 (cf. [**53**, Remarks 2.2]). *Let $\delta\colon A \to M(A \otimes C^*(G))$ be a nondegenerate homomorphism which satisfies conditions (i) and (ii) of Definition A.21 above. Let $1_G\colon G \to \mathbb{C}$ denote the trivial representation of G and let us identify A with $A \otimes \mathbb{C}$ in the canonical way. Then $\delta \circ (\mathrm{id}_A \otimes 1_G) \circ \delta = \delta$. In particular, δ is injective (hence a coaction) if and only if $(\mathrm{id}_A \otimes 1_G) \circ \delta\colon A \to A$ is the identity on A.*

PROOF. It follows from (i) that $(\mathrm{id}_A \otimes 1_G)\bigl(\delta(a)(1 \otimes z)\bigr) \in A$ for all $a \in A$, $z \in C^*(G)$, and hence that $(\mathrm{id}_A \otimes 1_G) \circ \delta(a) \in A$. Since $(\mathrm{id}_G \otimes 1_G) \circ \delta_G(s) = (\mathrm{id}_G \otimes 1_G)(s \otimes s) = s$ for all $s \in G$, we have $(\mathrm{id}_G \otimes 1_G) \circ \delta_G = \mathrm{id}_G$. This together with (ii) gives

$$\delta(a) = (\mathrm{id}_A \otimes \mathrm{id}_G \otimes 1_G) \circ (\mathrm{id}_A \otimes \delta_G) \circ \delta(a) = (\mathrm{id}_A \otimes \mathrm{id}_G \otimes 1_G) \circ (\delta_A \otimes \mathrm{id}_G) \circ \delta(a)$$
$$= (\delta \otimes 1_G) \circ \delta(a) = \delta\bigl((\mathrm{id}_A \otimes 1_G)(\delta(a))\bigr).$$

This completes the proof. $\square$

REMARK A.25. The map $\mathrm{id}_A \otimes 1_G\colon M(A \otimes C^*(G)) \to M(A)$ above is the first appearance of a slice map. We shall see later that more general slice maps play an important rôle in the theory.

EXAMPLE A.26. Let (A, G, α) be an action, let (i_A, i_G) denote the canonical maps from (A, G) into $M(A \times_\alpha G)$ and let u denote the canonical map from G into $M(C^*(G))$. The *dual coaction* of G on $A \times_\alpha G$ is defined as the integrated form
$$\widehat{\alpha} = (i_A \otimes 1) \times (i_G \otimes u)\colon A \times_\alpha G \to M\bigl((A \times_\alpha G) \otimes C^*(G)\bigr).$$
Let us check conditions (i) and (ii) of Definition A.21. For (i) we consider the map $\Psi\colon C_c(G \times G, A) \to (A \times_\alpha G) \otimes C^*(G)$ given by
$$\Psi(g) = \int_{G \times G} (i_A \otimes 1)(f(s,t))\bigl(i_G(s) \otimes u(t)\bigr) d(s,t).$$
If $z \in C_c(G, A)$ and $w \in C_c(G)$, let $z \diamond w(s, t) = z(s)w(s^{-1}t) \in C_c(G \times G, A)$. Then
$$\widehat{\alpha}(z)(1 \otimes u(w)) = \int_{G \times G} (i_A \otimes 1)(z(s)w(t))(i_G(s) \otimes u(st))\, d(s,t)$$
$$= \int_{G \times G} (i_A \otimes 1)(z(s)w(s^{-1}t))(i_G(s) \otimes u(t))\, d(s,t)$$
$$= \Psi(z \diamond w) \in (A \times_\alpha G) \otimes C^*(G).$$

Hence (i) follows from the fact that $C_c(G, A)$ and $C_c(G)$ are dense in $A \times_\alpha G$ and $C^*(G)$, respectively. Now we check (ii). Since $\widehat{\alpha}(i_A(a)) = i_A(a) \otimes 1$ we first get

$$(\widehat{\alpha} \otimes \mathrm{id}_G) \circ \widehat{\alpha}(i_A(a)) = i_A(a) \otimes 1 \otimes 1 = (\mathrm{id}_{A \times_\alpha G} \otimes \delta_G) \circ \widehat{\alpha}(i_A(a))$$

for all $a \in A$, and using $\widehat{\alpha}(i_G(s)) = i_G(s) \otimes s$ and $\delta_G(s) = s \otimes s \in M(C^*(G) \otimes C^*(G))$ we get

$$(\widehat{\alpha} \otimes \mathrm{id}_G) \circ \widehat{\alpha}(i_G(s)) = i_G(s) \otimes s \otimes s = (\mathrm{id}_{A \times_\alpha G} \otimes \delta_G) \circ \widehat{\alpha}(i_G(s))$$

for $s \in G$. Hence (ii) follows from integration. Finally, injectivity follows from Lemma A.24 by observing that $(\mathrm{id}_{A \times_\alpha G} \otimes 1_G) \circ \widehat{\alpha} = i_A \times i_G$.

Note that $\widehat{\alpha}$ is always nondegenerate. This follows from inductive-limit density of $C_c(G, A) \diamond C_c(G)$ in $C_c(G \times G, A)$, continuity of Ψ with respect to the inductive limit topology on $C_c(G \times G, A)$, and density of $\Psi(C_c(G \times G, A))$ in $(A \times_\alpha G) \otimes C^*(G)$ (since it contains $C_c(G, A) \odot C_c(G)$).

It is a good exercise for the reader to check that in case where G is abelian, the dual coaction $\widehat{\alpha}$ corresponds to the dual action of $\widehat{G}$ on $A \times_\alpha G$ under the correspondence given in Example A.23. The next example shows that we also have a dual coaction on the reduced crossed product $A \times_{\alpha,r} G$.

EXAMPLE A.27. Let (i_A^r, i_G^r) denote the canonical maps of (A, G) into $M(A \times_{\alpha,r} G)$, and let $u: G \to M(C^*(G))$ be as above. Then Lemma A.18 implies that $(i_A^r \otimes 1) \times (i_G^r \otimes u)$ factors through a faithful representation $\widehat{\alpha}^n$ of $A \times_{\alpha,r} G$ into $M((A \times_{\alpha,r} G) \otimes C^*(G))$ (write $i_A^r \times i_G^r = \mathrm{Ind}\,\pi$ for some regular representation $\mathrm{Ind}\,\pi$ and use part (i) of Lemma A.18 to deduce that $(i_A^r \otimes 1) \times (i_G^r \otimes u)$ is injective). Exactly the same arguments as in the preceding example show that $\widehat{\alpha}^n$ is a nondegenerate coaction of G on $A \times_{\alpha,r} G$, which is called the *dual coaction* of G on $A \times_{\alpha,r} G$

It turns out later that $\widehat{\alpha}^n$ is the *normalization* of the coaction $\widehat{\alpha}$ of Example A.26 (see Proposition A.61 below). This is the motivation for using the superscript "n" in our notation. However, we shall often skip the "n" if no confusion is possible—in particular, we shall always write $\widehat{\alpha}$ instead of $\widehat{\alpha}^n$ in the main body of this work.

The following example shows that coactions of G restrict to coactions of G/N for any closed normal subgroup N of G. In such a situation we shall always assume that Haar measures on G/N and N are normalized so that $\int_G f(s)\,ds = \int_{G/N} \int_N f(sn)\,dn\,ds N$ for all $f \in C_c(G)$.

EXAMPLE A.28. Let (A, G, δ) be a coaction and let N be a closed normal subgroup of G. Let $q: G \to G/N$ denote the quotient map. Identifying $sN \in G/N$ with its image in $M(C^*(G/N))$ we may view q as a unitary homomorphism from G into $UM(C^*(G/N))$ whose integrated form is the quotient map $q: C^*(G) \to C^*(G/N)$ (to see that it takes values in $C^*(G/N)$ and is surjective one checks that it restricts on $C_c(G)$ to the surjective map $C_c(G) \to C_c(G/N)$ given by $q(f)(sN) = \int_N f(sn)\,dn$). Then

$$\delta| = (\mathrm{id}_A \otimes q) \circ \delta \colon A \to M(A \otimes C^*(G/N))$$

is a coaction of G/N on A, which is called the *restriction* of δ to G/N. Note that Condition (i) of Definition A.21 follows from

(A.4) $\quad \delta|(a)(1 \otimes q(z)) = (\mathrm{id}_A \otimes q)(\delta(a))(1 \otimes q(z)) = (\mathrm{id}_A \otimes q)(\delta(a)(1 \otimes z))$

for $z \in C^*(G)$, and the coaction identity for $\delta|$ follows from applying $\mathrm{id}_A \otimes q \otimes q$ to both sides of the coaction identity for δ. Since $1_{G/N} \circ q = 1_G$, injectivity of $\delta|$

follows from Lemma A.24. Moreover, it is a direct consequence of Equation (A.4) that $\delta|$ is nondegenerate if δ is.

The next example shows that any coaction of a closed subgroup H of G inflates to a coaction of G.

EXAMPLE A.29. Let H be a closed subgroup of G. Let $\varphi\colon C^*(H) \to M(C^*(G))$ denote the integrated form of $u|_H\colon H \to M(C^*(G))$, where $u\colon G \to M(C^*(G))$ denotes the canonical map. If (A, H, ϵ) is a coaction, then $\operatorname{Inf}\epsilon = (\operatorname{id}_A \otimes \varphi) \circ \epsilon$ is a coaction of G on A: Condition (i) of the definition follows from factoring $z \in C^*(G)$ as $z = \varphi(v)w$ for some $v \in C^*(H)$ and $w \in C^*(G)$, and computing

$$(A.5) \qquad \operatorname{Inf}\epsilon(a)(1 \otimes z) = (\operatorname{id}_A \otimes \varphi)\big(\epsilon(a)(1 \otimes v)\big)(1 \otimes w) \in A \otimes C^*(G).$$

Condition (ii) follows from $(\varphi \otimes \varphi) \circ \delta_H = \delta_G \circ \varphi$, and injectivity of $\operatorname{Inf}\epsilon$ is a consequence of Lemma A.24 and the fact that $1_G|_H = 1_H$. We call $\operatorname{Inf}\epsilon$ the coaction *inflated* from ϵ. Note that (A.5) also implies that $\operatorname{Inf}\epsilon$ is nondegenerate if ϵ is nondegenerate.

A.4. Slice maps and nondegeneracy

Slicing in tensor products is one of the basic tools in the theory of coactions. In particular, slice maps are indispensable for the study of covariant representations of coactions. Thus we now provide the basic facts about slice maps which are needed in this work.

If B is a C^*-algebra and $\rho\colon B \to \mathcal{B}(\mathcal{H})$ is a nondegenerate representation, then $b \mapsto \langle \rho(b)\xi, \eta \rangle$ is a continuous linear functional on B for each $\xi, \eta \in \mathcal{H}$ (called a *matrix coefficient* of ρ), and every element of B^* can be written in this way. We denote by B^*_ρ the matrix coefficients of ρ. There are canonical left and right actions of $M(B)$ on B^* (respectively B^*_ρ) given by

$$(A.6) \qquad m \cdot f(b) = f(bm) \quad \text{and} \quad f \cdot m(b) = f(mb)$$

for $m \in M(B)$, $f \in B^*$ (respectively $f \in B^*_\rho$). For $f(b) = \langle \rho(b)\xi, \eta \rangle$ in B^*_ρ we can factor $\eta = \rho(c)\zeta$ and $\xi = \rho(c')\zeta'$ for some $c, c' \in B$ and $\zeta, \zeta' \in \mathcal{H}$. Thus, if $g, g' \in B^*_\rho$ are given by $g(b) = \langle \rho(b)\xi, \zeta \rangle$ and $g'(b) = \langle \rho(b)\zeta', \eta \rangle$ we see that $f = g \cdot c^* = c' \cdot g'$. The possibility of factoring elements $f \in B^*_\rho$ as above turns out to be essential.

If A and B are C^*-algebras and $f \in B^*$, then the *slice map* $S_f\colon A \odot B \to A$ defined by

$$S_f\left(\sum_{i=1}^n a_i \otimes b_i\right) = \sum_{i=1}^n a_i f(b_i)$$

extends to a bounded linear map of $A \otimes B$ into A of norm $\|f\|$. We shall frequently use the following lemma, which is essentially [**35**, Lemma 15]. We give an alternative and (probably) more elementary proof here.

LEMMA A.30 (cf. [**35**, Lemma 1.5]). S_f *extends to a strictly continuous linear map* $S_f\colon M(A \otimes B) \to M(A)$ *satisfying*

$$(A.7) \qquad S_{b \cdot f}(m)a = S_f\big(m(a \otimes b)\big) \quad \text{and} \quad aS_{f \cdot b}(m) = S_f\big((a \otimes b)m\big)$$

for all $a \in M(A)$, $f \in B^*$ *and* $b \in M(B)$.[3] *If* $m \in M(A \otimes B)$ *such that* $S_f(m) = 0$ *for all* $f \in B^*$, *then* $m = 0$. *Finally, if* $\rho\colon A \to M(D)$ *is a nondegenerate*

---
[3]Note that our definition of the action of B on B^* differs from that used in [**35**].

homomorphism then

(A.8) $$S_f \circ (\rho \otimes \mathrm{id}_B) = \rho \circ S_f.$$

PROOF. It is straightforward to check that (A.7) holds for all $m \in A \otimes B$. To see that S_f extends to $M(A \otimes B)$ write $f = c \cdot g \cdot c'$ for some $c, c' \in B$ and $g \in B^*$. Then, for $m \in M(A \otimes B)$ and $a \in A$, define

$$S_f(m)a = S_{g \cdot c'}(m(a \otimes c)) \quad \text{and} \quad aS_f(m) = S_{c \cdot g}((a \otimes c')m).$$

The maps $a \mapsto S_f(m)a$ and $a \mapsto aS_f(m)$ satisfy the double-centralizer property:

$$(aS_f(m))b = S_{c \cdot g}((a \otimes c')m)b = S_g\big((a \otimes c')m(b \otimes c)\big) = a(S_f(m)b)$$

for all $a, b \in A$, so $S_f(m) \in M(A)$. If $\{m_i\}_{i \in I}$ is a net in $M(A \otimes B)$ which converges strictly to $m \in M(A \otimes B)$, then

$$S_f(m_i)a = S_{g \cdot c'}(m_i(a \otimes c)) \to S_{g \cdot c'}(m(a \otimes c)) = S_f(m)a$$

for all $a \in A$ (since S_f is norm continuous on $A \otimes B$). Similarly, $aS_f(m_i) \to aS_f(m)$ for $a \in A$, which proves that S_f is strictly continuous. Since S_f extends the slice map on $A \otimes B$ and $A \otimes B$ is strictly dense in $M(A \otimes B)$ this also shows that the definition of $S_f(m)$ does not depend on the particular factorization $f = c \cdot g \cdot c'$. But this implies that (A.7) holds for all $a \in A$, $b \in B$ and $m \in M(A \otimes B)$. Further, if $a \in M(A)$, $b \in M(B)$ and $f = c \cdot g$, then

$$S_{b \cdot f}(m)ad = S_{(bc) \cdot g}(m)ad = S_{g \cdot c'}(m(ad \otimes bc)) = S_{c \cdot g}(m(a \otimes b))d = S_f(m(a \otimes b))d$$

for all $d \in A$, which implies that $S_{b \cdot f}(m)a = S_f(m(a \otimes b))$. Similarly, $aS_{f \cdot b}(m) = S_f((a \otimes b)m)$, so (A.7) holds for all $a \in M(A)$, $b \in M(B)$, and $m \in M(A \otimes B)$.

If $g \in A^*$, $f \in B^*$, then $g \otimes f \in (A \otimes B)^*$ extends to a strictly continuous functional on $M(A \otimes B)$. Moreover, the set $\{g \otimes f \mid g \in A^*, f \in B^*\}$ separates the points of $M(A \otimes B)$[4]. Since $m \mapsto g(S_f(m))$ and $m \mapsto (g \otimes f)(m)$ are both strictly continuous and clearly agree on $A \otimes B$, they also agree on $M(A \otimes B)$. Thus, if $S_f(m) = 0$ for all $f \in B^*$, then $(g \otimes f)(m) = g(S_f(m)) = 0$ for all $g \in A^*, f \in B^*$ which implies $m = 0$.

The final assertion now follows because both sides of Equation (A.8) are strictly continuous and coincide on $A \otimes B$. □

In this work we are mainly interested in the case $B = C^*(G)$ for a locally compact group G. In this case we can identify $C^*(G)^*$ with the *Fourier-Stieltjes algebra* $B(G)$ of G, which consists of all bounded continuous functions on G which can be expressed as matrix coefficients of unitary representations of G, *i.e.*, a function $f \in B(G)$ is of the form $f(s) = \langle V_s \xi, \eta \rangle$, where V is a unitary representation of G on a Hilbert space $\mathcal{H}$ and $\xi, \eta \in \mathcal{H}$. Of course, if $x \in C^*(G)$ (or $M(C^*(G))$) and $f \in B(G)$ with $f(s) = \langle V_s \xi, \eta \rangle$, then $f(x) = \langle V(x)\xi, \eta \rangle$. In particular, we have $f(u(s)) = f(s)$ if $u \colon G \to UM(C^*(G))$ denotes the canonical map. Note that $B(G)$ becomes an involutive Banach algebra when equipped with the dual norm, pointwise multiplication and the involution $f \mapsto \bar{f}$.

The *Fourier algebra* $A(G)$ is the (closed) $*$-subalgebra of $B(G)$ consisting of all matrix coefficients of the left regular representation λ, *i.e.*, $A(G) = C^*(G)^*_\lambda$. The nontrivial fact that $A(G)$ is an algebra can be deduced from the fact that $\lambda \otimes \lambda$

---

[4]To see this, choose faithful representations π and σ of A and B and observe that $(g \otimes f)(m) = 0$ for all $g \in A^*_\pi, f \in B^*_\sigma$ implies that $(\pi \otimes \sigma)(m) = 0$, and hence $m = 0$ since $\pi \otimes \sigma$ is faithful on $A \otimes B$.

is unitarily equivalent to $\lambda \otimes 1_{L^2(G)}$ (Lemma A.18) and that $\lambda \oplus \lambda$ is unitarily equivalent to $\lambda \otimes 1_{\mathbb{C}^2}$. One can show that $A(G)$ is dense in $C_0(G)$ with respect to the supremum-norm. Moreover, the subalgebra $A_c(G)$ of compactly supported elements is dense in $A(G)$ (which follows from the fact that the compactly supported elements are dense in $L^2(G)$) and for each compact subset C of G there exists an element $f \in A_c(G)$ with $f|_C \equiv 1$. For more details on $A(G)$ and $B(G)$ we refer to [**23**]. The following proposition indicates that any nondegenerate coaction turns A into a nondegenerate $A(G)$-module.

PROPOSITION A.31. *Let (A, G, δ) be a nondegenerate coaction. Then for all $f \in B(G)$ the composition $\delta_f = S_f \circ \delta$ maps A back into itself, and $\mathrm{span}\{\delta_f(a) \mid a \in A, f \in A(G)\}$ is dense in A.*

PROOF. Let $x \in C^*(G)$ and $f \in A(G)$ with $f(x) = 1$. Let $a \in A$ be given. Since $\delta(A)(1 \otimes C^*(G))$ is dense in $A \otimes C^*(G)$ we can approximate $a \otimes x$ by a finite sum $\sum_{i=1}^n \delta(a_i)(1 \otimes x_i)$ with $a_i \in A$ and $x_i \in C^*(G)$. Then, using Lemma A.30, we get

$$a = f(x)a = S_f(a \otimes x) \approx S_f\left(\sum_{i=1}^n \delta(a_i)(1 \otimes x_i)\right) = \sum_{i=1}^n S_{x_i \cdot f}(\delta(a_i)) = \sum_{i=1}^n \delta_{x_i \cdot f}(a_i).$$

□

Note that by [**47**, Corollary 1.5] the converse of the proposition above is also true, *i.e.*, δ is nondegenerate if and only if $\mathrm{span}\{\delta_f(a) \mid a \in A, f \in A(G)\}$ is dense in A.

A.5. Covariant representations and crossed products

We are now going to introduce covariant representations of coactions. Of course, if G is abelian, these should coincide with the covariant representations of the corresponding action $(A, \widehat{G}, \alpha)$ as discussed in Section A.3. So let us start our discussion with a covariant homomorphism (π, V) of an action $(A, \widehat{G}, \alpha)$ into $M(D)$, where G an abelian group. Identifying $C^*(G)$ with $C_0(\widehat{G})$ via Fourier transform we may view the integrated form of V as a homomorphism $\mu \colon C_0(\widehat{G}) \to M(D)$. Thus we may expect covariant homomorphisms of coactions (A, G, δ) to be pairs of nondegenerate maps (π, μ) with $\pi \colon A \to M(D), \mu \colon C_0(\widehat{G}) \to M(D)$.

Since $\delta^\alpha \colon A \to C_b(\widehat{G}, A) \subseteq M(A \otimes C_0(\widehat{G}))$ is given by the formula $\delta^\alpha(a)(\chi) = \alpha_\chi(a)$, the covariance condition on (π, V) can be expressed as

(A.9) $$(\pi \otimes \mathrm{id})(\delta(a)) = \widetilde{V}(\pi(a) \otimes 1)\widetilde{V}^*$$

in $M(D \otimes C_0(\widehat{G}))$, where $\widetilde{V} \in UM(D \otimes C_0(\widehat{G}))$ is given by the strictly continuous function $\chi \mapsto V_\chi$. Since the canonical embedding of $\widehat{G}$ into $M(C^*(\widehat{G})) \cong C_b(G)$ maps a character $\chi \in \widehat{G}$ to the function $s \mapsto \chi(s)$, and since the Fourier transform $\mathcal{F} \colon C^*(G) \to C_0(\widehat{G})$ maps $s \in M(C^*(G))$ to the function $\chi \mapsto \chi(s)$ in $C_b(\widehat{G})$, it follows that

$$\widetilde{V} = (\mu \otimes \mathcal{F})(w_G),$$

where w_G denotes the function $s \mapsto u(s)$ in $UM(C_0(G) \otimes C^*(G))$. Applying the inverse Fourier transform (or more precisely $\mathrm{id}_D \otimes \mathcal{F}^{-1}$) to both sides of Equation (A.9) we see that the covariance condition for (π, V) translates into the condition
$$(\pi \otimes \mathrm{id}_G) \circ \delta^\alpha(a) = (\mu \otimes \mathrm{id}_G)(w_G)(\pi(a) \otimes 1)(\mu \otimes \mathrm{id}_G)(w_G)^*$$
in $M(D \otimes C^*(G))$. Thus we are led to the following definition of covariant representations for coactions.

DEFINITION A.32. Let (A, G, δ) be a coaction and let $w_G \in UM(C_0(G) \otimes C^*(G))$ be given by the map $s \mapsto u(s) \colon G \to UM(C^*(G))$. A *covariant homomorphism* of (A, G, δ) into $M(D)$ is a pair (π, μ) of nondegenerate homomorphisms of $(A, C_0(G))$ into $M(D)$ satisfying the covariance condition
$$(\mathrm{A.10}) \qquad (\pi \otimes \mathrm{id}_G) \circ \delta(a) = (\mu \otimes \mathrm{id}_G)(w_G)(\pi(a) \otimes 1)(\mu \otimes \mathrm{id}_G)(w_G)^*.$$
If $D = \mathcal{K}(\mathcal{H})$ for some Hilbert space $\mathcal{H}$, then (π, μ) is called a *covariant representation* of (A, G, δ) on $\mathcal{H}$.

REMARK A.33. It is sometimes useful to work with pairs (π, μ) in which the homomorphism $\pi \colon A \to M(D)$ is degenerate, but the pair (π, μ) satisfies all other conditions on a covariant representation (including nondegeneracy of $\mu \colon C_0(G) \to M(D)$). We shall call such a pair a *degenerate* covariant representation of $(A, G, \mathcal{E})$. Note that the covariance condition (A.10) makes sense if π is degenerate, since $\delta(A) \in M_G(A \otimes C^*(G))$, and $\pi \otimes \mathrm{id}_G$ extends uniquely to $M_G(A \otimes C^*(G))$ by Proposition A.6.

However, if not explicitly stated otherwise, all covariant representations in this work are assumed to be nondegenerate as in the definition.

When working with covariant representations, it is often necessary to recover the nondegenerate homomorphism $\mu \colon C_0(G) \to M(A)$ from the unitary multiplier $(\mu \otimes \mathrm{id}_G)(w_G) \in UM(A \otimes C^*(G))$ and to know some further properties of this unitary. For notation: if $m \in M(A \otimes C)$ for some C^*-algebra C then $m_{12}, m_{13} \in M(A \otimes C \otimes C)$ are defined by $m_{12} = m \otimes 1$ and $m_{13} = (\mathrm{id}_A \otimes \Sigma)(m \otimes 1)$, where $\Sigma \colon C \otimes C \to C \otimes C$ denotes the flip isomorphism $c \otimes d \mapsto d \otimes c$.

PROPOSITION A.34 (*cf.* [**48**, Lemma A1, Lemma A2]). *Let A be a C^*-algebra and let $\mu \colon C_0(G) \to M(A)$ be a nondegenerate homomorphism. Then*
$$S_f\bigl((\mu \otimes \mathrm{id}_G)(w_G)\bigr) = \mu(f) \qquad \text{for all } f \in B(G) \subseteq C_b(G).$$
Moreover, $w = (\mu \otimes \mathrm{id}_G)(w_G)$ satisfies the equation $w_{12} w_{13} = (\mu \otimes \delta_G)(w_G) = (\mathrm{id}_A \otimes \delta_G)(w)$.

PROOF. First, we have $S_f(w_G) = f$ for all $f \in B(G)$. To see this let $s \in G$ and let $\chi_s \colon C_b(G) \to \mathbb{C}$ denote evaluation at s. Using Lemma A.30 we get
$$\chi_s(S_f(w_G)) = S_f\bigl((\chi_s \otimes \mathrm{id}_G)(w_G)\bigr) = S_f(1_\mathbb{C} \otimes u(s)) = f(s).$$
Again using Lemma A.30 this implies
$$S_f\bigl((\mu \otimes \mathrm{id}_G)(w_G)\bigr) = \mu(S_f(w_G)) = \mu(f).$$
From the definition of w_G it follows that $(w_G)_{12}(w_G)_{13} \in M(C_0(G) \otimes C^*(G) \otimes C^*(G))$ is given by the map $s \mapsto u(s) \otimes u(s)$, which clearly equals $(\mathrm{id}_{C_0(G)} \otimes \delta_G)(w_G)$. Applying $\mu \otimes \mathrm{id}_G \otimes \mathrm{id}_G$ to both sides gives $w_{12} w_{13} = (\mu \otimes \delta_G)(w_G) = (\mathrm{id}_A \otimes \delta_G)(w)$ for $w = (\mu \otimes \mathrm{id}_G)(w_G)$. $\square$

REMARK A.35. In fact, there is a certain converse to the above result: If $w \in UM(A \otimes C^*(G))$ satisfies $w_{12}w_{13} = (\mathrm{id}_A \otimes \delta_G)(w)$ in $M(A \otimes C^*(G) \otimes C^*(G))$, then $w = (\mu \otimes \mathrm{id}_G)(w_G)$ for some nondegenerate homomorphism $\mu \colon C_0(G) \to M(A)$ (see [**48**, Lemma A.1]). Of course, the above result then implies that $\mu(f) = S_f(w)$ for all $f \in B(G) \subseteq C_b(G)$.

PROPOSITION A.36 (cf. [**53**, Lemma 2.10]). *Let (π, μ) be a (possibly degenerate) covariant homomorphism of (A, G, δ) into $M(D)$ for some C^*-algebra D. Then*

$$C^*(\pi, \mu) = \overline{\pi(A)\mu(C_0(G))}$$

is a C^*-algebra.

PROOF. The result will follow as soon as it is clear that an element $\mu(f)\pi(a)$ can be approximated in norm by a finite sum $\sum_i \pi(a_i)\mu(f_i)$, where $f, f_i \in A(G)$, $a, a_i \in A$. So let $f \in A(G)$, and factor $f = g \cdot x$ for some $x \in C^*(G)$ and $g \in A(G)$. The covariance identity together with Lemma A.30 and Proposition A.34 gives

$$\mu(f)\pi(a) = S_f\Big(\big((\mu \otimes \mathrm{id}_G)(w_G)\big)(\pi(a) \otimes 1)\Big)$$
$$= S_f\Big(\big(\pi \otimes \mathrm{id}_G(\delta(a))\big)\big((\mu \otimes \mathrm{id}_G)(w_G)\big)\Big)$$
$$= S_g\Big(\big(\pi \otimes \mathrm{id}_G((1 \otimes x)\delta(a))\big)\big((\mu \otimes \mathrm{id}_G)(w_G)\big)\Big).$$

Approximating $(1 \otimes x)\delta(a) \in A \otimes C^*(G)$ by a finite sum $\sum_i a_i \otimes x_i$ then implies

$$\mu(f)\pi(a) \approx S_g\Big(\Big(\sum_i \pi(a_i) \otimes x_i\Big)\big((\mu \otimes \mathrm{id}_G)(w_G)\big)\Big)$$
$$= \sum_i \pi(a_i) S_{g \cdot x_i}\big((\mu \otimes \mathrm{id}_G)(w_G)\big) = \sum_i \pi(a_i)\mu(g \cdot x_i),$$

which completes the proof. □

The following proposition shows that covariant homomorphisms do exist.

PROPOSITION A.37 (cf. [**53**, Proposition 2.6]). *If π is a nondegenerate homomorphism of A into $M(D)$, then $((\pi \otimes \lambda) \circ \delta, 1 \otimes M)$ is a covariant homomorphism of (A, G, δ) into $M(D \otimes \mathcal{K}(L^2(G)))$.*

PROOF. We first establish the identity

(A.11) $\qquad (\lambda \otimes \mathrm{id}_G) \circ \delta_G = \mathrm{Ad}(M \otimes \mathrm{id}_G)(w_G) \circ (\lambda \otimes 1).$

Applied to $s \in G$ (regarded as an element of $UM(C^*(G))$) the left hand side becomes $\lambda_s \otimes s$. To compute the right hand side at s, assume that $C^*(G)$ is represented faithfully on a Hilbert space $\mathcal{H}$. Then for $\xi \in L^2(G, \mathcal{H})$ and $t \in G$ we get

$$\big((M \otimes \mathrm{id}_G)(w_G)(\lambda_s \otimes 1)(M \otimes \mathrm{id}_G)(w_G)^*\xi\big)(t) = t\big((\lambda_s \otimes 1)(M \otimes \mathrm{id}_G)(w_G)^*\xi\big)(t)$$
$$= t\big((M \otimes \mathrm{id}_G)(w_G)^*\xi\big)(s^{-1}t) = t(t^{-1}s)\xi(s^{-1}t) = \big((\lambda_s \otimes s)\xi\big)(t).$$

Thus (A.11) follows from integration. Together with the coaction identity this implies

$$\big((\pi \otimes \lambda) \circ \delta \otimes \mathrm{id}_G\big) \circ \delta(a) = (\pi \otimes \lambda \otimes \mathrm{id}_G) \circ (\delta \otimes \mathrm{id}_G) \circ \delta(a)$$
$$= (\pi \otimes \lambda \otimes \mathrm{id}_G) \circ (\mathrm{id}_A \otimes \delta_G) \circ \delta(a)$$
$$= (\pi \otimes \mathrm{id}_\mathcal{K} \otimes \mathrm{id}_G) \circ \big(\mathrm{id}_A \otimes(\lambda \otimes \mathrm{id}_G) \circ \delta_G\big) \circ \delta(a)$$
$$= (\pi \otimes \mathrm{id}_\mathcal{K} \otimes \mathrm{id}_G)$$
$$\Big((1 \otimes M \otimes \mathrm{id}_G)(w_G)\big((\mathrm{id}_A \otimes\lambda)(\delta(a)) \otimes 1\big)(1 \otimes M \otimes \mathrm{id}_G)(w_G)^*\Big)$$
$$= (1 \otimes M \otimes \mathrm{id}_G)(w_G)\big((\pi \otimes \lambda)(\delta(a)) \otimes 1\big)(1 \otimes M \otimes \mathrm{id}_G)(w_G)^*,$$

which completes the proof. $\square$

DEFINITION A.38. Let $\pi \colon A \to M(D)$ be a nondegenerate homomorphism. Then the covariant homomorphism $\mathrm{Ind}\,\pi = \big((\pi \otimes \lambda) \circ \delta, 1 \otimes M\big)$ of (A, G, δ) into $M(D \otimes \mathcal{K}(L^2(G)))$ is called the homomorphism of (A, G, α) *induced from* π. If π is faithful, then $\mathrm{Ind}\,\pi$ is called a *regular* homomorphism of (A, G, α).

Somewhat surprisingly, it turns out that for coactions there is no difference between full and reduced crossed products. In fact we are now going to define the crossed product by a coaction as the C^*-algebra generated by a certain regular representation (similarly to the definition of the reduced crossed product $A \times_{\alpha,r} G$ as given in Definition A.13) and then show that this crossed product enjoys universal properties similar to those enjoyed by the full crossed product $A \times_\alpha G$ of an action $\alpha \colon G \to \mathrm{Aut}(A)$. However, if we view the crossed product $A \times_\delta G$ as a skew tensor product of A with $C_0(G)$, this might be less surprising in view of the fact that any commutative C^*-algebra is nuclear.

DEFINITION A.39. Let (A, G, δ) be a coaction and let $(j_A, j_G) = \big((\mathrm{id}_A \otimes\lambda) \circ \delta, 1 \otimes M\big)$ be the regular homomorphism induced by $\mathrm{id}_A \colon A \to A$. Then $A \times_\delta G = C^*(j_A, j_G)$ is called the *crossed product* of the coaction (A, G, δ). The maps (j_A, j_G), viewed as maps of $(A, C_0(G))$ into $M(A \times_\delta G)$, are called the *canonical maps* of $(A, C_0(G))$ into $M(A \times_\delta G)$.

REMARK A.40. By definition, the crossed product $A \times_\delta G$ is a subalgebra of $M(A \otimes \mathcal{K}(L^2(G)))$. It is important to observe that it actually lies in the $\mathcal{K}(L^2(G))$-multiplier algebra $M_{\mathcal{K}(L^2(G))}(A \otimes \mathcal{K}(L^2(G)))$. To see this just observe that

$$j_A(A) = (\mathrm{id}_A \otimes\lambda)\circ\delta(A) \subseteq (\mathrm{id}_A \otimes\lambda)\big(M_G(A\otimes C^*(G))\big) \subseteq M_{\mathcal{K}(L^2(G))}(A\otimes\mathcal{K}(L^2(G))),$$

where the last inclusion follows from Proposition A.6. Since $j_G(C_0(G)) = 1 \otimes M(C_0(G)) \subseteq 1 \otimes M(\mathcal{K}(L^2(G)))$, it follows from item (iii) of Proposition A.5 that $A \times_\delta G = \overline{j_A(A)j_G(C_0(G))} \subseteq M_{\mathcal{K}(L^2(G))}(A \otimes \mathcal{K}(L^2(G)))$.

THEOREM A.41 (cf. [**53**, Theorem 4.1]). *The triple* $(A\times_\delta G, j_A, j_G)$ *satisfies the following universal property: if* (π, μ) *is any covariant homomorphism of* (A, G, δ) *into* $M(D)$ *for some* C^*-*algebra* D, *then there exists a unique nondegenerate homomorphism* $\pi \times \mu \colon A \times_\delta G \to M(D)$ *such that* $(\pi \times \mu) \circ j_A = \pi$ *and* $(\pi \times \mu) \circ j_G = \mu$.

Moreover, if (π, μ) *is a degenerate covariant homomorphism, then there exists a unique (degenerate) homomorphism* $\pi \times \mu \colon A \times_\delta G \to M(D)$ *such that* $\pi \times \mu(j_A(a)j_G(f)) = \pi(a)\mu(f)$.

DEFINITION A.42. The homomorphism $\pi \times \mu$ of Theorem A.41 is called the *integrated form* of the covariant homomorphism (π, μ) of (A, G, δ).

REMARK A.43. (1) It is not hard to check that if $\rho \colon A \times_\delta G \to M(D)$ is a nondegenerate homomorphism of $A \times_\delta G$, then $(\rho \circ j_A, \rho \circ j_G)$ is a covariant homomorphism. We then have $\rho = (\rho \circ j_A) \times (\rho \circ j_G)$. Thus, $(\pi, \mu) \mapsto \pi \times \mu$ gives a one-to-one correspondence between the (nondegenerate) covariant homomorphisms of (A, G, δ) and the nondegenerate homomorphisms of $A \times_\delta G$.

(2) As for actions, one could alternatively *define* the crossed product via universal properties, i.e., as a triple (C, l_A, l_G) satisfying

  (i) (l_A, l_G) is a covariant homomorphism of (A, G, δ) and $C = C^*(l_A, l_G)$; and
  (ii) for every covariant homomorphism (π, μ) of (A, G, δ) into $M(D)$ there exists a unique nondegenerate homomorphism $\pi \times \mu \colon C \to M(D)$ such that $\pi \times \mu \circ l_A = \pi$ and $\pi \times \mu \circ l_G = \mu$.

Of course, for any such triple (C, l_A, l_G), the C^*-algebra C is isomorphic to $A \times_\delta G$ via the integrated form $l_A \times l_G \colon A \times_\delta G \to C$ with inverse map $j_A \times j_G \colon C \to A \times_\delta G$, where $j_A \times j_G$ is the map associated to (j_A, j_G) by (ii).

(3) If $\operatorname{Ind} \pi = ((\pi \otimes \lambda) \circ \delta, 1 \otimes M)$ is any regular homomorphism of (A, G, δ) induced by the faithful homomorphism $\pi \colon A \to M(D)$, then its integrated form (also denoted $\operatorname{Ind} \pi$) is a faithful homomorphism of $A \times_\delta G$. For this one observes that
$$\operatorname{Ind} \pi = (\pi \otimes \operatorname{id}_\mathcal{K}) \circ \operatorname{Ind} \operatorname{id}_A = (\pi \otimes \operatorname{id}_\mathcal{K}) \circ (j_A \times j_G),$$
which is faithful since $\pi \otimes \operatorname{id}_\mathcal{K}$ is faithful.

(4) For abelian G, the reader can check without too much difficulty (using the Plancherel isomorphism $L^2(G) \to L^2(\widehat{G})$) that regular representations of $(A, \widehat{G}, \alpha)$ correspond to regular representations of (A, G, δ^α), which implies that $A \times_\alpha \widehat{G} \cong A \times_{\delta^\alpha} G$.

The proof of Theorem A.41 we give here is based on the proof of [**35**, Theorem 3.7]. We need the following lemma.

LEMMA A.44. *Let (π, μ) be a (possibly degenerate) covariant homomorphism of (A, G, δ) into $M(E)$ for some C^*-algebra E, let $W = (\mu \otimes \lambda)(w_G) \in M(E \otimes \mathcal{K}(L^2(G)))$ and let $\epsilon \colon C_0(G) \to M(C_0(G) \otimes C_0(G))$ be the comultiplication on $C_0(G)$, i.e., $\epsilon(f)(s, t) = f(st)$. Then*

(A.12) $\quad \operatorname{Ad} W^* \circ (\pi \otimes \lambda) \circ \delta = \pi \otimes 1 \quad and \quad \operatorname{Ad} W^* \circ (1 \otimes M) = (\mu \otimes M) \circ \epsilon.$

PROOF. The first equation follows from applying $\operatorname{id}_E \otimes \lambda$ to both sides of the covariance condition $\pi \otimes 1 = \operatorname{Ad}\bigl((\mu \otimes \operatorname{id}_G)(w_G)\bigr)^* \circ (\pi \otimes \operatorname{id}_G) \circ \delta$. For the proof of the second equation we first show that
$$(M \otimes \lambda(w_G))^*(1 \otimes M(f))(M \otimes \lambda(w_G)) = (M \otimes M)(\epsilon(f))$$
in $\mathcal{B}(L^2(G \times G))$. For this write $W_G = M \otimes \lambda(w_G)$. Then $(W_G \xi)(s, t) = \xi(s, s^{-1}t)$, from which it follows that
$$\bigl(W_G^*(1 \otimes M(f))W_G \xi\bigr)(s, t) = \bigl((1 \otimes M(f))W_G \xi\bigr)(s, st) = f(st)\bigl(W_G \xi\bigr)(s, st)$$
$$= f(st)\xi(s, t) = \bigl((M \otimes M)(\epsilon(f))\xi\bigr)(s, t).$$

Since M is faithful on $C_0(G)$ we get the equation

$$(\mathrm{id}\otimes\lambda(w_G))^*(1\otimes M(f))(\mathrm{id}\otimes\lambda(w_G)) = (\mathrm{id}\otimes M)(\epsilon(f))$$

for all $f \in C_0(G)$, and applying $\mu \otimes \mathrm{id}_{\mathcal{K}}$ to both sides of this equation gives the desired result. $\square$

PROOF OF THEOREM A.41. Let (π, μ) be a (possibly degenerate) covariant homomorphism of (A, G, δ) into $M(D)$ and let $W = (\mu \otimes \lambda)(w_G) \in UM(D \otimes \mathcal{K}(L^2(G)))$. Further let $\varphi\colon A \times_\delta G \to M(D \otimes \mathcal{K}(L^2(G)))$ denote the restriction to $A \times_\delta G$ of the homomorphism

$$\operatorname{Ad} W^* \circ (\pi \otimes \mathrm{id}_{\mathcal{K}})\colon M_{\mathcal{K}(L^2(G))}(A \otimes \mathcal{K}(L^2(G))) \to M(D \otimes \mathcal{K}(L^2(G))).$$

Using Lemma A.44, we get

$$\begin{aligned}\varphi(j_A(a)j_G(f)) &= \varphi\big((\mathrm{id}\otimes\lambda)(\delta(a))(1\otimes M_f)\big)\\&= W^*((\pi\otimes\lambda)(\delta(a))WW^*(1\otimes M_f)W = (\pi(a)\otimes 1)(\mu\otimes M)(\epsilon(f))\end{aligned}$$

for $a \in A$, $f \in C_0(G)$. It follows that $\varphi(A \times_\delta G)$ lies in the image of $M(D \otimes C_0(G))$ under the homomorphism $\mathrm{id}_D \otimes M\colon D \otimes C_0(G) \to M(D \otimes \mathcal{K}(L^2(G)))$. Since $\mathrm{id}_D \otimes M$ is faithful, we therefore obtain a homomorphism $\psi\colon A \times_\delta G \to M(D \otimes C_0(G))$ satisfying

$$\psi(j_A(a)j_G(f)) = (\pi(a) \otimes 1)(\mu \otimes \mathrm{id})(\epsilon(f)).$$

Let $\chi_e\colon C_0(G) \to \mathbb{C}$ denote evaluation at the identity $e \in G$ and define

$$\pi \times \mu = (\mathrm{id}_D \otimes \chi_e) \circ \psi\colon A \times_\delta G \to M(D).$$

We then have

$$\begin{aligned}\pi \times \mu(j_A(a)j_G(f)) &= (\mathrm{id}_D \otimes \chi_e)\big((\pi(a) \otimes 1)(\mu \otimes \mathrm{id})(\epsilon(f))\big)\\&= (\mathrm{id}_D \otimes \chi_e)\big((\pi(a) \otimes 1)\big)(\mathrm{id}_D \otimes \chi_e)\big((\mu \otimes \mathrm{id})(\epsilon(f))\big)\\&= \pi(a)(\mu \otimes \chi_e)(\epsilon(f)) = \pi(a)\mu(f)\end{aligned}$$

for all $a \in A$, $f \in C_0(G)$, where the identity $(\mu \otimes \chi_e)(\epsilon(f))) = \mu(f)$ follows from the equation $(\mathrm{id}_{C_0(G)} \otimes \chi_e)(\epsilon(f)) = f$ for $f \in C_0(G)$. If π is degenerate, then so is $\pi \times \mu$, since

$$\pi \times \mu(A \times_\delta G)D = \overline{\pi(A)\mu(C_0(G))}D = \pi(A)(\mu(C_0(G))D) = \pi(A)D \neq D.$$

If π is nondegenerate, then as a composition of nondegenerate homomorphisms, $\pi \times \mu$ is nondegenerate, too. We then have

$$(\pi \times \mu) \circ j_A(a) = (\mathrm{id}_D \otimes \chi_e)\big(\pi(a) \otimes 1\big) = \pi(a)$$

for $a \in A$, and

$$(\pi \times \mu) \circ j_G(f) = (\mu \otimes \chi_e)(\epsilon(f)) = \mu(f),$$

for all $f \in C_0(G)$. $\square$

DEFINITION A.45. Let (A, δ) and (B, ϵ) be coactions of G. A (possibly degenerate) homomorphism $\varphi\colon A \to M(B)$ is called $\delta - \epsilon$ *equivariant* if $(\varphi \otimes \mathrm{id}_G) \circ \delta = \epsilon \circ \varphi$. (Note that the composition on the left hand side is well-defined by Proposition A.6 since $\delta(A) \subseteq M_G(A \otimes C^*(G))$.)

LEMMA A.46. *Let (A, δ) and (B, ϵ) be coactions of G and let $\varphi \colon A \to M(B)$ be a $\delta - \epsilon$ equivariant homomorphism. Then there is a well-defined homomorphism*
$$\varphi \times G = (j_B \circ \varphi) \times j_G \colon A \times_\delta G \to M(B \times_\epsilon G),$$
where $(j_B, j_G) \colon (B, C_0(G)) \to M(B \times_\epsilon G)$ denotes the canonical pair of maps. Moreover, $\varphi \times G$ is nondegenerate if (and only if) φ is nondegenerate, and $\varphi \times G$ is faithful if φ is faithful.

PROOF. The covariance condition for (j_B, j_G) (except nondegeneracy) extends to $(j_B \circ \varphi, j_G)$. Thus $\varphi \times G$ is well-defined. If φ is nondegenerate, then so is $j_B \circ \varphi$ and hence $\varphi \times G = (j_B \circ \varphi) \times j_G$. So suppose finally that φ is faithful, and consider the composition
$$A \times_\delta G \xrightarrow{j_A \times j_G} M_{\mathcal{K}(L^2(G))}(A \otimes \mathcal{K}(L^2(G))) \xrightarrow{\varphi \otimes \mathrm{id}_\mathcal{K}} M(B \otimes \mathcal{K}(L^2(G))).$$
Using the definition of j_A, j_B and j_G, and the $\delta - \epsilon$ equivariance of φ we compute for all $a \in A$ and $f \in C_0(G)$:
$$\begin{aligned}
\varphi \otimes \mathrm{id}_\mathcal{K}(j_A(a)j_G(f)) &= \varphi \otimes \mathrm{id}_\mathcal{K}\left(((\mathrm{id}_A \otimes \lambda) \circ \delta(a))(1 \otimes M(f))\right) \\
&= \left((\mathrm{id}_B \otimes \lambda) \circ (\varphi \otimes \mathrm{id}_G) \circ \delta(a)\right)(1 \otimes M(f)) \\
&= \left((\mathrm{id}_B \otimes \lambda) \circ \epsilon(\varphi(a))\right)(1 \otimes M(f)) \\
&= j_B \circ \varphi(a)j_G(f),
\end{aligned}$$
which implies that $\varphi \times G$ coincides with the above composition of maps. Thus, if φ is faithful, the composition $(\varphi \otimes \mathrm{id}_\mathcal{K}) \circ (j_A \times j_G)$ is faithful by Proposition A.6. □

A.6. Dual actions and decomposition coactions

In Example A.26 we saw that for any action (A, G, α) there is a dual coaction $\widehat{\alpha}$ of G on $A \times_\alpha G$. Obversely, for any coaction (A, G, δ) there is a dual action of G on $A \times_\delta G$.

DEFINITION A.47. Let $A \times_\delta G = C^*(j_A, j_G)$ be the crossed product of the coaction (A, G, δ). For $s \in G$ let σ_s denote right translation on $C_0(G)$, i.e., $\sigma_s(f)(t) = f(ts)$. Then $\widehat{\delta} \colon G \to \mathrm{Aut}(A \times_\delta G)$ defined by $\widehat{\delta}_s = j_A \times (j_G \circ \sigma_s)$ is called the *dual action* of G on $A \times_\delta G$.

REMARK A.48. Note that $\widehat{\delta}_s$ is given on a typical element of the form $j_A(a)j_G(f)$ by
$$\widehat{\delta}_s(j_A(a)j_G(f)) = j_A(a)j_G(\sigma_s(f)).$$
Since right translation on $C_0(G)$ is continuous it follows that $\widehat{\delta} \colon G \to \mathrm{Aut}(A \times_\delta G)$ is strongly continuous.

In the main body of the paper we need to work with a certain canonical coaction, the *decomposition coaction* δ^{dec} of G on $A \times_{\delta|} G/N$, where δ is a given coaction of G on A and N is a closed normal subgroup of G. (The reason for this terminology is that, by [**44**, Theorem 3.1], $A \times_\delta G$ can be decomposed into a twisted crossed product of $A \times_{\delta|} G/N$ by δ^{dec}.) We are now going to describe this coaction.

LEMMA A.49. *Let (A, G, δ) and N be as above. The formula*
(A.13)
$$\delta^{\mathrm{dec}}(j_A(a)j_{G/N}(f)) = (j_A \otimes \mathrm{id}) \circ \delta(a)(j_{G/N}(f) \otimes 1) \quad \text{for } a \in A, f \in C_0(G/N)$$
defines a coaction of G on $A \times_{\delta|} G/N$, which is nondegenerate if δ is.

PROOF. First, it is easy to see that $(j_A \otimes \mathrm{id}) \circ \delta$ and $j_{G/N} \otimes 1$ are nondegenerate homomorphisms of A and $C_0(G)$, respectively, into $M((A \times_{\delta|} G/N) \otimes C^*(G))$. We show that $((j_A \otimes \mathrm{id}) \circ \delta, j_{G/N} \otimes 1)$ is a covariant pair: for $a \in A$ we have

$$\mathrm{Ad}(j_{G/N} \otimes 1 \otimes \mathrm{id})(w_{G/N})\big((j_A \otimes \mathrm{id}) \circ \delta(a) \otimes 1\big)$$
$$= \mathrm{Ad}(\mathrm{id} \otimes \sigma)\big((j_{G/N} \otimes \mathrm{id})(w_{G/N}) \otimes 1\big)\big((j_A \otimes \mathrm{id}) \circ \delta(a) \otimes 1\big)$$
$$= (\mathrm{id} \otimes \sigma) \circ \mathrm{Ad}\big((j_{G/N} \otimes \mathrm{id})(w_{G/N}) \otimes 1\big) \circ (\mathrm{id} \otimes \sigma)\big((j_A \otimes \mathrm{id}) \circ \delta(a) \otimes 1\big)$$
$$= (\mathrm{id} \otimes \sigma) \circ \big(\mathrm{Ad}(j_{G/N} \otimes \mathrm{id})(w_{G/N}) \otimes \mathrm{id}\big) \circ (j_A \otimes 1 \otimes \mathrm{id}) \circ \delta(a)$$
$$= (\mathrm{id} \otimes \sigma) \circ \big(\mathrm{Ad}(j_{G/N} \otimes \mathrm{id})(w_{G/N}) \circ (j_A \otimes 1) \otimes \mathrm{id}\big) \circ \delta(a)$$
$$= (\mathrm{id} \otimes \sigma) \circ \big((j_A \otimes \mathrm{id}) \circ \delta| \otimes \mathrm{id}\big) \circ \delta(a)$$
$$= (j_A \otimes \mathrm{id} \otimes \mathrm{id}) \circ (\mathrm{id} \otimes \sigma) \circ (\delta| \otimes \mathrm{id}) \circ \delta(a)$$
$$= (j_A \otimes \mathrm{id} \otimes \mathrm{id}) \circ (\delta \otimes \mathrm{id}) \circ \delta|(a)$$
$$= \big((j_A \otimes \mathrm{id}) \circ \delta \otimes \mathrm{id}\big) \circ \delta|(a).$$

Thus there is a nondegenerate homomorphism $\delta^{\mathrm{dec}} \colon A \times_{\delta|} G/N \to M((A \times_{\delta|} G/N) \otimes C^*(G))$. This homomorphism is injective, because (letting 1_G denote the trivial character on G) $(\mathrm{id} \otimes 1_G) \circ (j_A \otimes \mathrm{id}) \circ \delta = j_A \circ (\mathrm{id} \otimes 1_G) \circ \delta = j_A$ and $(\mathrm{id} \otimes 1_G) \circ (j_{G/N} \otimes 1) = j_{G/N}$. The coaction identity holds because

$$(\delta^{\mathrm{dec}} \otimes \mathrm{id}) \circ \delta^{\mathrm{dec}} \circ j_A = (\delta^{\mathrm{dec}} \otimes \mathrm{id}) \circ (j_A \otimes \mathrm{id}) \circ \delta = (\delta^{\mathrm{dec}} \circ j_A \otimes \mathrm{id}) \circ \delta$$
$$= \big((j_A \otimes \mathrm{id}) \circ \delta \otimes \mathrm{id}\big) \circ \delta = (j_A \otimes \mathrm{id} \otimes \mathrm{id}) \circ (\delta \otimes \mathrm{id}) \circ \delta$$
$$= (j_A \otimes \mathrm{id} \otimes \mathrm{id}) \circ (\mathrm{id} \otimes \delta_G) \circ \delta = (\mathrm{id} \otimes \delta_G) \circ (j_A \otimes \mathrm{id}) \circ \delta$$
$$= (\mathrm{id} \otimes \delta_G) \circ \delta^{\mathrm{dec}} \circ j_A$$

and

$$(\delta^{\mathrm{dec}} \otimes \mathrm{id}) \circ \delta^{\mathrm{dec}} \circ j_{G/N} = (\delta^{\mathrm{dec}} \otimes \mathrm{id}) \circ (j_{G/N} \otimes 1) = (\delta^{\mathrm{dec}} \circ j_{G/N}) \otimes 1$$
$$= j_{G/N} \otimes 1 \otimes 1 = (\mathrm{id} \otimes \delta_G) \circ (j_{G/N} \otimes 1) = (\mathrm{id} \otimes \delta_G) \circ \delta^{\mathrm{dec}} \circ j_{G/N}.$$

Also, for $a \in A$, $f \in C_0(G/N)$, and $c \in C^*(G)$ we have

$$\delta^{\mathrm{dec}}\big(j_A(a)j_{G/N}(f)\big)(1 \otimes c) = (j_A \otimes \mathrm{id}) \circ \delta(a)(j_{G/N}(f) \otimes c)$$
$$= (j_A \otimes \mathrm{id})\big(\delta(a)(1 \otimes c)\big)(j_{G/N}(f) \otimes 1) \in \big(j_A(A) \otimes C^*(G)\big)(j_{G/N}(f) \otimes 1)$$
$$\subseteq (A \times_{\delta|} G/N) \otimes C^*(G).$$

Hence δ^{dec} is a coaction.

Now assume δ is nondegenerate. We must show δ^{dec} is nondegenerate, too. If $a \in A$, $f \in C_0(G/N)$, and $c \in C^*(G)$ we have

$$\delta^{\mathrm{dec}}\big(j_A(a)j_{G/N}(f)\big)(1 \otimes c) = (j_A \otimes \mathrm{id}) \circ \delta(a)(j_{G/N}(f) \otimes 1)(1 \otimes c)$$
$$= (j_A \otimes \mathrm{id}) \circ \delta(a)(1 \otimes c)(j_{G/N}(f) \otimes 1).$$

The latter elements densely span $(j_A \otimes \mathrm{id})\big(A \otimes C^*(G/N)\big)\big(j_{G/N}(C_0(G)) \otimes 1\big)$, which of course densely spans $(A \times_{\delta|} G/N) \otimes C^*(G)$. $\square$

A.7. Normal coactions and normalizations

Normal coactions and normalizations of coactions will play a fundamental rôle in this work. We start the discussion with:

DEFINITION A.50. A coaction (A, G, δ) is *normal* if $j_A \colon A \to M(A \times_\delta G)$ is injective.

REMARK A.51. Since for every covariant homomorphism (π, μ) of (A, G, δ) the homomorphism π of A factors through j_A (which follows from $(\pi \times \mu) \circ j_A = \pi$) we see that a coaction δ is normal if and only if it has a covariant homomorphism (or representation) (π, μ) with π faithful.

EXAMPLE A.52. Recall from the preceding section that if (A, G, δ) is a coaction and N is a closed normal subgroup of G, then the decomposition coaction δ^{dec} of G on $A \times_{\delta|} G/N$ is defined on the generators by

$$\delta^{\text{dec}}(j_A(a) j_{G/N}(f)) = (j_A \otimes \text{id}) \circ \delta(a)(j_{G/N}(f) \otimes 1).$$

Suppose δ is normal. We will now show that δ^{dec} is normal, too. It suffices to show that there is a covariant homomorphism (π, μ) of $(A \times_{\delta|} G/N, G, \delta^{\text{dec}})$ with π faithful. We take $\pi = j_A^G \times j_G|$ and $\mu = j_G$; as we mention in Appendix B, our hypotheses on δ guarantee that π is a faithful nondegenerate homomorphism of $A \times_{\delta|} G/N$ into $M(A \times_\delta G)$. So, it remains to verify that the pair (π, μ) is covariant: we have

$$\text{Ad}(j_G \otimes \text{id})(w_G) \circ (j_A \times j_G|) \circ (j_A \otimes 1) = \text{Ad}(j_G \otimes \text{id})(w_G) \circ (j_A^G \otimes 1)$$
$$= (j_A^G \otimes \text{id}) \circ \delta = ((j_A^G \times j_G|) \circ j_A^{G/N} \otimes \text{id}) \circ \delta$$
$$= ((j_A^G \otimes j_G|) \otimes \text{id}) \circ (j_A^{G/N} \otimes \text{id}) \circ \delta = ((j_A^G \times j_G|) \otimes \text{id}) \circ \delta^{\text{dec}} \circ j_A,$$

and

$$\text{Ad}(j_G \otimes \text{id})(w_G) \circ (j_A \times j_G|) \circ (j_{G/N} \otimes 1) = \text{Ad}(j_G \otimes \text{id})(w_G) \circ (j_G| \otimes 1)$$
$$= j_G| \otimes 1 = (j_A \times j_G|) \circ j_{G/N} \otimes 1$$
$$= ((j_A \times j_G|) \otimes \text{id}) \circ (j_{G/N} \otimes 1) = ((j_A \times j_G|) \otimes \text{id}) \circ \delta^{\text{dec}} \circ j_{G/N}.$$

In what follows next we show that for every coaction δ there exists a canonically-defined normal coaction δ^n such that δ and δ^n have the same representation theory and crossed products.

DEFINITION A.53. Let (A, δ) and (B, ϵ) be coactions of G. If $\varphi \colon A \to B$ is a $\delta - \epsilon$ equivariant isomorphism, then we say that φ is a *conjugacy* between δ and ϵ. We say that δ and ϵ are *conjugate* if there exists a conjugacy between δ and ϵ.

REMARK A.54. It is straightforward to check that conjugacy defines an equivalence relation, and that if $\varphi \colon A \to B$ is a conjugacy, then the homomorphism $\varphi \times G \colon A \times_\delta G \to B \times_\epsilon G$ of Lemma A.46 is an isomorphism with inverse $\varphi^{-1} \times G$.

LEMMA A.55 (*cf.* [**45**, Proposition 2.4]). *Let (π, μ) be a covariant homomorphism of the coaction (A, G, δ) into $M(D)$. Then the map $\delta^\mu \colon \pi(A) \to M\bigl(\pi(A) \otimes C^*(G)\bigr)$ defined by*

$$\pi(a) \mapsto \text{Ad}\bigl((\mu \otimes \text{id}_G)(w_G)\bigr)\bigl(\pi(a) \otimes 1\bigr)$$

is a normal coaction of G on $\pi(A)$ which is nondegenerate if δ is nondegenerate. Moreover, $\pi \colon A \to \pi(A)$ is $\delta - \delta^\mu$ equivariant.

PROOF. The covariance condition for (π, μ) implies that

(A.14) $$\delta^\mu(\pi(a)) = (\pi \otimes \mathrm{id}_G) \circ \delta(a).$$

This shows that δ^μ takes values in $M(\pi(A) \otimes C^*(G))$ and

$$\delta^\mu(\pi(A))(1 \otimes C^*(G)) = \pi \otimes \mathrm{id}_G \left(\delta(A)(1 \otimes C^*(G))\right) \subseteq \pi(A) \otimes C^*(G).$$

Of course we have equality if δ is nondegenerate. The coaction identity for δ^μ follows from applying $\pi \otimes \mathrm{id}_G \otimes \mathrm{id}_G$ to both sides of the coaction identity of δ. Finally, observe that $(\mathrm{id}_{\pi(A)}, \mu)$ is covariant for δ^μ, so δ^μ is normal by Remark A.51. □

Of course, by identifying $\pi(A)$ with $A/\ker \pi$ we may also view δ^μ as a coaction on $A/\ker \pi$.

DEFINITION A.56. Let (A, G, δ) be a coaction and let $A^n = j_A(A) \cong A/\ker j_A$. Then $\delta^n = \delta^{j_G} : A^n \to M(A^n \otimes C^*(G))$ is called the *normalization* of δ.

REMARK A.57. It follows from Lemma A.55 that δ^n is nondegenerate if δ is nondegenerate. The converse is shown in [**47**, Proposition 2.5].

Since $j_A : A \to A^n$ is $\delta - \delta^n$ equivariant we see that if (π, μ) is a covariant homomorphism of (A^n, G, δ^n), then $(\pi \circ j_A, \mu)$ is a covariant homomorphism of (A, G, δ) and $C^*(\pi \circ j_A, \mu) = C^*(\pi, \mu)$. Conversely, since $\ker j_A \subseteq \ker \rho$ for every covariant homomorphism (ρ, ν) of (A, G, δ) (see Remark A.51) it follows that every covariant homomorphism of (A, G, δ) arises this way.

The following proposition is now completely straightforward.

PROPOSITION A.58 (*cf.* [**47**, Proposition 2.6]). *For any coaction (A, G, δ), the map*

$$j_A \times G : A \times_\delta G \to A^n \times_{\delta^n} G$$

is a $\widehat{\delta} - \widehat{\delta^n}$ equivariant isomorphism.

As an interesting corollary we get:

COROLLARY A.59 (*cf.* [**47**, Corollary 2.7]). *Let (A, G, δ) be a coaction and let $\pi : A \to M(D)$ be any nondegenerate homomorphism of A such that $\ker \pi \subseteq \ker j_A$. Then*

$$\mathrm{Ind}\,\pi = (\pi \otimes \lambda) \circ \delta \times (1 \otimes M) : A \times_\delta G \to M(D \otimes \mathcal{K}(L^2(G)))$$

is faithful.

PROOF. Let $\varphi : \pi(A) \to j_A(A)$ be such that $\varphi \circ \pi = j_A$. Since $\delta^n \circ j_A = (j_A \otimes \mathrm{id}_G) \circ \delta$ we have

$$(\varphi \circ \pi \otimes \lambda) \circ \delta = (j_A \otimes \lambda) \circ \delta = (\mathrm{id}_{A^n} \otimes \lambda) \circ \delta^n \circ j_A.$$

Thus it follows from the proposition and Theorem A.41 that $(\varphi \otimes \mathrm{id}) \circ (\pi \otimes \lambda) \circ \delta \times (1 \otimes M) = (j_A \otimes \lambda) \circ \delta \times (1 \otimes M)$ is faithful on $A \times_\delta G = A^n \times_{\delta^n} G$. □

We can even strengthen the above result as follows:

PROPOSITION A.60. *Let (A, G, δ) be a coaction and let $\pi : A \to M(D)$ be a nondegenerate homomorphism such that $\ker(\pi \otimes \lambda) \circ \delta = \ker j_A$. Then $\mathrm{Ind}\,\pi : A \times_\delta G \to M(D \otimes \mathcal{K}(L^2(G)))$ is faithful.*

PROOF. Let us assume that D is represented faithfully on the Hilbert space $\mathcal{H}$. By Corollary A.59 we know that
$$\operatorname{Ind}\bigl((\pi \otimes \lambda) \circ \delta\bigr) = \bigl((\pi \otimes \lambda \otimes \lambda) \circ (\delta \otimes \operatorname{id}_G) \circ \delta\bigr) \times (1 \otimes 1 \otimes M)$$
is a faithful representation of $A \times_\delta G$ on $\mathcal{H} \otimes L^2(G) \otimes L^2(G)$. Hence the result will follow if we can show that $\operatorname{Ind}\bigl((\pi \otimes \lambda) \circ \delta\bigr)$ is unitarily equivalent to $(\operatorname{Ind} \pi) \otimes 1$. For this let $U \in U(L^2(G \times G))$ be defined by $(U\xi)(s,t) = \Delta(t)^{-1/2}\xi(st^{-1}, s)$. A quick calculation shows that U^* is given by $(U^*\xi)(s,t) = \Delta(s^{-1}t)^{1/2}\xi(t, s^{-1}t)$. Using this, another straightforward computation shows that

(A.15) $$U\bigl((\lambda \otimes \lambda) \circ \delta_G(s)\bigr)U^* = U(\lambda_s \otimes \lambda_s)U^* = \lambda_s \otimes 1$$

for all $s \in G$ and

(A.16) $$U(1 \otimes M(f))U^* = M(f) \otimes 1$$

for all $f \in C_0(G)$.[5] Using this and the coaction identity for δ, we get
$$(1 \otimes U)\bigl((\pi \otimes \lambda \otimes \lambda) \circ (\delta \otimes \operatorname{id}_G) \circ \delta(a)\bigr)(1 \otimes U^*)$$
$$= (1 \otimes U)\bigl((\pi \otimes \lambda \otimes \lambda) \circ (\operatorname{id}_A \circ \delta_G) \circ \delta(a)\bigr)(1 \otimes U^*)$$
$$= (1 \otimes U)\bigl((\pi \otimes ((\lambda \otimes \lambda) \circ \delta_G)) \circ \delta(a)\bigr)(1 \otimes U^*)$$
$$\stackrel{(A.15)}{=} (\pi \otimes (\lambda \otimes 1)) \circ \delta(a) = \bigl((\pi \otimes \lambda) \circ \delta(a)\bigr) \otimes 1$$
for $a \in A$, and (A.16) clearly implies that
$$(1 \otimes U)(1 \otimes 1 \otimes M(f))(1 \otimes U^*) = 1 \otimes M(f) \otimes 1$$
for $f \in C_0(G)$. This completes the proof. $\square$

The most enlightening example of normalizations is possibly given in the case of dual coactions.

PROPOSITION A.61. *Let (A, G, α) be an action and let $\widehat{\alpha}$ be the dual coaction of G on $A \times_\alpha G$. Then the normalization $\widehat{\alpha}^n$ of $\widehat{\alpha}$ is the dual coaction of G on $A \times_{\alpha,r} G$. Hence $\widehat{\alpha}$ is normal if and only if $i_A^r \times i_G^r \colon A \times_\alpha G \to A \times_{\alpha,r} G$ is an isomorphism. Moreover, the double crossed products $A \times_\alpha G \times_{\widehat{\alpha}} G$ and $A \times_{\alpha,r} G \times_{\widehat{\alpha}^n} G$ are canonically isomorphic.*

PROOF. Let $\pi \times U$ be any faithful representation of $A \times_\alpha G$ on a Hilbert space $\mathcal{H}$. Then Theorem A.41 implies that we can write $j_{A \times_\alpha G} = (\pi \times U \otimes \lambda) \circ \widehat{\alpha}$. By the definition of $\widehat{\alpha}$ it follows that $j_{A \times_\alpha G} \circ i_A(a) = \pi(a) \otimes 1$ and $j_{A \times_\alpha G} \circ i_G(s) = U_s \otimes \lambda_s$, where (i_A, i_G) denotes the canonical map of (A, G) into $M(A \times_\alpha G)$. Hence $j_{A \times_\alpha G} = (\pi \otimes 1) \times (U \otimes \lambda)$, which by Lemma A.18 is unitarily equivalent to a regular representation of $A \times_\alpha G$. Thus, identifying $A \times_{\alpha,r} G$ with $j_{A \times_\alpha G}(A \times_\alpha G)$, we get $i_A^r = \pi \otimes 1, i_G^r = U \otimes \lambda$.

By definition (using $j_{A \times_\alpha G} = i_A^r \times i_G^r$), α^n is determined by the equation
$$\alpha^n \circ (i_A^r \times i_G^r) = \bigl((i_A^r \times i_G^r) \otimes \operatorname{id}_G\bigr) \circ \widehat{\alpha}.$$
But the right hand side applied to the elements $i_A(a), i_G(s) \in M(A \times_\alpha G)$ gives $i_A^r(a) \otimes 1$ and $i_G^r(s) \otimes s$, respectively. Hence, $\widehat{\alpha}^n$ is precisely the dual coaction of

---

[5]Notice that U is the composition $U = V_G^* W_G^*$, where $(W_G \xi)(s,t) = \xi(s, s^{-1}t)$ and $(V_G \xi)(s,t) = \Delta(t)^{1/2}\xi(st, t)$. $\operatorname{Ad} W_G^*$ moves $1 \otimes M$ to $M \otimes M \circ \epsilon$ and $(\lambda \otimes \lambda) \circ \delta_G$ to $\lambda \otimes 1$, where ϵ is the comultiplication on $C_0(G)$, and $\operatorname{Ad} V_G^*$ moves $M \otimes M \circ \epsilon$ to $M \otimes 1$.

G on $A \times_{\alpha,r} G$ as defined in Example A.27. The final assertion now follows from Proposition A.58. □

EXAMPLE A.62. As an interesting application of the above results we now show that $(\mathcal{K}(L^2(G)), \lambda, M)$ is a crossed product for the coaction $(C^*(G), G, \delta^G)$. By Proposition A.61 we know that the normalization δ_G^n is the coaction on $C_r^*(G)$ determined by the map $\lambda_s \mapsto \lambda_s \otimes u(s)$ for $s \in G$ (note that δ_G is just the dual coaction on $C^*(G) = \mathbb{C} \times_{\mathrm{id}} G$). Consider the trivial representation $1_G \colon G \to \mathbb{C}$ of G. Then $\mathrm{Ind}\, 1_G = (1_G \otimes \lambda) \circ \delta_G \times (1_{\mathbb{C}} \otimes M) = \lambda \times M$. Thus $\ker(1_G \otimes \lambda) \circ \delta_G = \ker \lambda$, which by Proposition A.61 coincides with $j_{C^*(G)}$, and it follows from Proposition A.60 that $\lambda \times M$ is a faithful representation of $C^*(G) \times_{\delta_G} G$. Since we already observed in Remark A.11 that $C^*(\lambda, M) = \lambda(C^*(G))M(C_0(G)) = \mathcal{K}(L^2(G))$, the result follows.

Example A.62 should be compared with Example A.10, where we mentioned that $(\mathcal{K}(L^2(G)), M, \lambda)$ is a crossed product for the action $(C_0(G), G, \tau)$. Note that for the latter result we had to refer to a deep theorem of Mackey (the imprimitivity theorem), while the result on $(C^*(G), G, \delta_G)$ is a natural outcome of the theory of coactions. We shall see below (Theorem A.64) that there is a natural isomorphism between $C^*(G) \times_{\delta_G} G$ and $C_0(G) \times_\tau G$, so the result for $(C_0(G), G, \tau)$ will actually be a consequence of the above example.

A.8. The duality theorems of Imai-Takai and Katayama

In this section we want to deduce the classical duality theorem of Imai and Takai for actions and the duality theorem of Katayama for coactions, *i.e.*, the analogues of the Takesaki-Takai duality theorem for nonabelian groups. We start with a result which is actually more general than what we need here, but which will be important later.

Let (A, G, α) be an action and let N be a closed normal subgroup of G. Recall from Examples A.26 and A.28 that the restriction $\widehat{\alpha}|$ of the dual coaction $\widehat{\alpha}$ to G/N is the integrated form of the covariant homomorphism $(i_A \otimes 1, i_G \otimes q)$ of (A, G, α) on $M(A \times_\alpha G \otimes C^*(G/N))$, where $q \colon G \to G/N \subseteq M(C^*(G/N))$ denotes the quotient map. In what follows next we first want to show that $(A \times_\alpha G) \times_{\widehat{\alpha}|} G/N$ is canonically isomorphic to the crossed product $(A \otimes C_0(G/N)) \times_{\alpha \otimes \tau} G$, where $\tau \colon G \to \mathrm{Aut}(C_0(G/N))$ is given by $\tau_s(f)(tN) = f(s^{-1}tN)$, as in Example A.10.

Recall that if (π, μ) is a pair of commuting nondegenerate homomorphisms $\pi \colon A \to M(D)$, $\mu \colon C_0(G/N) \to M(D)$, then (since $C_0(G/N)$ is nuclear) there is a nondegenerate homomorphism

$$\pi \otimes \mu \colon A \otimes C_0(G/N) \to M(D)$$

which is determined on elementary tensors by $\pi \otimes \mu(a \otimes f) = \pi(a)\mu(f)$, and every nondegenerate homomorphism of $A \otimes C_0(G)$ arises this way. Since the canonical maps $A \to M(A \otimes C_0(G/N))$, $C_0(G/N) \to M(A \otimes C_0(G/N))$ are G-equivariant, it follows that a triple (π, μ, V) of nondegenerate homomorphisms $\pi \colon A \to M(D)$, $\mu \colon C_0(G/N) \to M(D)$, $V \colon C^*(G) \to M(D)$ determines a covariant homomorphism $(\pi \otimes \mu, V)$ of $(A \otimes C_0(G/N), G, \alpha \otimes \tau)$ if and only if

(i) π and μ have commuting ranges in $M(D)$,
(ii) (π, V) is covariant for α, and
(iii) (μ, V) is covariant for τ.

PROPOSITION A.63 (*cf.* [53, Examples 2.9]). *The assignment* $(\pi \otimes \mu, V) \mapsto (\pi \times V, \mu)$ *is a one-to-one correspondence between the covariant homomorphisms of* $(A \otimes C_0(G/N), G, \alpha \otimes \tau)$ *and the covariant homomorphisms of* $(A \times_\alpha G, G/N, \widehat{\alpha}|)$. *Moreover,*

$$((\pi \otimes \mu) \times V)\big((A \otimes C_0(G/N) \times_{\alpha \otimes \tau} G\big) = \big((\pi \times V) \times \mu\big)\big(A \times_\alpha G \times_{\widehat{\alpha}|} G/N\big).$$

PROOF. We only have to check that conditions (i)–(iii) above are equivalent to the covariance condition for $(\pi \times V, \mu)$ with respect to $\widehat{\alpha}|$. First note that, by Lemma A.30, the covariance condition for $(\pi \times V, \mu)$ is equivalent to

$$\begin{aligned}\text{(A.17)} \quad S_f\big((\pi \times V \otimes \mathrm{id}_{G/N}) \circ \widehat{\alpha}|(c)((\mu \otimes \mathrm{id}_G)(w_{G/N}))\big) \\ = S_f\big((\mu \otimes \mathrm{id}_{G/N})(w_{G/N})(\pi \times V(c) \otimes 1)\big)\end{aligned}$$

for all $c \in A \times_\alpha G$, $f \in B(G/N)$. By integration it is enough to check this equation on all elements $i_A(a), i_G(s) \in M(A \times_\alpha G)$. For $c = i_A(a)$, using Lemma A.30 and Proposition A.34, the left-hand side of (A.17) becomes

$$S_f\big((\pi(a) \otimes 1)(\mu \otimes \mathrm{id}_G)(w_G)\big) = \pi(a)S_f\big((\mu \otimes \mathrm{id}_G)(w_G)\big) = \pi(a)\mu(f),$$

and a similar computation shows that the right hand side of (A.17) becomes $\mu(f)\pi(a)$. Thus, since $A(G/N)$ is norm-dense in $C_0(G/N)$ and weak*-dense in $B(G/N)$ we see that having (A.17) hold for all $c = i_A(a) \in i_A(A)$ and $f \in B(G/N)$ is equivalent to (i). For $c = i_G(s)$ the left hand side of (A.17) becomes

$$\begin{aligned}S_f\big((V_s \otimes q(s))(\mu \otimes \mathrm{id}_{G/N})(w_{G/N})\big) &= V_s S_{f \cdot q(s)}\big((\mu \otimes \mathrm{id}_{G/N})(w_{G/N})\big) \\ &= V_s \mu(f \cdot q(s)) = V_s \mu(\tau_{s^{-1}}(f)),\end{aligned}$$

while the right hand side becomes $\mu(f)V_s$. Thus, having (A.17) hold for all $c = i_G(s)$ and $f \in B(G/N)$ is equivalent to (iii).

The last assertion follows from the equality

$$\pi(A)\mu(C_0(G/N))V(C^*(G)) = \pi(A)V(C^*(G))\mu(C_0(G/N)).$$

□

Let us write $i_A \otimes i_{C_0(G/N)}$ for the canonical map

$$i_{A \otimes C_0(G/N)} \colon A \otimes C_0(G/N) \to M\big((A \otimes C_0(G/N)) \times_{\alpha \otimes \tau} G\big).$$

We define an action $\beta \colon G/N \to \mathrm{Aut}((A \otimes C_0(G/N)) \times_{\alpha \otimes \tau} G)$ by

$$\beta_{sN}\big(i_A(a)i_{C_0(G/N)}(f)i_G(z)\big) = i_A(a)i_{C_0(G/N)}(\sigma_{sN}(f))i_G(z),$$

where σ_{sN} denotes right translation by sN. The following theorem is now a consequence of Proposition A.63 and the universal properties of the crossed products.

THEOREM A.64. *Let* (A, G, α) *be an action and let* N *be a closed normal subgroup of* G. *Then there is a* $\beta - \widehat{\alpha}|$ *equivariant isomorphism*

$$(A \otimes C_0(G/N)) \times_{\alpha \otimes \tau} G \cong (A \times_\alpha G) \times_{\widehat{\alpha}|} G/N$$

taking $i_{A \otimes C_0(G/N)}(a \otimes f)i_G(c)$ *to* $j_{G/N}(f)j_{A \times G}(i_A(a)i_G(c))$ *for* $a \in A$, $f \in C_0(G/N)$, *and* $c \in C^*(G)$.

We also present a reduced version of Theorem A.64.

THEOREM A.65. *Let (A, G, α) be an action and let N be a closed normal subgroup of G. Then there is a $\beta - \widehat{\alpha}|$ equivariant isomorphism*

$$(A \otimes C_0(G/N)) \times_{\alpha \otimes \tau, r} G \cong (A \times_{\alpha, r} G) \times_{\widehat{\alpha}|} G/N$$

which takes $i^r_{A \otimes C_0(G/N)}(a \otimes f) i^r_G(c)$ to $j_{G/N}(f) j_{A \times_r G}(i^r_A(a) i^r_G(c))$ for $a \in A$, $f \in C_0(G/N)$, and $c \in C^*(G)$.

PROOF. We show that, up to unitary equivalence, there are regular representations of $(A \otimes C_0(G/N), G, \alpha \otimes \tau)$ and $(A \times_{\alpha, r} G, G/N, \widehat{\alpha}|)$ which match up as in Proposition A.63. This will give the result.

We start with a faithful representation $\pi \colon A \to \mathcal{B}(\mathcal{H})$. Let $M^{G/N} \colon C_0(G/N) \to \mathcal{B}(L^2(G/N))$ and $M^G \colon C_0(G) \to \mathcal{B}(L^2(G))$ denote the respective representations by multiplication operators. Consider the regular representation

$$\Pi = \mathrm{Ind}(\pi \otimes M^{G/N}) = (\pi \otimes M^{G/N} \otimes M^G) \circ (\alpha \otimes \tau) \times (1 \otimes 1 \otimes \lambda)$$

of $(A \otimes C_0(G/N)) \times_{\alpha \otimes \tau} G$ on $\mathcal{H} \otimes L^2(G/N) \otimes L^2(G)$ (and recall that in this formula we view $\alpha \otimes \tau$ as a homomorphism of $A \otimes C_0(G/N)$ to $M(A \otimes C_0(G/N) \otimes C_0(G))$ as in Definition A.13), and the regular representation

$$\Lambda = \mathrm{Ind}(\mathrm{Ind}\,\pi) = \big((\pi \otimes M^G) \circ \alpha \times (1 \otimes \lambda)\big) \circ \widehat{\alpha}| \times \big(1 \otimes 1 \otimes M^{G/N}\big)$$

of $(A \times_{\alpha, r} G) \times_{\widehat{\alpha}|} G/N$ on $\mathcal{H} \otimes L^2(G) \otimes L^2(G/N)$. Define a unitary operator $U \colon \mathcal{H} \otimes L^2(G/N) \otimes L^2(G) \to \mathcal{H} \otimes L^2(G) \otimes L^2(G/N)$ by

$$(U\xi)(t, sN) = \xi(t^{-1}sN, t).$$

for $\xi \in L^2(G/N \times G, \mathcal{H}) = \mathcal{H} \otimes L^2(G/N) \otimes L^2(G)$. Moreover, let's write $i^r_A \otimes i^r_{C_0(G/N)}$ for $i^r_{A \otimes C_0(G/N)}$ and $j_A \times j_G$ for $j_{A \times_r G}$. We claim that

(i) $\mathrm{Ad}\, U \circ \Pi \circ i^r_A = \Lambda \circ j_A$,
(ii) $\mathrm{Ad}\, U \circ \Pi \circ i^r_{C_0(G/N)} = \Lambda \circ j_{C_0(G/N)}$, and
(iii) $\mathrm{Ad}\, U \circ \Pi \circ i^r_G = \Lambda \circ j_G$.

Since $\mathrm{Ad}\, U \circ \Pi$ and Λ are faithful on the respective (reduced) crossed products, the result will follow from Proposition A.63. For (i), for $a \in A$ and $\xi \in L^2(G \times G/N, \mathcal{H}) \cong \mathcal{H} \otimes L^2(G) \otimes L^2(G/N)$ we compute

$$\big(\mathrm{Ad}\, U \circ \Pi(i^r_A(a))\xi\big)(t, sN) = \big(U(\pi \otimes M^{G/N} \otimes M^G) \circ (\alpha \otimes \tau)(a \otimes 1)U^*\xi\big)(t, sN)$$
$$= \big((\pi \otimes M^G \otimes M^{G/N}) \circ (\alpha \otimes \tau)(a \otimes 1)U^*\xi\big)(t^{-1}sN, t)$$
$$= \pi(\alpha_{t^{-1}}(a))\big(U^*\xi\big)(t^{-1}sN, t) = \pi(\alpha_{t^{-1}}(a))\xi(t, sN)$$
$$= \big((\pi \otimes M^G) \circ \alpha(a) \otimes 1)\xi\big)(t, sN) = \big(\Lambda \circ j_A(a)\xi\big)(t, sN).$$

For (ii) (writing $(\pi \otimes M)^\sim$ instead of $(\pi \otimes M^{G/N} \otimes M^G) \circ (\alpha \otimes \tau)$), for $f \in C_0(G/N)$ compute

$$\big(\mathrm{Ad}\, U \circ \Pi(i^r_{C_0(G/N)}(f))\xi\big)(t, sN) = \big(U(\pi \otimes M)^\sim(1 \otimes f)U^*\xi\big)(t, sN)$$
$$= \big((\pi \otimes M)^\sim(1 \otimes f)U^*\xi\big)(t^{-1}sN, t) = M(\tau_{t^{-1}}(f))\big(U^*\xi\big)(t^{-1}sN, t)$$
$$= \tau_{t^{-1}}(f)(t^{-1}sN)(U\xi)(t^{-1}sN, t) = f(sN)\xi(t, sN)$$
$$= \big((1 \otimes 1 \otimes M)(f)\xi\big)(t, sN) = \big(\Lambda \circ j_{C_0(G/N)}(f)\xi\big)(t, sN),$$

and for (iii), for $r \in G$ we compute

$$\begin{aligned}
\big(\operatorname{Ad} U \circ \Pi(i_G^r(r))\xi\big)(t, sN) &= \big(U(1 \otimes 1 \otimes \lambda(r))U^*\xi\big)(t, sN) \\
&= \big((1 \otimes 1 \otimes \lambda(r))U^*\xi\big)(t^{-1}sN, t) = \big(U^*\xi\big)(t^{-1}sN, r^{-1}t) \\
&= \xi(r^{-1}t, r^{-1}sN) = \big((1 \otimes \lambda(r) \otimes \lambda^{G/N}(rN))\xi\big)(t, sN) \\
&= \big(\Lambda \circ j_G(r)\xi\big)(t, sN).
\end{aligned}$$
□

In case $N = \{e\}$ and $A = \mathbb{C}$ we get, as a corollary of Theorem A.64 and Example A.62, the complete proof of the statement of Example A.10.

COROLLARY A.66. $(\mathcal{K}(L^2(G)), M, \lambda)$ *is a crossed product for* $(C_0(G), G, \tau)$.

Another (more general) consequence of the special case $N = \{e\}$ together with Example A.12 is the Imai-Takai duality theorem for actions.

THEOREM A.67. *Let* (A, G, α) *be an action. Then*

$$(A \times_\alpha G) \times_{\widehat{\alpha}} G \cong A \otimes \mathcal{K}(L^2(G)),$$

equivariantly for the double dual action $\widehat{\widehat{\alpha}}$ *and for the action* $\alpha \otimes \operatorname{Ad}\rho$, *where* ρ *denotes the right regular representation of* G *on* $L^2(G)$.

The only statement in the above theorem we haven't explicitly shown so far is the statement about the action $\widehat{\widehat{\alpha}}$. But this can be done by checking carefully what the isomorphism of Example A.12 does to the action β of the theorem. Since the dual coaction of G on $A \times_{\alpha,r} G$ is the normalization of $\widehat{\alpha}$, we get from Proposition A.58 a reduced version of the above result

THEOREM A.68. *There exists an* $\widehat{\widehat{\alpha}} - \alpha \otimes \operatorname{Ad}\rho$ *equivariant isomorphism*

$$(A \times_{\alpha,r} G) \times_{\widehat{\alpha}} G \cong A \otimes \mathcal{K}(L^2(G)).$$

We now turn our attention to the duality theorem of Katayama, where we consider the double crossed product $(A \times_\delta G) \times_{\widehat{\delta}} G$ of a coaction (A, G, δ). Unfortunately the situation is a bit more complicated in this case, since the crossed product $A \times_\delta G$ doesn't seem to see the full algebra A but rather only the quotient $j_A(A)$ of A. So in principle we cannot expect to recover A from the double crossed product. However, that this reasoning is somewhat too naive follows from [**45**, Theorem 3.7] which shows that full duality does work for all dual coactions on full crossed products, which are very often not normal (see also the discussion in [**19**, §3]).

THEOREM A.69 (*cf.* [**31**, Theorem 8]). *Let* (A, G, δ) *be a nondegenerate coaction. Then the reduced double crossed product* $A \times_\delta G \times_{\widehat{\delta},r} G$ *is isomorphic to* $(A/\ker j_A) \otimes \mathcal{K}(L^2(G))$.

For the proof we need

LEMMA A.70 (*cf.* [**31**, Theorem 5, Theorem 8]). *Let* $\delta \colon A \to M(A \otimes C^*(G))$ *be a nondegenerate coaction, and let* $W_G = M \otimes \lambda(w_G)$. *Then*

(A.18) $\qquad (\operatorname{id}_A \otimes \lambda) \circ \delta(A)(1 \otimes \mathcal{K}(L^2(G))) = A \otimes \mathcal{K}(L^2(G))$

and

$$(A.19) \quad (1 \otimes W_G)\big((\mathrm{id}_A \otimes \lambda) \circ \delta(A) \otimes 1\big)(1 \otimes W_G^*)(1 \otimes 1 \otimes \mathcal{K}(L^2(G))) \\ = \big((\mathrm{id}_A \otimes \lambda) \circ \delta(A)\big) \otimes \mathcal{K}(L^2(G)).$$

PROOF. It follows from Remark A.11 that $\lambda(C^*(G))M(C_0(G)) = \mathcal{K}(L^2(G))$. Since δ is nondegenerate we have $(\mathrm{id}_A \otimes \lambda) \circ \delta(A)(1 \otimes C_r^*(G)) = A \otimes C_r^*(G)$. Multiplying both sides of the equation with $1 \otimes M(C_0(G))$ from the right we get (A.18).

Since $\mathrm{Ad}\, W_G \circ (\lambda \otimes 1) = (\lambda \otimes \lambda) \circ \delta_G$ (which follows from applying $\mathrm{id} \otimes \lambda$ to Equation (A.11)), we get

$$(1 \otimes W_G)\big((\mathrm{id}_A \otimes \lambda) \circ \delta(A) \otimes 1\big)(1 \otimes W_G^*) = (\mathrm{id}_A \otimes \lambda \otimes \lambda)\big((\mathrm{id}_A \otimes \delta_G) \circ \delta(A)\big)$$
$$= (\mathrm{id}_A \otimes \lambda \otimes \lambda)\big((\delta \otimes \mathrm{id}_G) \circ \delta(A)\big) = (\mathrm{id}_A \otimes \lambda \otimes \mathrm{id}_\mathcal{K})\big((\delta \otimes \lambda) \circ \delta(A)\big).$$

Thus, the left hand side of (A.19) becomes

$$(\mathrm{id}_A \otimes \lambda \otimes \mathrm{id}_\mathcal{K}) \circ (\delta \otimes \mathrm{id}_\mathcal{K})\big((\mathrm{id}_A \otimes \lambda)(\delta(A))(1 \otimes \mathcal{K}(L^2(G)))\big)$$
$$\stackrel{(A.18)}{=} (\mathrm{id}_A \otimes \lambda \otimes \mathrm{id}_\mathcal{K}) \circ (\delta \otimes \mathrm{id}_\mathcal{K})\big(A \otimes \mathcal{K}(L^2(G))\big)$$
$$= \big((\delta \otimes \lambda) \circ \delta(A)\big) \otimes \mathcal{K}(L^2(G)).$$

$\square$

PROOF OF THEOREM A.69. Let us assume that A is faithfully represented on a Hilbert space $\mathcal{H}$. By Theorem A.41, $\pi = (\mathrm{id}_A \otimes \lambda) \circ \delta \times (1 \otimes M)$ is a faithful representation of $A \times_\delta G$ on $\mathcal{H} \otimes L^2(G)$, and then $\mathrm{Ind}\,\pi$ (the regular representation of $(A \times_\delta G, G, \widehat{\delta})$ induced from π) is a faithful representation of $A \times_\delta G \times_{\widehat{\delta},r} G$ on $\mathcal{H} \otimes L^2(G) \otimes L^2(G)$. Let us write $k_A, k_{C(G)}$ and k_G for the compositions $\mathrm{Ind}\,\pi \circ i^r_{A \times_\delta G} \circ j_A$, $\mathrm{Ind}\,\pi \circ i^r_{A \times_\delta G} \circ j_G$ and i^r_G, respectively, where $(i^r_{A \times_\delta G}, i^r_G)$ denote the canonical maps of $(A \times_\delta G, G)$ into $M(A \times_\delta G \times_{\widehat{\delta},r} G)$. By definition of $\mathrm{Ind}\,\pi$ we have

$$k_A(a) = \big((\mathrm{id}_A \otimes \lambda) \circ \delta(a)\big) \otimes 1, \qquad k_{C(G)}(f) = 1 \otimes \big((M \otimes M) \circ \nu(f)\big)$$
$$\text{and} \quad k_G(s) = 1 \otimes 1 \otimes \lambda_s,$$

where $\nu: C_0(G) \to C_b(G \times G)$ is defined by $\nu(f)(s,t) = f(st^{-1})$. We define $V = (1 \otimes W_G)(1 \otimes S)$, where $W_G = M \otimes \lambda(w_G)$ and $S: L^2(G \times G) \to L^2(G \times G)$ is the self-adjoint unitary operator $(S\xi)(s,t) = \Delta(t)^{-1/2}\xi(s, t^{-1})$. Since $S(\lambda \otimes 1)S = \lambda \otimes 1$ we have $\mathrm{Ad}(1 \otimes S) \circ k_A = k_A$. For $\xi \in L^2(G \times G)$, we compute

$$\big(W_G S((M \otimes M) \circ \nu(f)) S W_G^* \xi\big)(s,t)$$
$$= \Delta(t^{-1}s)^{1/2}\big((M \otimes M) \circ \nu(f)) S W_G^* \xi\big)(s, t^{-1}s)$$
$$= \Delta(t^{-1}s)^{1/2} f(t)\big(S W_G^* \xi\big)(s, t^{-1}s) = f(t)\xi(s,t) = \big((1 \otimes M(f))\xi\big)(s,t),$$

which implies that $\mathrm{Ad}\,V \circ k_{C(G)} = 1 \otimes 1 \otimes M$, and the identity $W_G S(1 \otimes \lambda) S W_G^* = 1 \otimes \rho$ implies that $\mathrm{Ad}\,V \circ k_G = 1 \otimes 1 \otimes \rho$, where $\rho: G \to U(L^2(G))$ denotes the right regular representation of G. Thus, since $C_0(G)\rho(C^*(G)) = \mathcal{K}(L^2(G))$ by

Remark A.11, it follows that

$$A \times_\delta G \times_{\widehat{\delta},r} G \cong \operatorname{Ad} V(k_A(A)) \operatorname{Ad} V(k_{C(G)}(C_0(G))) \operatorname{Ad} V(k_G(C^*(G)))$$
$$= \bigl((1 \otimes W_G)\bigl((\operatorname{id}_A \otimes \lambda) \circ \delta(A) \otimes 1\bigr)(1 \otimes W_G^*)\bigr)$$
$$\cdot \bigl(1 \otimes 1 \otimes M(C_0(G))\bigr)\bigl(1 \otimes 1 \otimes \rho(C^*(G))\bigr)$$
$$= \Bigl((1 \otimes W_G)\bigl((\operatorname{id}_A \otimes \lambda) \circ \delta(A) \otimes 1\bigr)(1 \otimes W_G^*)\Bigr)\bigl(1 \otimes 1 \otimes \mathcal{K}(L^2(G))\bigr)$$
$$\stackrel{(A.18)}{=} \bigl((\operatorname{id}_A \otimes \lambda) \circ \delta(A)\bigr) \otimes \mathcal{K}(L^2(G)).$$

Finally, $(\operatorname{id}_A \otimes \lambda) \circ \delta = j_A$ by definition, so the proof is complete. $\square$

A.9. Other definitions of coactions

As mentioned before, full coactions were first introduced by the fourth author in [**53**] using maximal tensor products instead of the minimal ones we use here. Although we shall see below that this definition has some disadvantages, we should point out that the approach of [**53**] inspired much of the theory we use in this work.

Let us denote by $\delta_G^{\max}\colon C^*(G) \to M(C^*(G) \otimes_{\max} C^*(G))$ the integrated form of the unitary homomorphism $s \mapsto u(s) \otimes_{\max} u(s)$. The fourth author defined a coaction of G on a C^*-algebra A as a nondegenerate homomorphism $\delta\colon A \to M(A \otimes_{\max} C^*(G))$ satisfying

 (i) $\delta(a)(1 \otimes z) \in A \otimes_{\max} C^*(G)$ for all $a \in A$, $z \in C^*(G)$.
 (ii) $(\delta \otimes \operatorname{id}_G) \circ \delta = (\operatorname{id}_A \otimes \delta_G^{\max}) \circ \delta$ as maps from A to $M(A \otimes_{\max} C^*(G) \otimes_{\max} C^*(G))$.

Note that there is no assumption on injectivity here, but it was pointed out in [**53**, Remarks 2.2] that if δ is not injective, then it factors through an injective coaction $\delta_1\colon A/\ker\delta \to M(A/\ker\delta \otimes_{\max} C^*(G))$. Covariant representations of a full coaction (A, G, δ) (in this sense) on a Hilbert space $\mathcal{H}$ are defined completely analogous to Definition A.32: They are pairs (π, μ) of nondegenerate representations $\pi\colon A \to \mathcal{B}(\mathcal{H}), \mu\colon C_0(G) \to \mathcal{B}(\mathcal{H})$ satisfying

$$(\pi \otimes \operatorname{id}_G) \circ \delta = \operatorname{Ad}(\mu \otimes \operatorname{id}_G)(w_G) \circ (\pi \otimes 1)$$

in $M(\mathcal{K}(\mathcal{H}) \otimes C^*(G))$.

If $\delta\colon A \to M(A \otimes_{\max} C^*(G))$ is an injective coaction as above, then $\delta^{\min} = \Phi \circ \delta$ is a coaction in our sense (i.e., in the sense of [**47**]), where $\Phi\colon A \otimes_{\max} C^*(G) \to A \otimes C^*(G)$ denotes the quotient map (see [**53**, Remarks 2.2 (4)]). Moreover, by [**53**, Remarks 2.5] a pair (π, μ) is a covariant representation for δ if and only if it is a covariant representation for $\delta^{\min}$, which in particular shows (since both satisfy the same universal properties) that the crossed products of δ and $\delta^{\min}$ are the same.

One of the main defects of the theory of full coactions with maximal tensor products is the fact that, if (π, μ) is a covariant representation of (A, G, δ), then $\delta^\mu(\pi(a)) = \operatorname{Ad}\bigl((\mu \otimes_{\max} \operatorname{id}_G)(w_G)\bigr)(\pi(a) \otimes 1)$ does not define a coaction on $\pi(A)$ in general. So one of the basic features of the theory (see Section A.7) is not available anymore.

EXAMPLE A.71 (*cf.* [**47**, Example 1.15]). We consider the coaction $\delta_G^{\max}$ of G on $C^*(G)$. Since $(\delta_G^{\max})^{\min} = \delta_G$ it follows from the above discussion that $C^*(G) \times_{\delta_G^{\max}} G = C^*(G) \times_{\delta_G} G$. Thus, Example A.62 implies that $(\mathcal{K}(L^2(G)), \lambda, M)$ is a crossed product for $(C^*(G), G, \delta_G^{\max})$. Now let G be a discrete non-amenable group. We claim that $\delta^M = \mathrm{Ad}(M\otimes_{\max}\mathrm{id}_G(w_G))\circ(\lambda(\cdot)\otimes 1)$ is not a full coaction on $\lambda(C^*(G)) = C_r^*(G)$ in the sense of [**53**], *i.e.*, we show that it doesn't factor through $C_r^*(G) = C^*(G)/\ker\lambda$. To see this, first observe that if it did, then it would be determined by the map $\lambda_s \mapsto \lambda_s\otimes_{\max} u(s)$ for $s \in G$. Let ρ denote the right regular representation of G, and define a representation π of $C_r^*(G)\otimes_{\max} C^*(G)$ on $\ell^2(G)$ by $\pi(\lambda(x_i)\otimes y_i) = \lambda(x_i)\rho(y_i)$, $x_i, y_i \in C^*(G)$ (since the left regular representation commutes with the right regular representation, π is well-defined). For $s \in G$ it follows that $\pi \circ \delta^M(\lambda_s) = \lambda_s\rho_s$, hence $\pi\circ\delta^M(\lambda_s)\chi_{\{e\}} = \chi_{\{e\}}$, where $\chi_{\{e\}} \in \ell^2(G)$ denotes the characteristic function on the trivial element e. Thus, $\ker\lambda \subseteq \ker\pi\circ\delta^M\circ\lambda \subseteq \ker 1_G$, where 1_G is the trivial representation of G, which contradicts the fact that G is non-amenable.

Note that the (non existent) coaction δ^M in the above example corresponds to the dual coaction δ_G^n of $C^*(G)$ on the reduced group algebra $C_r^*(G)$ in our theory. Thus there seems to be no natural definition of a dual coaction of $C^*(G)$ on the reduced crossed product $A \times_{\alpha,r} G$ in the theory of [**53**]. Of course this is another serious drawback.

We want to use this opportunity to point out a gap in [**47**, Example 1.15] (from where we actually have extracted the above example). There it was stated that there exists no full coaction δ of G on $C_r^*(G)$ in the sense of [**53**] such that $\delta^{\min} = \delta_G^n$, arguing that if such a coaction did exist, it would have to be given by δ^M as in the example above. But of course, there is no guarantee that, if I is the kernel of the quotient map $C_r^*(G)\otimes_{\max} C^*(G) \to C_r^*(G)\otimes C^*(G)$ in $M(C_r^*(G)\otimes_{\max} C^*(G))$, we could not modify δ^M by adding suitable elements of I so that the new map would be a coaction of G on $C_r^*(G)$ in the sense of [**53**]. However, we do not see how this can be done.

The other different approach (and the one which is actually most established in the literature [**34, 31, 35**]) is to define coactions of G as coactions of the reduced group C^*-algebra $C_r^*(G)$ with comultiplication given by $\delta_G^r\colon \lambda_s \mapsto \lambda_s\otimes\lambda_s$.

DEFINITION A.72. A *reduced coaction* of G on A is an injective and nondegenerate homomorphism $\delta\colon A \to M(A\otimes C_r^*(G))$ satisfying

(i) $\delta(a)(1\otimes\lambda(x)) \in A\otimes C_r^*(G)$ for all $x \in C^*(G)$ and
(ii) $(\delta\otimes\mathrm{id}_G^r)\circ\delta = (\mathrm{id}_A\otimes\delta_G^r)\circ\delta$ as maps from A into $M(A\otimes C_r^*(G)\otimes C_r*(G))$.

We also call the triple (A, G, δ) a reduced coaction.

Again, the definition of covariant representations of reduced coactions is completely analogous to the definition we use here: They are pairs (π, μ) of nondegenerate representations of $(A, C_0(G))$ satisfying

$$(\pi\otimes\mathrm{id}_G^r)\circ\delta = \mathrm{Ad}\big((\mu\otimes\lambda)(w_g)\big)(\pi(\cdot)\otimes 1).$$

If δ is a coaction in our sense, then $\delta^r = (\mathrm{id}_A\otimes\lambda)\circ\delta$ factors through a reduced coaction on $A^r = A/\ker j_A$ [**47**, Corollary 3.4], which is called the *reduction* of δ. δ^r is nondegenerate (in the appropriate sense) if and only if δ is nondegenerate. In particular, we see that if δ is normal, then δ^r coacts an A itself and we have $(\delta^n)^r = \delta^r$. Thus we see that taking the assignment $\delta \mapsto \delta^r$ is not a one-to-one correspondence between (conjugacy classes of) coactions in our sense and (conjugacy classes of) reduced coactions. However, it is if we restrict to nondegenerate normal coactions [**47**, Theorem 4.7].

The covariant homomorphisms of δ and δ^r coincide by Remark A.57 and [**47**, Proposition 3.7]. To be more precise: If (π, μ) is a covariant homomorphism of (A^r, G, δ^r), then $(\pi \circ j_A, \mu)$ is a covariant homomorphism of (A, G, δ) and every covariant homomorphism of (A, G, δ) arises this way. In particular, it follows from the universal properties that the crossed products of (A^r, G, δ^r) and (A, G, δ) are the same (more precisely, $(A^r \times_{\delta^r} G, j_{A^r} \circ j_A, j_G)$ is a crossed product for (A, G, δ)).

The fact that the reductions of a coaction δ and its normalization δ^n are the same shows that the theory of coactions we use here is potentially richer than the theory of reduced coactions. One major drawback in the theory of reduced coactions is given by the fact that, in case where N is a non-amenable normal subgroup of G, there is no well-defined quotient map $C_r^*(G) \to C_r^*(G/N)$, so it would be difficult to define the restriction of a reduced coaction of G to the quotient G/N in the realm of reduced coactions (see Example A.27). Since restriction of coactions is one of the most basic concepts in this work, the use of reduced coactions would be inadequate for our purposes. However, the above discussion clearly shows that results on reduced coactions can often be applied (with care) to problems of coactions in our sense (in particular to problems concerning the representation theory and the crossed products of coactions (A, G, δ)). Of course, if G is amenable, all the different notions of coactions do coincide.

APPENDIX B

The Imprimitivity Theorems of Green and Mansfield

In this appendix we want to recall the main ingredients of the imprimitivity theorems of Green and Mansfield. The main idea behind those imprimitivity theorems is to show that certain C^*-algebras which appear in the study of crossed products by actions and coactions, and in particular in the study of induced representations, are Morita equivalent in the sense that there are canonical imprimitivity bimodules linking those algebras together.

B.1. Imprimitivity theorems for actions

In what follows we recall the main results of [51]. We refer to [54] (and Chapter 1 of this work) for the necessary background on (pre-)imprimitivity bimodules and Morita equivalence.

Consider a C^*-algebra D, two locally compact groups K and H, and a locally compact space P. Suppose that K acts freely and properly on the left of P, and that H acts likewise on the right such that these actions commute (i.e., $k(ph) = (kp)h$). Suppose also that we have commuting actions β of K and α of H on D. For the left action of K we define the induced C^*-algebra $\operatorname{Ind}_K^P(D, \beta)$ (or just $\operatorname{Ind}_K^P D$, if confusions seems unlikely) to be the set of bounded continuous functions $f \colon P \to D$ such that $f(kp) = \beta_k(f(p))$ for all $k \in K$ and $p \in P$, and such that the function $Kp \mapsto \|f(p)\|$ vanishes at infinity on $K\backslash P$. For the right action of H we define the *induced C^*-algebra* $\operatorname{Ind}_H^P(D, \alpha)$ (or just $\operatorname{Ind}_H^P D$) as the set of bounded continuous functions $g \colon P \to D$ such that $g(ph) = \alpha_{h^{-1}}(g(p))$ for all $p \in P$ and $h \in H$, and such that $pH \mapsto \|g(p)\|$ vanishes at infinity on P/H. Note that it follows from the properness of the actions of K and H on P that the quotient spaces $K\backslash P$ and P/H are locally compact Hausdorff spaces.

The induced algebras $\operatorname{Ind}_K^P D$ and $\operatorname{Ind}_H^P D$ are C^*-algebras with pointwise operations, and carry actions $\sigma \colon H \to \operatorname{Aut}(\operatorname{Ind}_K^P D)$, $\rho \colon K \to \operatorname{Aut}(\operatorname{Ind}_H^P D)$ given by

(B.1) $$\sigma_h(f)(p) = \alpha_h(f(ph)) \quad \text{and} \quad \rho_k(g)(p) = \beta_k(g(k^{-1}p)).$$

Theorem 1.1 of [51] states that $C_c(P, D)$ can be made into a $C_c(K, \operatorname{Ind}_H^P D)$ – $C_c(H, \operatorname{Ind}_K^P D)$ pre-imprimitivity bimodule which completes to an imprimitivity bimodule, Z, for the full crossed products $\operatorname{Ind}_H^P D \times_\rho K$ and $\operatorname{Ind}_K^P D \times_\sigma H$. The actions and inner products are given for $b \in C_c(K, \operatorname{Ind}_H^P D) \subseteq \operatorname{Ind}_H^P D \times K$, $x, y \in$

$C_c(P, D)$, and $c \in C_c(H, \operatorname{Ind}_K^P D) \subseteq \operatorname{Ind}_K^P D \times H$ as follows:

$$b \cdot x(p) = \int_K b(k, p) \beta_k(x(k^{-1}p)) \Delta_K(k)^{1/2} dk$$

$$x \cdot c(p) = \int_H \alpha_h \big(x(ph) c(h^{-1}, ph)\big) \Delta_H(h)^{-1/2} dh$$

(B.2)

$$_{C_c(K, \operatorname{Ind}_H^P D)} \langle x, y \rangle (k, p) = \Delta_K(k)^{-1/2} \int_H \alpha_h \big(x(ph) \beta_k(y(k^{-1}ph)^*)\big) dh$$

$$\langle x, y \rangle_{C_c(H, \operatorname{Ind}_K^P D)} (h, p) = \Delta_H(h)^{-1/2} \int_K \beta_k \big(x(k^{-1}p)^* \alpha_h(y(k^{-1}ph))\big) dk.$$

By some abuse of notation, we shall often regard an element in $C_c(K, \operatorname{Ind}_H^P D)$ as a continuous function of two variables via $b(k, p) = b(k)(p)$, and similarly for $C_c(H, \operatorname{Ind}_K^P D)$.

Note that the formulas above are not precisely those given in [**51**]. This results from the fact that we are working with left *and* right actions on P, while in [**51**] all actions were on the left. To convert to the two-left-actions situation of [**51**], just put $hp = ph^{-1}$. Note also that a similar symmetric imprimitivity theorem has been deduced independently by Kasparov in [**30**, Theorem 3.15], which also gives a Morita equivalence for the reduced crossed products. However, the construction of Kasparov's bimodule is given by a composition of several other bimodules, and does not really fit our needs in this work. Nevertheless, we do need to know that the imprimitivity bimodule constructed above factors through an imprimitivity bimodule for the reduced crossed products, *i.e.*, we want to have:

PROPOSITION B.1. *If we view $C_c(K, \operatorname{Ind}_H^P D)$ and $C_c(H, \operatorname{Ind}_K^P D)$ as dense subalgebras of the reduced crossed products $\operatorname{Ind}_H^P D \times_{\rho,r} K$ and $\operatorname{Ind}_K^P D \times_{\sigma,r} H$, respectively, then the $C_c(K, \operatorname{Ind}_H^P D) - C_c(H, \operatorname{Ind}_K^P D)$ bimodule $C_c(P, D)$ with actions and inner products as given in Equation B.2 completes to an $(\operatorname{Ind}_H^P D \times_{\rho,r} K) - (\operatorname{Ind}_K^P D \times_{\sigma,r} H)$ imprimitivity bimodule.*

PROOF. The hard work for this proof has been done in [**49**]; we show how the proof follows from [**49**, Lemma 4.1]. Using [**54**, Proposition 3.24], the content of [**49**, Lemma 4.1] can be translated into the statement:

If $I \subseteq \operatorname{Ind}_K^P D \times H$ is the kernel of a regular representation, say $\operatorname{Ind} \pi$, of $\operatorname{Ind}_K^P D \times H$, then the ideal Z-$\operatorname{Ind} I$ of $\operatorname{Ind}_H^P D \times K$ induced from I via Z contains the kernel, say J, of the regular representation(s) of $\operatorname{Ind}_H^P D \times K$.

By symmetry, we also get $\widetilde{Z}$-$\operatorname{Ind} J \supseteq I$, where $\widetilde{Z}$ denotes the conjugate bimodule. Since by [**54**, Corollary 3.31] Z-$\operatorname{Ind} \circ \widetilde{Z}$-$\operatorname{Ind}$ is the identity map on the set of closed ideals of $\operatorname{Ind}_H^P D \times K$, and since induction via Z and $\widetilde{Z}$ preserves inclusion of ideals by [**54**, Theorem 3.22 and Proposition 3.24], we get

$$Z\text{-}\operatorname{Ind} I \supseteq J = Z\text{-}\operatorname{Ind} \circ \widetilde{Z}\text{-}\operatorname{Ind} J \supseteq Z\text{-}\operatorname{Ind} I,$$

and hence $J = Z$-$\operatorname{Ind} I$. But it follows then from [**54**, Proposition 3.25] that Z factors through an imprimitivity bimodule for the quotients $\operatorname{Ind}_H^P D \times_r K \cong (\operatorname{Ind}_H^P D \times K)/J$ and $\operatorname{Ind}_K^P D \times_r H \cong (\operatorname{Ind}_K^P D \times H)/I$. Applying the respective quotient maps on the dense subspaces $C_c(K, \operatorname{Ind}_H^P D)$, $C_c(P, D)$ and $C_c(H, \operatorname{Ind}_K^P D)$ then completes the proof. □

We are now going to derive a one-sided version of the above symmetric imprimitivity theorem. The full-crossed-product version of this result has first been deduced by Green (it follows from [25, Theorem 17]), so we shall call it *Green's imprimitivity theorem for induced algebras*.

Let G be a locally compact group and let $\alpha\colon H \to \operatorname{Aut} A$ be an action of a closed subgroup H of G. We let G act on itself by left translation, we let H act on G by right translation, and we let G act trivially on A. Then, if we put $P = G$, $K = G$, and $D = A$ in the setting of the symmetric imprimitivity theorem, we obtain a Morita equivalence between the crossed products $\operatorname{Ind}_H^G A \times_\rho G$ on the left and $\operatorname{Ind}_G^G A \times_\sigma H$ on the right, and another between the respective reduced crossed products. Moreover, it is easy to check that $\operatorname{Ind}_G^G A \to A$ defined by $f \mapsto f(e)$ is an isomorphism which transforms σ to α. Hence, the map $\Phi\colon C_c(H, \operatorname{Ind}_G^G A) \to C_c(H, A)$ defined by $\Phi(F)(h) = F(h, e)$ extends to an isomorphism $\operatorname{Ind}_G^G A \times_\sigma H \cong A \times_\alpha H$ (and similarly for the reduced crossed products).

In this special situation we shall always denote the action $\rho\colon G \to \operatorname{Aut}(\operatorname{Ind}_H^G A)$ of Equation B.1 by $\operatorname{Ind}\alpha$. Since we start with the trivial action of G on A, $\operatorname{Ind}\alpha$ is given by the formula

(B.3) $$\operatorname{Ind}\alpha_s(f)(t) = f(s^{-1}t)$$

for $f \in \operatorname{Ind}_H^G A$ and $s, t \in G$. Using the map $\Phi\colon C_c(H, \operatorname{Ind}_G^G A) \to C_c(H, A)$ described above, it follows from the formulas given in Equation B.2 that $C_c(G, A)$ becomes a $C_c(G, \operatorname{Ind}_H^G A) - C_c(H, A)$ pre-imprimitivity bimodule with actions and inner products given by the formulas

(B.4)
$$b \cdot x(s) = \int_G b(t, s) x(t^{-1}s) \Delta_G(t)^{1/2} dt$$
$$x \cdot c(s) = \int_H \alpha_h\bigl(x(sh) c(h^{-1})\bigr) \Delta_H(h)^{-1/2} dh$$
$$_{C_c(G, \operatorname{Ind}_H^G A)}\langle x, y\rangle(s, t) = \Delta_G(s)^{-1/2} \int_H \alpha_h\bigl(x(th) y(s^{-1}th)^*\bigr) dh$$
$$\langle x, y\rangle_{C_c(H, A)}(h) = \Delta_H(h)^{-1/2} \int_G x(t^{-1})^* \alpha_h(y(t^{-1}h)) \, dt.$$

More precisely, we obtain:

THEOREM B.2. *Regard $C_c(G, \operatorname{Ind}_H^G A)$ and $C_c(H, A)$ as dense subalgebras of the full crossed products $\operatorname{Ind}_H^G A \times_{\operatorname{Ind}\alpha} G$ and $A \times_\alpha H$, respectively. Then $C_c(G, A)$ with actions and inner products as given in Equation B.4 completes to an $(\operatorname{Ind}_H^G A \times_{\operatorname{Ind}\alpha} G) - (A \times_\alpha H)$ imprimitivity bimodule $V_H^G(A)^f$ (the f stands for "full crossed products").*

Similarly, if we regard $C_c(G, \operatorname{Ind}_H^G A)$ and $C_c(H, A)$ as dense subalgebras of the respective reduced crossed products, $C_c(G, A)$ completes to give a $(\operatorname{Ind}_H^G A \times_{\operatorname{Ind}\alpha, r} G) - (A \times_{\alpha, r} H)$ imprimitivity bimodule $V_H^G(A)^r$.

We usually suppress the superscripts in the notation of the bimodules, if confusion seems unlikely. In fact, in the main body of our work we use *reduced* crossed products almost exclusively.

As an important special case of Green's imprimitivity theorem for induced actions, we shall now derive Green's original imprimitivity theorem, which we call *Green's imprimitivity theorem for induced representations*, since it was used by

Green to develop a very strong version of Mackey's imprimitivity theorem for induced representations of locally compact groups. Thus, the original motivation to develop this theorem was to solve the following problem:

Suppose that $\alpha\colon G \to \operatorname{Aut} A$ is an action, and H is a closed subgroup of G. Assume further that $\pi \times V$ is a representation of $A \times_\alpha G$ on a Hilbert space $\mathcal{H}$. When is this representation equivalent to a representation $\operatorname{Ind}_H^G(\sigma \times W)$ induced from a representation $\sigma \times W$ of $A \times_{\alpha|} H$?

Of course, before one can solve this problem, one has to make clear what the construction of the induced representation $\operatorname{Ind}_H^G(\sigma \times W)$ should be. Green's approach to this problem was to use the modern bimodule techniques of Rieffel [56] to define induction via a certain natural right-Hilbert $(A \times_\alpha G) - (A \times_{\alpha|} H)$ bimodule. We now derive this bimodule as a special case of Theorem B.2 above.

For this assume that (A, G, α) is an action and H is a closed subgroup of G. We observe that the map $\varphi\colon \operatorname{Ind}_H^G(A, \alpha|) \to C_0(G/H, A) \cong A \otimes C_0(G/H)$ defined by $\varphi(f)(sH) = \alpha_s(f(s))$ is an $\operatorname{Ind}\alpha - (\alpha \otimes \tau)$ equivariant isomorphism, where $\tau\colon G \to \operatorname{Aut} C_0(G/N)$ is given by $\tau_s(f)(tN) = f(s^{-1}tN)$, as in Example A.10. Moreover, if we define $\Psi\colon C_c(G, A) \to C_c(G, A)$ by $\Psi(x)(s) = \alpha_s(x(s))$, then the formulas given in Equation B.4 transform into

(B.5)
$$b \cdot x(s) = \int_G b(t, sH)\alpha_t(x(t^{-1}s))\Delta_G(t)^{1/2}dt$$
$$x \cdot c(s) = \int_H x(sh)\alpha_{sh}(c(h^{-1}))\Delta_H(h)^{-1/2}dh$$
$${}_{C_c(G,C_0(G/H,A))}\langle x, y\rangle(s, tH) = \Delta_G(s)^{-1/2}\int_H x(th)\alpha_s(y(s^{-1}th)^*)\, dh$$
$$\langle x, y\rangle_{C_c(H,A)}(h) = \Delta_H(h)^{-1/2}\int_G \alpha_t\bigl(x(t^{-1})^*y(t^{-1}h)\bigr)\, dt,$$

and as a direct consequence of Theorem B.2 we obtain:

THEOREM B.3. *With the above formulas for the actions and inner products, $C_c(G, A)$ completes to give $((A \otimes C_0(G/H)) \times_{\alpha \otimes \tau} G) - (A \times_{\alpha|} H)$ and $((A \otimes C_0(G/H)) \times_{\alpha \otimes \tau, r} G) - (A \times_{\alpha|, r} H)$ imprimitivity bimodules $X_H^G(A)^f$ and $X_H^G(A)^r$, respectively.*

In practice, the dense subalgebra $C_c(G, C_0(G/H, A))$ of $(A \otimes C_0(G/H)) \times_{\alpha \otimes \tau, r} G$ is actually replaced with the smaller (but still dense) subalgebra $C_c(G \times G/H, A)$, and $C_c(G, A)$ is regarded as a $C_c(G \times G/H, A) - C_c(H, A)$ pre-imprimitivity bimodule. In the (important) special case $H = \{e\}$, it takes a little effort to remember that it is the second variable that comes from $C_0(G, A)$ (so that for $b \in C_c(G \times G, A) \subseteq C_0(G, A) \times_r G$ we have $b(s, t) = b(s)(t)$, where $b(s)$ is an element of the C^*-algebra $C_0(G, A)$).

Let us briefly explain what this theorem has to do with the problem mentioned above. For this let $(k_A \otimes k_{C_0(G/H)}, k_G)$ denote the canonical maps of $(A \otimes C_0(G/H), G)$ into $M((A \otimes C_0(G/H)) \times_{\alpha \otimes \tau} G)$. Then we get a nondegenerate $*$-homomorphism

$$k_A \times k_G \colon A \times_\alpha G \to M\bigl((A \times C_0(G/H)) \times_{\alpha \otimes \tau} G\bigr).$$

Identifying $M\bigl((A \otimes C_0(G/H)) \times_{\alpha \otimes \tau} G\bigr)$ with $\mathcal{L}(X_H^G(A)^f)$ (see, for example, [54, Corollary 2.54]), this makes $X_H^G(A)^f$ into a right-Hilbert $(A \times_\alpha G) - (A \times_{\alpha|} H)$

bimodule, which we can use to induce representations from $A \times_{\alpha|} H$ to $A \times_\alpha G$. On the other side, since $X_H^G(A)^f$ is a $((A \otimes C_0(G/H)) \times_{\alpha \otimes \tau} G) - (A \times_{\alpha|} H)$ imprimitivity bimodule, induction via $X_H^G(A)^f$ provides an *equivalence* between the representation spaces of $A \times_{\alpha|} H$ and $(A \otimes C_0(G/H)) \times_{\alpha \otimes \tau} G$. It follows from the definition of the action of $A \times_\alpha G$ on $X_H^G(A)^f$, that if $\sigma \times W$ is a representation of $A \times_{\alpha|} H$ and $(\pi \otimes \mu) \times V$ is the representation of $(A \otimes C_0(G/H)) \times_{\alpha \otimes \tau} G$ induced from $\sigma \times W$ via $X_H^G(A)^f$, then $\pi \times V$ is the representation of $A \times_\alpha G$ induced from $A \times_{\alpha|} H$ via $X_H^G(A)^f$. Thus we arrive at the following answer to the above-stated problem:

THEOREM B.4 (Mackey-Green). *Let $\pi \times V$ be a representation of $A \times_\alpha G$ on a Hilbert space $\mathcal{H}$. Then $\pi \times V$ is induced from a representation $\sigma \times W$ of $A \times_{\alpha|} H$ (via $X_H^G(A)^f$) if and only if there exists a nondegenerate representation $\mu \colon C_0(G/H) \to \mathcal{B}(\mathcal{H})$ such that μ and π have commuting images in $\mathcal{B}(\mathcal{H})$ and such that $(\pi \otimes \mu, V)$ is a covariant representation of $(A \otimes C_0(G/H)) \times_{\alpha \otimes \tau} G$.*

We finally want to give an interpretation of the above imprimitivity theorems in terms of duality theory. For this we specialize even further to the case where $N = H$ is *normal* in G. If α is an action of G, we can form the dual coaction $\widehat{\alpha}$ of G on the crossed product $A \times_\alpha G$, and the dual coaction $\widehat{\alpha}^n$ of G on $A \times_{\alpha,r} G$. As described in Example A.28, we may restrict the coactions $\widehat{\alpha}$ and $\widehat{\alpha}^n$ to coactions $\widehat{\alpha}|$ and $\widehat{\alpha}^n|$ of G/N, respectively. Theorem A.64 provides a canonical isomorphism

$$(A \otimes C_0(G/N)) \times_{\alpha \otimes \tau} G \cong (A \times_\alpha G) \times_{\widehat{\alpha}|} G/N$$

and Theorem A.65 provides an isomorphism

$$(A \otimes C_0(G/N)) \times_{\alpha \otimes \tau, r} G \cong (A \times_{\alpha, r} G) \times_{\widehat{\alpha}^n|} G/N.$$

Thus, replacing stable isomorphism by Morita equivalence, the above-derived Morita equivalences should be regarded as a generalization of the Imai-Takai duality theorems (see Theorem A.67 and Theorem A.68):

THEOREM B.5. *Suppose that (A, G, α) is an action and N is a closed normal subgroup of G. Then the above constructions provide (natural)*

$$((A \times_\alpha G) \times_{\widehat{\alpha}|} G/N) - (A \times_{\alpha|} N) \quad \text{and} \quad ((A \times_{\alpha, r} G) \times_{\widehat{\alpha}^n|} G/N) - (A \times_{\alpha|, r} N)$$

imprimitivity bimodules, still denoted $X_N^G(A)^f$ and $X_N^G(A)^r$, respectively.

The main purpose of this paper is to study how "natural" these bimodules are, and to study various actions and coactions on these bimodules, which fit with certain canonical actions and coactions on the algebras involved.

B.2. Mansfield's imprimitivity bimodule

Starting with a coaction (A, G, δ), in [38] Mansfield provided a dual mirror to Theorem B.5, at least if N is an *amenable* closed normal subgroup of G. Later, in [28], Mansfield's result was generalized to arbitrary closed normal subgroups N of G, provided the coaction δ satisfies certain extra conditions. Since those conditions are always satisfied if δ is *normal*, we may formulate:

THEOREM B.6 (*cf.* [**38**, Theorem 27],[**28**, Theorem 3.3]). *Assume that (A, G, δ) is a nondegenerate normal coaction, and let N be a closed normal subgroup of G. There exists an $(A \times_\delta G) \times_{\widehat{\delta}, r} N - A \times_{\delta|} G/N$ imprimitivity bimodule $Y_{G/N}^G(A)$.*

Of course, in the same way as Theorem B.5 should be viewed as a generalization of the Imai-Takai duality theorem, this theorem should be viewed as a generalization of Katayama's duality theorem (see Theorem A.69).

Let us be a bit more precise about how the above-stated result follows from [28, Theorem 3.3]. For this let (j_A, j_G) denote the canonical maps of $(A, C_0(G))$ into $M(A \times_\delta G)$. Restricting j_G to $C_C(G/N) \subseteq M(C_0(G))$ gives a homomorphism $j_G|: C_0(G/N) \to M(A \times_\delta G)$, and one can check without too much pain that the pair $(j_A, j_G|)$ is then a covariant homomorphism of $(A, G/N, \delta|)$ into $M(A \times_\delta G)$. Now, [28, Theorem 3.3] states that the constructions of Mansfield provide a $(A \times_\delta G \times_{\hat\delta|,r} N) - (j_A \times j_G|)(A \times_{\delta|} G/N)$ imprimitivity bimodule $Y_{G/N}^G$. On the other hand, if δ is normal, then it follows from [28, Lemma 3.2] that $\delta|$ is normal, too, and that $j_A \times j_G|: A \times_{\delta|} G/N \to M(A \times_\delta G)$ is faithful. Thus in this situation we may identify $(j_A \times j_G|)(A \times_{\delta|} G/N)$ with $A \times_{\delta|} G/N$. The following example shows that Theorem B.6 does not hold for arbitrary coactions.

EXAMPLE B.7. Consider the coaction δ_G of G on $C^*(G)$, and let $N = G$. The restriction of δ_G to G/G is of course the trivial coaction of the trivial group, so we get $C^*(G) \times_{\delta_G|} G/G = C^*(G)$. On the other hand, we have $C^*(G) \times_{\delta_G} G \cong C_r^*(G) \times_{\delta_G^n} G \cong \mathcal{K}(L^2(G))$ by Example A.62 and Proposition A.61. Applying the Imai-Takai duality theorem (see Theorem A.68) to the trivial action of G on $\mathbb{C}$, we see that the dual action $\widehat{\delta_G}$ corresponds to the unitary action $\operatorname{Ad}\rho$ of G on $\mathcal{K}(L^2(G))$. But this implies that $C^*(G) \times_{\delta_G} G \times_{\widehat{\delta_G},r} G \cong \mathcal{K}(L^2(G)) \times_{\operatorname{Ad}\rho,r} G \cong \mathcal{K}(L^2(G)) \otimes C_r^*(G)$. Thus, if it were true for δ_G, Theorem B.6 would provide a Morita equivalence between $C^*(G)$ and $C_r^*(G)$. Such an equivalence should rarely exist for any non-amenable group, and it certainly does not exist if $G = \mathbb{F}_2$, the free group on two generators, since $C_r^*(\mathbb{F}_2)$ is simple but $C^*(\mathbb{F}_2)$ is not.

Note that it is this kind of problem which makes us stick to reduced crossed products and normal coactions in the body of the paper.

In the rest of this section we want to briefly recall the construction of the bimodule $Y_{G/N}^G(A)$ of Theorem B.6. As in the other imprimitivity theorems we have to work with certain dense subalgebras, but, unfortunately, the constructions which have to be done here are much more complicated than the corresponding constructions in the action case (however, see [18] for an easier construction in the case of a dual coaction).

For a compact subset $E \subseteq G$ let $C_E(G) = \{f \in C_c(G) \mid \operatorname{supp} f \subseteq E\}$, and let $A_c(G) = A(G) \cap C_c(G)$. Let $\varphi: C_c(G) \to C_c(G/N)$ denote the surjection

$$\varphi(f)(sN) = \int_N f(sn)\, dn,$$

with the usual convention on the Haar measures such that $\int_{G/N} \circ \int_N = \int_G$. Recall from Proposition A.31 that if $u \in A(G)$, then $\delta_u: A \to A$ denotes the composition $S_u \circ \delta$, where $S_u: M(A \otimes C^*(G)) \to M(A)$ is the slice map corresponding to u.

DEFINITION B.8. For fixed $u \in A_c(G)$ and compact $E \subseteq G$ we put

$$\mathcal{D}_{(u,E,N)} = \overline{j_A(\delta_u(A))j_G|(\varphi(C_E(G)))} \subseteq M(A \times_\delta G),$$

and we define

$$\mathcal{D}_N = \bigcup \{\mathcal{D}_{(u,E,N)} \mid u \in A_c(G), E \subseteq G \text{ compact}\}.$$

Moreover, we write $\mathcal{D}_{(u,E)}$ for $\mathcal{D}_{(u,E,\{e\})}$ and $\mathcal{D}$ for $\mathcal{D}_{\{e\}}$.

It follows from Mansfield's computations in [**38**] that $\mathcal{D}$ is a dense $*$-subalgebra of $A \times_\delta G$ and that $\mathcal{D}_N$ is a dense $*$-subalgebra of $(j_A \times j_G|)(A \times_{\delta|} G/N)$. Thus, if δ is normal, and if we identify $A \times_{\delta|} G/N$ with its image in $M(A \times_\delta G)$, we can regard $\mathcal{D}_N$ as a dense $*$-subalgebra of $A \times_{\delta|} G/N$. It is also a consequence of Mansfield's computations that there is a linear map $\Psi \colon \mathcal{D} \to \mathcal{D}_N$ such that

$$\Psi\big(j_A(a) j_G(f)\big) = j_A(a) j_G|(\varphi(f)) \quad \text{for } a \in \delta_{A_c(G)}(A), f \in C_c(G).$$

Indeed it follows from [**38**, Lemma 18] that for all $x, y \in \mathcal{D}$ the maps $s \mapsto \hat{\delta}_s(x) y$ and $s \mapsto y \hat{\delta}_s(x)$ are norm continuous with compact supports and that, as an element of $M(A \times_\delta G)$, $\Psi(x)$ is characterized by

$$(\text{B.6}) \qquad \Psi(x) y = \int_N \hat{\delta}_n(x) y \, dn \quad \text{and} \quad y \Psi(x) = \int_N y \hat{\delta}_n(x) \, dn \qquad \text{for } y \in \mathcal{D}.$$

So, although $n \mapsto \hat{\delta}_n(x)$ rarely has compact support, we shall often write

$$\Psi(x) = \int_N \hat{\delta}_n(x) \, dn.$$

Actually, the following proposition shows that the latter expression exists as a *weak* integral:

PROPOSITION B.9. *For each $x \in \mathcal{D}$ and $\omega \in (A \times_\delta G)^*$ the function $n \mapsto \omega(\hat{\delta}_n(x))$ is integrable, and the element $\Psi(x) \in M(A \times_\delta G)$ is uniquely determined by the equations*

$$\omega(\Psi(x)) = \int_N \omega(\hat{\delta}_n(x)) \, dn.$$

PROOF. It follows from [**38**, Lemma 9] that $\mathcal{D}$ is closed under multiplication on either side by $j_G(C_b(G))$, and in particular by $j_G(C_c(G))$. Hence, [**46**, Lemmas 3.5, 3.8, and 3.10 and Corollary 3.6] (which are based upon [**41**, Section 2] and [**42**, note added in proof]) tell us that for each $x \in \mathcal{D}$ there is a unique element $\int_N \hat{\delta}_n(x) \, dn$ of $M(A \times G)$ such that

$$\omega\left(\int_N \hat{\delta}_n(x) \, dn \right) = \int_N \omega(\hat{\delta}_n(x)) \, dn \qquad \text{for } \omega \in (A \times G)^*.$$

To see that this integral coincides with $\Psi(x)$, we use Equation B.6 to see that for all $y \in \mathcal{D}$ and all $\omega \in (A \times_\delta G)^*$ we have

$$\omega(\Psi(x) y) = \omega\left(\int_N \hat{\delta}_n(x) y \, dn \right) = \int_N \omega(\hat{\delta}_n(x) y) \, dn = \int_N y \cdot \omega(\hat{\delta}_n(x)) \, dn$$
$$= y \cdot \omega\left(\int_N \hat{\delta}_n(x) \, dn \right) = \omega\left(\int_N \hat{\delta}_n(x) \, dn \, y \right)$$

and similarly $\omega(y \Psi(x)) = \omega\big(y \int_N \hat{\delta}_n(x) \, dn \big)$. Since $\mathcal{D}$ is dense in $A \times_\delta G$, this finishes the proof. □

If we view $C_c(N, \mathcal{D}) \subseteq C_c(N, A \times_\delta G)$ as a dense subalgebra of $A \times_\delta G \times_{\hat{\delta}|, r} N$, then $\mathcal{D}$ becomes a $C_c(N, \mathcal{D}) - \mathcal{D}_N$ pre-imprimitivity bimodule with module actions and inner products given for $g \in C_c(N, \mathcal{D})$, $x, y \in \mathcal{D}$, and $c \in \mathcal{D}_N$ by

(B.7)
$$g \cdot x = \int_N g(n)\hat{\delta}_n(x)\Delta(n)^{1/2} dn$$
$$x \cdot c = xc \quad \text{(product in } M(A \times_\delta G))$$
$$_{C_c(N,\mathcal{D})}\langle x, y\rangle(n) = x\hat{\delta}_n(y^*)\Delta(n)^{-1/2}$$
$$\langle x, y\rangle_{\mathcal{D}_N} = \Psi(x^*y) \quad \left(= \int_N \hat{\delta}_n(x^*y)\, dn\right).$$

The $(A\times_\delta G\times_{\hat{\delta}|,r} N) - (A\times_{\delta|} G/N)$ imprimitivity bimodule $Y^G_{G/N}(A)$ of Theorem B.6 is the completion of the $C_c(N, \mathcal{D}) - \mathcal{D}_N$ pre-imprimitivity bimodule $\mathcal{D}$.

APPENDIX C

Function Spaces

In this appendix we develop machinery allowing us to work with certain function spaces in the various multiplier algebras of crossed products and multiplier bimodules.

C.1. The spaces $C_c(T, \mathcal{X})$ for locally convex spaces $\mathcal{X}$

We start with some general remarks on inductive limit topologies on functions with compact support taking values in locally convex spaces. Assume that T is a locally compact space and $\mathcal{X}$ is a Hausdorff locally convex vector space over $\mathbb{C}$. Let $\mathcal{P}$ be a set of continuous seminorms on $\mathcal{X}$ which generate the topology on $\mathcal{X}$. Let $C_c(T, \mathcal{X})$ denote the space of all continuous functions from T to $\mathcal{X}$ with compact support. If $K \subseteq T$ is compact, then $C_K(T, \mathcal{X})$ denotes the space of all $f \in C_c(T, \mathcal{X})$ with $\mathrm{supp}\, f \subseteq K$. For each $p \in \mathcal{P}$ we define a seminorm $p_K \colon C_K(T, \mathcal{X}) \to \mathbb{R}$ by

$$p_K(f) = \sup_{t \in T} p(f(t)),$$

and we topologize $C_K(T, \mathcal{X})$ via the family of seminorms $\mathcal{P}_K = \{p_K \mid p \in \mathcal{P}\}$. It is clear that $f_i \to 0$ in $C_K(T, \mathcal{X})$ if and only if $p \circ f_i \to 0$ uniformly for all $p \in \mathcal{P}$. It is also clear that if $L \subseteq T$ is a compact subset containing K, then the inclusion $C_K(T, \mathcal{X}) \to C_L(T, \mathcal{X})$ is a homeomorphism onto its image. Let $\mathcal{C}$ denote the set of all compact subsets of T ordered by inclusion. Then $C_c(T, \mathcal{X}) = \bigcup_{K \in \mathcal{C}} C_K(T, \mathcal{X})$ and we may equip $C_c(T, \mathcal{X})$ with the inductive limit topology (see [**58**, II section 6] for the precise definitions). The following useful properties of the inductive limit topology on $C_c(T, \mathcal{X})$ shall be used without reference:

  (i) For all $K \in \mathcal{C}$ the inclusion $C_K(T, \mathcal{X}) \to C_c(T, \mathcal{X})$ is a homeomorphism onto its image.
  (ii) If $\mathcal{Y}$ is any locally convex space and $\Phi \colon C_c(T, \mathcal{X}) \to \mathcal{Y}$ is a linear map, then Φ is continuous if and only if its restriction to $C_K(T, \mathcal{X})$ is continuous for all $K \in \mathcal{C}$.
  (iii) If $\mathcal{Y}$ is any locally convex space and $\Psi \colon \mathcal{X} \to \mathcal{Y}$ is a continuous linear map, then the linear map $f \mapsto \Psi \circ f \colon C_c(T, \mathcal{X}) \to C_c(T, \mathcal{Y})$ is continuous.

PROPOSITION C.1. *Suppose that S and T are locally compact spaces and $\mathcal{X}$ is a locally convex space. Then the natural embedding $\Phi \colon C_c(S \times T, \mathcal{X}) \to C_c(S, C_c(T, \mathcal{X}))$ given by $\Phi(f)(s) = f(s, \cdot)$ is continuous.*

PROOF. We only have to check that the restriction of Φ to $C_{L \times K}(S \times T, \mathcal{X})$ for any compact subset of the form $L \times K$ of $S \times T$ is continuous. The image of $C_{L \times K}(S \times T, \mathcal{X})$ clearly lies in $C_L(S, C_K(T, \mathcal{X}))$, and if $q \colon C_c(T, \mathcal{X}) \to \mathbb{R}$ is any continuous seminorm, its restriction to $C_K(T, \mathcal{X})$ is dominated by a finite sum of seminorms of the form p_K as above. So the result follows from the fact that $p_K(\Phi(f)) = p_{L \times K}(f)$ for all $f \in C_{L \times K}(S \times T, \mathcal{X})$. $\square$

In the sequel we need to know that $C_c(T, \mathcal{X})$ is complete whenever $\mathcal{X}$ is. Unfortunately, it seems to be not clear whether this is true in general, but the following result shows that it is true whenever T is a locally compact group.

PROPOSITION C.2. *Suppose that T is a disjoint union of a collection $(T_\lambda)_{\lambda \in \Lambda}$ of σ-compact open subsets of T. Then $C_c(T, \mathcal{X})$ is complete whenever $\mathcal{X}$ is. In particular, if T is a locally compact group, or a quotient of a locally compact group by a closed subgroup, then $C_c(T, \mathcal{X})$ is complete whenever $\mathcal{X}$ is.*

PROOF. If T is a disjoint union of a family of open subsets T_λ, $\lambda \in \Lambda$, then the natural map $C_c(T, \mathcal{X}) \to \bigoplus_{\lambda \in \Lambda} C_c(T_\lambda, \mathcal{X})$ is an isomorphism. Hence by [**58**, page 55] $C_c(T, \mathcal{X})$ is complete if and only if $C_c(T_\lambda, \mathcal{X})$ is complete for all $\lambda \in \Lambda$. Thus we have reduced to the case where T is σ-compact. But in this case there exists an increasing sequence $(K_n)_{n \in \mathbb{N}}$ such that $T = \cup_{n \in \mathbb{N}} K_n$ and $C_c(T, \mathcal{X})$ is the inductive limit of the sequence $C_{K_n}(T, \mathcal{X})$. Using [**58**, page 59] it follows that $C_c(T, \mathcal{X})$ is complete if $C_{K_n}(T, \mathcal{X})$ is complete for all $n \in \mathbb{N}$. But standard arguments show that each $C_{K_n}(T, \mathcal{X})$ is complete whenever $\mathcal{X}$ is.

If T is a locally compact group and V is a compact neighborhood of the identity in T, then the subgroup generated by V is open and σ-compact. Thus T is a disjoint union of open σ-compact subsets and the last assertion follows from the above. Finally, if T is a quotient of a locally compact group by a closed subgroup, the result follows from the fact that the quotient map is open and continuous. $\square$

We need to integrate functions in $C_c(T, \mathcal{X})$ with respect to a Radon measure μ on T. For $f \in C_c(T, \mathcal{X})$ the integral $\int_T f(t) d\mu(t)$ is defined as the unique element $y \in \mathcal{X}$ (if it exists) such that

$$x'(y) = \int_T x'(f(t)) \, d\mu(t)$$

for every continuous linear functional x' on $\mathcal{X}$. By [**57**, Theorem 3.27] such a y always exists if the convex hull of $f(T)$ has compact closure in $\mathcal{X}$. But if $\mathcal{X}$ is complete, this latter property of $f(T)$ follows from [**57**, Theorem 3.4] and the fact that the closure of a set $E \subseteq \mathcal{X}$ is compact if and only if E is totally bounded [**32**, page 198]. Moreover, if p is any continuous seminorm on $\mathcal{X}$, then similar arguments as used in [**57**, Theorem 3.29] show that

$$p\left(\int_T f(t) \, d\mu(t)\right) \leq \int_T p(f(t)) \, d\mu(t)$$

for all $f \in C_c(T, \mathcal{X})$.

We are now collecting some important properties of integration:

PROPOSITION C.3. *Assume that $\mathcal{X}$ is a complete locally convex space, μ is a Radon measure on T and S is a locally compact space which satisfies the conditions of Proposition C.2. Then*
  (i) *The linear map $\int : C_c(T, \mathcal{X}) \to \mathcal{X}$ given by $f \mapsto \int_T f(t) d\mu(t)$ is continuous;*
  (ii) *$\int_T : C_c(S \times T, \mathcal{X}) \to C_c(S, \mathcal{X})$ given by $f \mapsto \int_T f(\cdot, t) \, d\mu(t)$ is well-defined and continuous. Moreover,*

$$\left(\int_T f(\cdot, t) \, d\mu(t)\right)(s) = \int_T f(s, t) d\mu(t)$$

for all $s \in S$.

PROOF. Since $\int$ is linear, it is enough to check continuity on $C_K(T, \mathcal{X})$ for all compact $K \subseteq T$. So let $f_i \to 0$ in $C_K(T, \mathcal{X})$ and let p be a continuous seminorm on $\mathcal{X}$. By definition of the topology on $C_K(T, \mathcal{X})$ it follows that $p \circ f_i \to 0$ uniformly, which implies that

$$p\left(\int_T f(t)\, d\mu(t)\right) \leq \int_T p(f(t))\, d\mu(t) \to 0.$$

This proves (i). In order to see (ii) note that by Proposition C.1 the map $C_c(S \times T, \mathcal{X}) \to C_c(T, C_c(S, \mathcal{X}))$ which maps $f \in C_c(S \times T, \mathcal{X})$ to the function $t \mapsto f(\cdot, t)$ is continuous. By the assumption on S it follows from Proposition C.2 that $C_c(S, \mathcal{X})$ is complete. Therefore the integral $\int_T f(\cdot, t)\, d\mu(t)$ makes sense and takes its value in $C_c(S, \mathcal{X})$. The continuity of $\int_T$ follows then from (i) and Proposition C.1. Finally, since evaluation at $s \in S$ is continuous on $C_c(S, \mathcal{X})$ the last equality follows from the definition of the integral. $\square$

C.2. Functions in multiplier algebras and multiplier bimodules

In what follows let A be a C^*-algebra and let $M(A)$ denote the multiplier algebra of A. We shall always write $M^\beta(A)$ if we want to consider $M(A)$ with the strict topology. While some of the results we present here could be proven in much greater generality (involving things like separately continuous bilinear maps among complete locally convex spaces), we choose to emphasize the C^*-techniques rather than follow a more Bourbaki-esque approach.

LEMMA C.4. *Let A be a C^*-algebra and T a locally compact space. For maps $f, g\colon T \to M(A)$ let $L_f(g) = fg$ and $R_f(g) = gf$ denote left and right pointwise multiplication of g by f.*

(i) *If $f\colon T \to A$ is norm continuous, then*

$$L_f(g), R_f(g) \in C_c(T, A)$$

for every $g \in C_c(T, M^\beta(A))$, and the maps $L_f, R_f\colon C_c(T, M^\beta(A)) \to C_c(T, A)$ are continuous.

(ii) *If $f\colon T \to M^\beta(A)$ is (strictly) continuous, then*

$$L_f(g), R_f(g) \in C_c(T, A)$$

for every $g \in C_c(T, A)$, and $L_f, R_f\colon C_c(T, A) \to C_c(T, A)$ are continuous.

(iii) *If $f\colon T \to M^\beta(A)$ is (strictly) continuous, then*

$$L_f(g), R_f(g) \in C_c(T, M^\beta(A))$$

for every $g \in C_c(T, M^\beta(A))$, and $L_f, R_f\colon C_c(T, M^\beta(A)) \to C_c(T, M^\beta(A))$ are continuous.

PROOF. Since the natural pairings $M^\beta(A) \times A \to A$ and $A \times M^\beta(A) \to A$ are continuous on bounded sets and multiplication on $M^\beta(A)$ is strictly continuous on bounded sets, it follows that all linear maps in (i)–(iii) are well-defined. In order to prove continuity in (i) we have to prove continuity on $C_K(T, M^\beta(A))$ for all compact sets $K \subseteq T$. Let $h = f|_K$. By Cohen's factorization theorem we may write $h = a_1 h_1 = h_2 a_2$ with $h_1, h_2 \in C(K, A)$ and $a_1, a_2 \in A$, where, for example, $(a_1 h_1)(t) = a_1 h_1(t)$. If $g_i \to 0$ in $C_K(T, M^\beta(A))$, then by definition of the topology on $C_K(T, M^\beta(A))$ it follows that $g_i a_1 \to 0$ and $a_2 g_i \to 0$ uniformly on K. But by continuity of multiplication in $C(K, A)$ this implies that $g_i = h_2 a_2 g_i \to 0$ and $g_i f = g_i a_1 h_1 \to 0$ uniformly on K. This proves (i).

Suppose now that f is as in (ii). If $K \subseteq T$ is compact, $f|_K$ is a bounded strictly continuous function on K with values in $M(A)$ and hence an element in the multiplier algebra of $C(K, A)$. Thus, if $g_i \to 0$ in $C_K(T, A)$, then $g_i f \to 0$ and $f g_i \to 0$ in $C(K, A)$ and hence in $C_c(T, A)$.

Now (iii) follows from the fact that if $g_i \to 0$ in $C_K(T, M^\beta(A))$, then $g_i fa \to 0$ uniformly on K by (i) for all $a \in A$ since $fa \in C(T, A)$, and $a g_i f \to 0$ by (ii) since $a g_i \to 0$ in $C_K(T, A)$. Similar arguments show that $a f g_i \to 0$ and $f g_i a \to 0$ uniformly on K. □

We shall need the following easy consequence of Lemma C.4.

COROLLARY C.5. *If A is a C^*-algebra, then $C_c(T, A)$ is dense in $C_c(T, M^\beta(A))$.*

PROOF. Let $\{e_i\}_{i \in I}$ be an approximate unit in A with $\|e_i\| = 1$ for all $i \in I$. We claim that $\{f e_i\}_{i \in I} \subseteq C_c(T, A)$ converges to f in $C_c(T, M^\beta(A))$ for all $f \in C_c(T, M^\beta(A))$. Since all $f e_i$ have support in a single compact subset $K \subseteq T$ it is enough to show that if $a \in A$ then $(f e_i a)_{i \in I}$ and $(a f e_i)_{i \in I}$ converge to fa and af uniformly on K, respectively. For this let $\psi \in C_c(T)$ with $\psi|_K \equiv 1$. Then convergence of $(f e_i a)_{i \in I}$ to fa follows from Lemma C.4 and the fact that the functions $\psi_i(t) = \psi(t) e_i a$ converge to ψa in $C_c(T, A)$, and convergence of $(a f e_i)_{i \in I}$ follows from Lemma C.4 and the fact that $af \in C_c(T, A)$ and $\tilde{\psi}_i(t) = \psi(t) e_i$ converges to $\psi 1$ in $C_c(T, M^\beta(A))$. □

PROPOSITION C.6. *Let (A, G, α) be an action.*

(i) *The formulas*

$$f * g(s) = \int_G f(t) \alpha_t(g(t^{-1}s)) \, dt \quad \text{and}$$

$$f^*(s) = \Delta_G(s^{-1}) \alpha_s(f(s^{-1}))^*$$

for convolution and involution on $C_c(G, A)$ extend to $C_c(G, M^\beta(A))$ in such a way that $C_c(G, M^\beta(A))$ becomes a locally convex $*$-algebra with separately continuous multiplication.

(ii) *The pairing $C_c(G, M^\beta(A)) \times C_c(G, M^\beta(A)) \to C_c(G, M^\beta(A))$ given by convolution restricts to separately continuous pairings*

$$C_c(G, M^\beta(A)) \times C_c(G, A) \to C_c(G, A)$$

and

$$C_c(G, A) \times C_c(G, M^\beta(A)) \to C_c(G, A).$$

(iii) *If (π, u) is a covariant homomorphism of (A, G, α) into $M(D)$ for some C^*-algebra D, then the formula*

$$(\pi \times u)(f) = \int_G \pi(f(s)) v_s \, ds \quad \text{for } f \in C_c(G, M^\beta(A))$$

determines the unique continuous $*$-homomorphism

$$\pi \times u \colon C_c(G, M^\beta(A)) \to M^\beta(D)$$

which extends the usual integrated form $\pi \times u$ on $C_c(G, A)$. Moreover, $\pi \times u$ is faithful on $C_c(G, M^\beta(A))$ if and only if it is faithful on $C_c(G, A)$.

PROOF. Let $f, g \in C_c(G, M^\beta(A))$. Then $(s,t) \mapsto f(t)\alpha_t(g(t^{-1}s)) \in M(A)$ is clearly strictly continuous with compact support, so that by Proposition C.3 the formula
$$f * g(s) = \int_G f(t)\alpha_t(g(t^{-1}s))\, dt$$
makes sense and delivers a function $f * g \in C_c(G, M^\beta(A))$. Since the $*$-operation is continuous in $M^\beta(A)$ it is also clear that involution carries over to a continuous involution on $C_c(G, M^\beta(A))$.

We claim that convolution on $C_c(G, M^\beta(A))$ is separately continuous. For this let $g \in C_c(G, M^\beta(A))$ be fixed. Then by linearity of $f \mapsto f * g$ it is enough to show continuity on $C_K(G, M^\beta(A))$ for any compact subset $K \subseteq G$. Choose a compact subset $C \subseteq G \times G$ such that $(s,t) \mapsto f(t)\alpha_t(g(t^{-1}s))$ has support in C for all $f \in C_K(G, M^\beta(A))$, and choose $\psi \in C_c(G \times G)$ such that $\psi|_C \equiv 1$. Then $f \mapsto \psi f$ with $\psi f(s,t) = \psi(s,t)f(t)$ is a continuous mapping from $C_K(G, M^\beta(A))$ into $C_c(G \times G, M^\beta(A))$; hence it follows from Lemma C.4 that $f \mapsto \psi_f$, where
$$\psi_f(s,t) = \psi(s,t) f(t) \alpha_t(g(t^{-1}s)) = f(t)\alpha_t(g(t^{-1}s))$$
is also a continuous map from $C_K(G, M^\beta(A))$ into $C_c(G \times G, M^\beta(A))$. Now the continuity of $f \mapsto f * g$ follows from Proposition C.3. In order to see that $g \mapsto f * g$ is also continuous we use the same trick: chose a compact subset $C \subseteq G \times G$ such that $(s,t) \mapsto f(t)\alpha_t(g(t^{-1}s))$ has support in C for all $g \in C_K(G, M^\beta(A))$, and then look at the map $g \mapsto g\psi$ with ψ as above, and $g\psi(s,t) = \alpha_t(g(t^{-1}s))\psi(s,t)$. This proves (i).

Arguments similar to the above together with part (i) of Lemma C.4 show that $f * g$ and $g * f$ are in $C_c(G, A)$ for all $f \in C_c(G, M^\beta(A))$ and $g \in C_c(G, A)$, and that the resulting pairings $C_c(G, M^\beta(A)) \times C_c(G, A) \to C_c(G, A)$ and $C_c(G, A) \times C_c(G, M^\beta(A)) \to C_c(G, A)$ are separately continuous. This proves (ii).

In order to verify (iii) we first observe that the integrand $\psi_f(s) = \pi(f(s))v_s$ is the pointwise product of $\pi \circ f$ in $C_c(G, M^\beta(D))$ and the strictly continuous function u, so by Lemma C.4 the map $f \mapsto \psi_f \colon C_c(G, M^\beta(A)) \to C_c(G, M^\beta(D))$ is continuous. Thus by Proposition C.3 the linear map $\pi \times u \colon C_c(G, M^\beta(A)) \to M^\beta(D)$ is continuous. Since $C_c(G, A)$ is dense in $C_c(G, M^\beta(A))$, it follows that this extension is unique, and the separate continuity of multiplication and the continuity of involution imply that this extension is a $*$-homomorphism.

Assume now that $\pi \times u$ is faithful on $C_c(G, A)$. If $f \in C_c(G, M^\beta(A))$ with $(\pi \times u)(f) = 0$, it follows that $0 = \pi(a)(\pi \times u)(f) = (\pi \times u)(af)$ for all $a \in A$. Thus $af(s) = 0$ for all $a \in A, s \in G$, which implies that $f(s) = 0$ for all $s \in G$. □

Recall that the canonical embedding of $C_c(G, A)$ into $A \times_\alpha G$ (respectively $A \times_{\alpha,r} G$) is given by the integrated form of the canonical maps (i_A, i_G) (respectively (i_A^r, i_G^r)) of (A, G) into $M(A \times_\alpha G)$ (respectively $M(A \times_{\alpha,r} G)$). Thus, applying part (iii) of the above proposition to $i_A \times i_G$ (respectively $i_A^r \times i_G^r$), gives:

COROLLARY C.7. *There are unique continuous injective $*$-homomorphisms*
$$i_A \times i_G \colon C_c(G, M^\beta(A)) \to M^\beta(A \times_\alpha G) \quad \text{and}$$
$$i_A^r \times i_G^r \colon C_c(G, M^\beta(A)) \to M^\beta(A \times_{\alpha,r} G)$$
extending the usual embeddings of $C_c(G, A)$ into $A \times_\alpha G$ and $A \times_{\alpha,r} G$, respectively. Moreover, if we view $C_c(G, M^\beta(A))$ as a subalgebra of $M(A \times_\alpha G)$ via the above embedding, and if (π, u) is a covariant homomorphism of (A, G, α) into $M(D)$,

then the integrated form $\pi \times u\colon C_c(G, M^\beta(A)) \to M^\beta(D)$ of Proposition C.6 (iii) coincides with the restriction to $C_c(G, M^\beta(A))$ of the usual extension of $\pi \times u$ to $M(A \times_\alpha G)$. If $\pi \times u$ factors through $A \times_{\alpha,r} G$, a similar result is true for the reduced crossed product.

PROOF. The first part is a direct consequence of Proposition C.6 (iii) and the second part follows from the uniqueness of the continuous extension of $\pi \times u\colon C_c(G, A) \to M^\beta(D)$ to all of $C_c(G, M^\beta(A))$. $\square$

In what follows we want to show how the above approach may be used to get similar embeddings of the appropriate function spaces into various multiplier bimodules. The first and easiest example of this kind of result is given by the situation where we have a right-Hilbert bimodule action $_{(A,\alpha)}(X,\gamma)_{(B,\beta)}$ of G. Then there is a right-Hilbert $(A \times_{\alpha,r} G)-(B \times_{\beta,r} G)$ bimodule $X \times_{\gamma,r} G$ (and similarly for the full crossed products), where $X \times_{\gamma,r} G$ is the completion of the pre-right-Hilbert $C_c(G, A) - C_c(G, B)$ bimodule $C_c(G, X)$, with operations given by Equation (3.1) (see Chapter 3 for more details).

In the following we denote by $M^\beta(X)$ the multiplier bimodule of a right-Hilbert bimodule $_A X_B$ equipped with the strict topology (see Definition 1.25).

LEMMA C.8. *Suppose that (X, γ) is as above.*
  (i) *All pairings among $C_c(G, A)$, $C_c(G, X)$, and $C_c(G, B)$ as given by Equation (3.1) extend uniquely to separately continuous pairings among*
  $$C_c(G, M^\beta(A)), \quad C_c(G, M^\beta(X)), \quad \text{and} \quad C_c(G, M^\beta(B)),$$
  given by the same formulas.
  (ii) *The natural bimodule homomorphism*
  $$_{C_c(G,A)}C_c(G, X)_{C_c(G,B)} \to X \times_{\gamma,r} G$$
  extends uniquely to a continuous bimodule homomorphism
  $$_{C_c(G,M^\beta(A))}C_c(G, M^\beta(X))_{C_c(G,M^\beta(B))} \to M^\beta(X \times_{\gamma,r} G)$$
  which also preserves the inner products. Similar results hold for the full crossed products.

PROOF. First assume X is a right-partial $A - B$ imprimitivity bimodule (see Definition 1.5). Let $L = L(X)$ be the linking algebra, with associated action ν of G. We have
$$L \times_{\nu,r} G = \begin{pmatrix} A \times_{\eta,r} G & X \times_{\gamma,r} G \\ * & B \times_{\beta,r} G \end{pmatrix}$$
by Lemma 3.3, and it follows from Proposition 1.51 that
$$M(L \times_{\nu,r} G) = \begin{pmatrix} M(A \times_{\eta,r} G) & M(X \times_{\gamma,r} G) \\ * & M(B \times_{\beta,r} G) \end{pmatrix},$$
where the strict topology on $M(L \times G)$ agrees with the product of the strict topologies on the corners by Proposition 1.51. Since convolution in $C_c(G, L)$ and $C_c(G, M^\beta(L))$ are given by the above pairings among the corners, the desired conclusions follow in this case from the continuous embedding of $C_c(G, M^\beta(L)) = \begin{pmatrix} C_c(G, M^\beta(A)) & C_c(G, M^\beta(X)) \\ * & C_c(G, M^\beta(B)) \end{pmatrix}$ into $M^\beta(L \times_{\nu,r} G)$ as in Corollary C.7.

For the general case, let $K = \mathcal{K}_B(X)$ be the algebra of compact operators on X, with associated action μ. Then from the above we get a pairing $C_c(G, M^\beta(K)) \times$

$C_c(G, M^\beta(X)) \to C_c(G, M^\beta(X))$ satisfying the required properties. Letting $\varphi \colon A \to M(K)$ be the associated nondegenerate homomorphism, we get a continuous homomorphism
$$\varphi \times_r G \colon C_c(G, M^\beta(A)) \to C_c(G, M^\beta(K)).$$
It follows that the pairing
$$C_c(G, M^\beta(A)) \times C_c(G, M^\beta(X)) \to C_c(G, M^\beta(X))$$
has the required properties also. This proves the result for the case of reduced crossed products; exactly the same arguments (replacing all reduced crossed products by full crossed products) give the corresponding result for full crossed products. □

Another result of this kind is the embedding of $C_c(G, M^\beta(A))$ into the multiplier bimodule $M(V_H^G(A))$ of the $(\operatorname{Ind}_H^G A \times_{\operatorname{Ind}\alpha, r} G) - (A \times_{\alpha, r} H)$ imprimitivity bimodule $V_H^G(A)$ of Theorem B.2, where (A, α) is an action of a closed subgroup H of G. Recall that $V_H^G(A)$ is the completion of the $C_c(G, \operatorname{Ind}_H^G A) - C_c(H, A)$ preimprimitivity bimodule $C_c(G, A)$ with operations given by Equation (B.4). Note that the embedding $C_c(G, A) \hookrightarrow V_H^G(A)$ is continuous from the inductive limit topology to the norm topology.

It follows from [**48**, Corollary 3.2] that elements in $C_c(G, M^\beta(\operatorname{Ind}_H^G A))$ can be viewed as continuous functions of $G \times G$ into $M^\beta(A)$. We only have to check that for $f \in C_c(G, M^\beta(\operatorname{Ind}_H^G A))$ the functions $(s, t) \mapsto af(s, t)$ and $(s, t) \mapsto f(s, t)a$ of $G \times G$ into A are continuous for every $a \in A$, which follows from the fact that for each $t \in G$ and $a \in A$ there exists a $g \in \operatorname{Ind}_H^G A$ such that $g(t) = a$.

PROPOSITION C.9. *Let G and (A, H, α) be as above.*

(i) *All pairings among*
$$C_c(G, \operatorname{Ind}_H^G A), \quad C_c(G, A), \quad \text{and} \quad C_c(H, A)$$
given in Equation (B.4) *extend uniquely to separately continuous pairings among*
$$C_c(G, M^\beta(\operatorname{Ind}_H^G A)), \quad C_c(G, M^\beta(A)), \quad \text{and} \quad C_c(H, M^\beta(A)),$$
given by the same formulas.

(ii) *The canonical bimodule homomorphism*
$$_{C_c(G, \operatorname{Ind}_H^G A)} C_c(G, A)_{C_c(H, A)} \to V_H^G(A)$$
extends uniquely to a continuous bimodule homomorphism
$$_{C_c(G, M^\beta(\operatorname{Ind}_H^G A))} C_c(G, M^\beta(A))_{C_c(H, M^\beta(A))} \to M^\beta(V_H^G(A))$$
which also preserves the inner products.

PROOF. Arguments similar to those used in Proposition C.6 show that the formulas for the pairings extend uniquely to separately continuous pairings among $C_c(G, M^\beta(\operatorname{Ind}_H^G A))$, $C_c(G, M^\beta(A))$, and $C_c(H, M^\beta(A))$.

Note that by Corollary C.7 we have continuous embeddings
$$C_c(G, M^\beta(\operatorname{Ind}_H^G A)) \to M^\beta(\operatorname{Ind}_H^G A \times_{\operatorname{Ind}\alpha, r} G)$$
and $C_c(H, M^\beta(A)) \to M^\beta(A \times_{\alpha, r} H)$ which uniquely extend the canonical embeddings of $C_c(G, \operatorname{Ind}_H^G A)$ and $C_c(H, A)$ into $\operatorname{Ind}_H^G A \times_{\operatorname{Ind}\alpha, r} G$ and $A \times_{\alpha, r} H$, respectively.

We claim that the embedding $C_c(G, A) \hookrightarrow V_H^G(A)$ is continuous with respect to the topology on $C_c(G, A)$ inherited from $C_c(G, M^\beta(A))$ and the strict topology on $V_H^G(A)$. Since $C_c(G, A)$ is dense in $C_c(G, M^\beta(A))$ we then obtain a unique continuous extension $C_c(G, M^\beta(A)) \to M^\beta(V_H^G(A))$, which by the separate continuity of the above pairings preserves all actions and inner products.

The claim can be shown by factoring elements $b \in A \times_{\alpha, r} H$ as $i_A^r(a)c$ for some $c \in A \times_{\alpha, r} H$ and $a \in A$ and by factoring elements $d \in \operatorname{Ind}_H^G A \times_{\operatorname{Ind}\alpha, r} G$ as $e(i_{\operatorname{Ind} A}^r(g))$, for some $e \in \operatorname{Ind}_H^G A \times_{\operatorname{Ind}\alpha, r} G$ and $g \in \operatorname{Ind}_H^G A$. Then, for $x \in C_c(G, M^\beta(A))$ the function $x \cdot i_A^r(a)$ given by $x \cdot i_A^r(a)(s) = x(s)\alpha_s(a)$ is the pointwise product of x with the norm-continuous function $s \mapsto \alpha_s(a)$, hence by Lemma C.4 the linear map $x \mapsto x \cdot i_A^r(a)$ of $C_c(G, M^\beta(A))$ into $C_c(G, A)$ is continuous. Since $C_c(G, A) \hookrightarrow V_H^G(A)$ is continuous, it follows that the map $C_c(G, A) \to V_H^G(A)$ given by $x \mapsto (x \cdot i_A^r(a)) \cdot c = x \cdot b$ is continuous with respect to the topology on $C_c(G, A)$ inherited from $C_c(G, M^\beta(A))$. Similarly, $i_{\operatorname{Ind} A}^r(g) \cdot x(s) = g(s)x(s)$, which is an element of $C_c(G, A)$, and this implies that $x \mapsto e \cdot (i_{\operatorname{Ind} A}^r(g) \cdot x) = d \cdot x$ is continuous for all d. $\square$

Throughout this work we freely use strictly continuous multiplier-valued functions inside multiplier bimodules. In every case the computations can be justified by closely following the lines of Propositions C.6–C.9, so we omit any further details.

Bibliography

1. C. A. Akemann, G. K. Pedersen and J. Tomiyama, *Multipliers of C^*-algebras*, J. Funct. Anal. **13**, (1973), 277–301.
2. S. Baaj and G. Skandalis, *C^*-algèbres de Hopf et théorie de Kasparov équivariante*, K-Theory **2** (1989), 683–721.
3. _____, *Unitaires multiplicatifs et dualité pour les produits croisés de C^*-algèbres*, Ann. Sci. École Norm. Sup. (4) **26** (1993), 425–488.
4. L. Brown, P. Green, and M. Rieffel, *Stable isomorphism and strong Morita equivalence of C^*-algebras*, Pacific J. Math. **71** (1977), 349–363.
5. L. Brown, , J. A. Mingo, and N.-T. Shen, *Quasi-multipliers and embeddings of Hilbert C^*-bimodules*, Canad. J. Math **46** (1994), no. 6, 1150–1174.
6. H. H. Bui, *Morita equivalence of twisted crossed products by coactions*, J. Funct. Anal. **123** (1994), 59–98.
7. _____, *Morita equivalence of twisted crossed products*, Proc. Amer. Math. Soc. **123** (1995), 2771–2776.
8. J. Chabert and S. Echterhoff, *Permanence properties of the Baum-Connes conjecture*, Doc. Math. **6** (2001), 127–183.
9. J. Chabert and S. Echterhoff, *Twisted equivariant KK-theory and the Baum-Connes conjecture for group extensions*, K-Theory **23** (2001), 157–200.
10. J. Chabert, S. Echterhoff, and R. Nest, *The Connes-Kasparov conjecture for almost connected groups*, Publ. Math. Inst. Hautes Études Sci. No. 97 (2003), 239–278.
11. F. Combes, *Crossed products and Morita equivalence*, Proc. London Math. Soc. **49** (1984), 289–306.
12. R. Curto, P. Muhly, and D. Williams, *Cross products of strongly Morita equivalent C^*-algebras*, Proc. Amer. Math. Soc. **90** (1984), 528–530.
13. S. Echterhoff, *On induced covariant systems*, Proc. Amer. Math. Soc. **108** (1990), 703–706.
14. _____, *Duality of induction and restriction for abelian twisted covariant systems*, Math. Proc. Camb. Phil. Soc. **116** (1994), 301–315.
15. _____, *Morita equivalent twisted actions and a new version of the Packer-Raeburn stabilization trick*, J. London Math. Soc. **50** (1994), 170–186.
16. _____, *Crossed products with continuous trace*, Mem. Amer. Math. Soc. **123** (1996), no. 586, 1–134.
17. S. Echterhoff, S. Kaliszewski, J. Quigg, and I. Raeburn, *Naturality and induced representations*, Bull. Austral. Math. Soc. **61** (2000), 415–438.
18. S. Echterhoff, S. Kaliszewski, and I. Raeburn, *Crossed products by dual coactions of groups and homogeneous spaces*, J. Operator Theory **39** (1998), 151–176.
19. S. Echterhoff and J. Quigg, *Induced coactions of discrete groups on C^*-algebras*, Canad. J. Math. **51** (1999), 745–770.
20. S. Echterhoff and I. Raeburn, *Multipliers of imprimitivity bimodules and Morita equivalence of crossed products*, Math. Scand. **76** (1995), 289–309.
21. _____, *The stabilisation trick for coactions*, J. Reine Angew. Math. **470** (1996), 181–215.
22. _____, *Induced C^*-algebras, coactions, and equivariance in the symmetric imprimitivity theorem*, Math. Proc. Cambridge Philos. Soc. **128** (2000), 327–342.
23. P. Eymard, *L'algèbre de Fourier d'un groupe localement compact*, Bull. Soc. Math. France **92** (1964), 181–236.
24. J. M. G. Fell and R. S. Doran, *Representations of *-algebras, locally compact groups, and Banach *-algebraic bundles*, Academic Press, San Diego, 1988.
25. P. Green, *The local structure of twisted covariance algebras*, Acta Math. **140** (1978), 191–250.

26. S. Imai and H. Takai, *On a duality for C^*-crossed products by a locally compact group*, J. Math. Soc. Japan **30** (1978), 495–504.
27. K.K. Jensen and K. Thomsen, *Elements of KK-theory*, Birkhäuser, 1991.
28. S. Kaliszewski and J. Quigg, *Imprimitivity for C^*-coactions of non-amenable groups*, Math. Proc. Cambridge Philos. Soc. **123** (1998), 101–118.
29. S. Kaliszewski, J. Quigg, and I. Raeburn, *Duality of restriction and induction for C^*-coactions*, Trans. Amer. Math. Soc. **349** (1997), 2085–2113.
30. G. G. Kasparov, *Equivariant KK-theory and the Novikov conjecture*, Invent. Math. **91** (1988), 147–201.
31. Y. Katayama, *Takesaki's duality for a non-degenerate co-action*, Math. Scand. **55** (1985), 141–151.
32. J. L. Kelley, *General topology*, Springer-Verlag, 1991.
33. E. C. Lance, *Hilbert C^*-modules*, London Math. Soc. Lecture Note Ser., vol. 210, Cambridge University Press, 1995.
34. M. B. Landstad, *Duality theory for covariant systems*, Trans. Amer. Math. Soc. **248** (1979), 223–267.
35. M. B. Landstad, J. Phillips, I. Raeburn, and C. E. Sutherland, *Representations of crossed products by coactions and principal bundles*, Trans. Amer. Math. Soc. **299** (1987), 747–784.
36. N.P. Landsman, *Bicategories of operator algebras and Poisson manifolds*, Mathematical physics in mathematics and physics: Quantum and Operator Algebraic Aspects (Siena, 2000) (Providence, RI) (R. Longo, ed.), Fields Inst. Commun., vol. 30, Amer. Math. Soc., 2001, pp. 271–286.
37. N.P. Landsman, *Quantized reduction as a tensor product*, Quantization of Singular Symplectic Quotients (Basel) (M. Pflaum, N.P. Landsman, and M. Schlichenmaier, eds.), Birkhäuser, 2001, pp. 137–180.
38. K. Mansfield, *Induced representations of crossed products by coactions*, J. Funct. Anal. **97** (1991), 112–161.
39. P. S. Muhly and B. Solel, *Tensor algebras over C^*-correspondences: representations, dilations, and C^*-envelopes*, J. Funct. Anal. **158** (1998), 389–457.
40. C. K. Ng, *Coactions and crossed products of Hopf C^*-algebras II: Hilbert C^*-modules*, preprint, 1995.
41. D. Olesen and G. K. Pedersen, *Applications of the Connes spectrum to C^*-dynamical systems*, J. Funct. Anal. **30** (1978), 179–197.
42. _____, *Applications of the Connes spectrum to C^*-dynamical systems*, II, J. Funct. Anal. **36** (1980), 18–32.
43. G. K. Pedersen, *C^*-algebras and their automorphism groups*, Academic Press, 111 Fifth Avenue, New York, NY 10003, 1979.
44. J. Phillips and I. Raeburn, *Twisted crossed products by coactions*, J. Austral. Math. Soc. Ser. A **56** (1994), 320–344.
45. J. Quigg, *Full C^*-crossed product duality*, J. Austral. Math. Soc. Ser. A **50** (1991), 34–52.
46. _____, *Landstad duality for C^*-coactions*, Math. Scand. **71** (1992), 277–294.
47. _____, *Full and reduced C^*-coactions*, Math. Proc. Cambridge Philos. Soc. **116** (1994), 435–450.
48. J. Quigg and I. Raeburn, *Induced C^*-algebras and Landstad duality for twisted coactions*, Trans. Amer. Math. Soc. **347** (1995), 2885–2915.
49. J. Quigg and J. Spielberg, *Regularity and hyporegularity in C^*-dynamical systems*, Houston J. Math. **18** (1992), 139–152.
50. I. Raeburn, *A duality theorem for crossed products by nonabelian groups*, Proc. Cent. Math. Anal. Austral. Nat. Univ. **15** (1987), 214–227.
51. _____, *Induced C^*-algebras and a symmetric imprimitivity theorem*, Math. Ann. **280** (1988), 369–387.
52. I. Raeburn, *Crossed products of C^*-algebras by coactions of locally compact groups*, Operator algebras and quantum field theory (S. Doplicher, R. Longo, J. E. Roberts, and L.Zsido, eds.), International Press, January 1998, pp. 74–84.
53. _____, *On crossed products by coactions and their representation theory*, Proc. London Math. Soc. **64** (1992), 625–652.
54. I. Raeburn and D. P. Williams, *Morita equivalence and continuous-trace C^*-algebras*, American Mathematical Society, 1998.

55. M. A. Rieffel, *On the uniqueness of the Heisenberg commutation relations*, Duke Math. J. **39** (1972), 745–751.
56. ———, *Induced representations of C^*-algebras*, Adv. Math. **13** (1974), 176–257.
57. W. Rudin, *Functional analysis*, McGraw-Hill, New York, 1973.
58. H. H. Schaefer, *Topological vector spaces*, Springer-Verlag, 1971.
59. J. Schweizer, *Crossed products by equivalence bimodules*, preprint, 1999.
60. J. Schweizer, *Crossed products by C^*-correspondences and Cuntz-Pimsner algebras*, C^*-algebras (Münster, 1999), 203–226, Springer, Berlin, 2000.
61. J. Schweizer, *Dilations of C^*-correspondences and the simplicity of Cuntz-Pimsner algebras*, J. Funct. Anal. **180** (2001), 404–425.
62. N. Sieben, *Morita equivalence of C^*-crossed products by inverse semigroup actions*, Rocky Mountain J. Math. **31** (2001), 661–686.

Editorial Information

To be published in the *Memoirs*, a paper must be correct, new, nontrivial, and significant. Further, it must be well written and of interest to a substantial number of mathematicians. Piecemeal results, such as an inconclusive step toward an unproved major theorem or a minor variation on a known result, are in general not acceptable for publication. Papers appearing in *Memoirs* are generally at least 80 and not more than 200 published pages in length. Papers less than 80 or more than 200 published pages require the approval of the Managing Editor of the Transactions/Memoirs Editorial Board.

As of November 30, 2005, the backlog for this journal was approximately 15 volumes. This estimate is the result of dividing the number of manuscripts for this journal in the Providence office that have not yet gone to the printer on the above date by the average number of monographs per volume over the previous twelve months, reduced by the number of volumes published in four months (the time necessary for preparing a volume for the printer). (There are 6 volumes per year, each containing at least 4 numbers.)

A Consent to Publish and Copyright Agreement is required before a paper will be published in the *Memoirs*. After a paper is accepted for publication, the Providence office will send a Consent to Publish and Copyright Agreement to all authors of the paper. By submitting a paper to the *Memoirs*, authors certify that the results have not been submitted to nor are they under consideration for publication by another journal, conference proceedings, or similar publication.

Information for Authors

Memoirs are printed from camera copy fully prepared by the author. This means that the finished book will look exactly like the copy submitted.

The paper must contain a *descriptive title* and an *abstract* that summarizes the article in language suitable for workers in the general field (algebra, analysis, etc.). The *descriptive title* should be short, but informative; useless or vague phrases such as "some remarks about" or "concerning" should be avoided. The *abstract* should be at least one complete sentence, and at most 300 words. Included with the footnotes to the paper should be the 2000 *Mathematics Subject Classification* representing the primary and secondary subjects of the article. The classifications are accessible from www.ams.org/msc/. The list of classifications is also available in print starting with the 1999 annual index of *Mathematical Reviews*. The Mathematics Subject Classification footnote may be followed by a list of *key words and phrases* describing the subject matter of the article and taken from it. Journal abbreviations used in bibliographies are listed in the latest *Mathematical Reviews* annual index. The series abbreviations are also accessible from www.ams.org/publications/. To help in preparing and verifying references, the AMS offers MR Lookup, a Reference Tool for Linking, at www.ams.org/mrlookup/. When the manuscript is submitted, authors should supply the editor with electronic addresses if available. These will be printed after the postal address at the end of the article.

Electronically prepared manuscripts. The AMS encourages electronically prepared manuscripts, with a strong preference for $\mathcal{AMS}$-LaTeX. To this end, the Society has prepared $\mathcal{AMS}$-LaTeX author packages for each AMS publication. Author packages include instructions for preparing electronic manuscripts, the *AMS Author Handbook*, samples, and a style file that generates the particular design specifications of that publication series. Though $\mathcal{AMS}$-LaTeX is the highly preferred format of TeX, author packages are also available in $\mathcal{AMS}$-TeX.

Authors may retrieve an author package from e-MATH starting from www.ams.org/tex/ or via FTP to ftp.ams.org (login as anonymous, enter username as password, and type cd pub/author-info). The *AMS Author Handbook* and the *Instruction Manual* are available in PDF format following the author packages link from www.ams.org/tex/. The author package can be obtained free of charge by sending email to pub@ams.org (Internet) or from the Publication Division, American Mathematical Society, 201 Charles St., Providence, RI 02904, USA. When requesting an author package, please specify $\mathcal{AMS}$-LaTeX or $\mathcal{AMS}$-TeX, Macintosh or IBM (3.5) format, and the publication in which your paper will appear. Please be sure to include your complete mailing address.

Sending electronic files. After acceptance, the source file(s) should be sent to the Providence office (this includes any TeX source file, any graphics files, and the DVI or PostScript file).

Before sending the source file, be sure you have proofread your paper carefully. The files you send must be the EXACT files used to generate the proof copy that was accepted for publication. For all publications, authors are required to send a printed copy of their paper, which exactly matches the copy approved for publication, along with any graphics that will appear in the paper.

TeX files may be submitted by email, FTP, or on diskette. The DVI file(s) and PostScript files should be submitted only by FTP or on diskette unless they are encoded properly to submit through email. (DVI files are binary and PostScript files tend to be very large.)

Electronically prepared manuscripts can be sent via email to pub-submit@ams.org (Internet). The subject line of the message should include the publication code to identify it as a Memoir. TeX source files, DVI files, and PostScript files can be transferred over the Internet by FTP to the Internet node e-math.ams.org (130.44.1.100).

Electronic graphics. Comprehensive instructions on preparing graphics are available at www.ams.org/jourhtml/graphics.html. A few of the major requirements are given here.

Submit files for graphics as EPS (Encapsulated PostScript) files. This includes graphics originated via a graphics application as well as scanned photographs or other computer-generated images. If this is not possible, TIFF files are acceptable as long as they can be opened in Adobe Photoshop or Illustrator. No matter what method was used to produce the graphic, it is necessary to provide a paper copy to the AMS.

Authors using graphics packages for the creation of electronic art should also avoid the use of any lines thinner than 0.5 points in width. Many graphics packages allow the user to specify a "hairline" for a very thin line. Hairlines often look acceptable when proofed on a typical laser printer. However, when produced on a high-resolution laser imagesetter, hairlines become nearly invisible and will be lost entirely in the final printing process.

Screens should be set to values between 15% and 85%. Screens which fall outside of this range are too light or too dark to print correctly. Variations of screens within a graphic should be no less than 10%.

Inquiries. Any inquiries concerning a paper that has been accepted for publication should be sent directly to the Electronic Prepress Department, American Mathematical Society, 201 Charles St., Providence, RI 02904, USA.

Editors

This journal is designed particularly for long research papers, normally at least 80 pages in length, and groups of cognate papers in pure and applied mathematics. Papers intended for publication in the *Memoirs* should be addressed to one of the following editors. In principle the Memoirs welcomes electronic submissions, and some of the editors, those whose names appear below with an asterisk (*), have indicated that they prefer them. However, editors reserve the right to request hard copies after papers have been submitted electronically. Authors are advised to make preliminary email inquiries to editors about whether they are likely to be able to handle submissions in a particular electronic form.

*Algebra to ALEXANDER KLESHCHEV, Department of Mathematics, University of Oregon, Eugene, OR 97403-1222; email: `ams@noether.uoregon.edu`

Algebra and its application to MINA TEICHER, Emmy Noether Research Institute for Mathematics, Bar-Ilan University, Ramat-Gan 52900, Israel; email: `teicher@macs.biu.ac.il`

Algebraic geometry to DAN ABRAMOVICH, Department of Mathematics, Brown University, Box 1917, Providence, RI 02912; email: `amsedit@math.brown.edu`

*Algebraic number theory to V. KUMAR MURTY, Department of Mathematics, University of Toronto, 100 St. George Street, Toronto, ON M5S 1A1, Canada; email: `murty@math.toronto.edu`

*Algebraic topology to ALEJANDRO ADEM, Department of Mathematics, University of British Columbia, Room 121, 1984 Mathematics Road, Vancouver, British Columbia, Canada V6T 1Z2; email: `adem@math.ubc.ca`

Combinatorics to JOHN R. STEMBRIDGE, Department of Mathematics, University of Michigan, Ann Arbor, Michigan 48109-1109; email: `jrs@umich.edu`

Complex analysis and harmonic analysis to ALEXANDER NAGEL, Department of Mathematics, University of Wisconsin, 480 Lincoln Drive, Madison, WI 53706-1313; email: `nagel@math.wisc.edu`

*Differential geometry and global analysis to LISA C. JEFFREY, Department of Mathematics, University of Toronto, 100 St. George St., Toronto, ON Canada M5S 3G3; email: `jeffrey@math.toronto.edu`

Dynamical systems and ergodic theory to AMIE WILKINSON, Department of Mathematics, Northwestern University, 2033 Sheridan Road, Evanston, IL 60208-2730; email: `wilkinso@math.northwestern.edu`

*Functional analysis and operator algebras to MARIUS DADARLAT, Department of Mathematics, Purdue University, 150 N. University St., West Lafayette, IN 47907-2067; email: `mdd@math.purdue.edu`

*Geometric analysis to TOBIAS COLDING, Courant Institute, New York University, 251 Mercer St., New York, NY 10012; email: `traneditor@cims.nyu.edu`

*Geometric analysis to MLADEN BESTVINA, Department of Mathematics, University of Utah, 155 South 1400 East, JWB 233, Salt Lake City, Utah 84112-0090; email: `bestvina@math.utah.edu`

Harmonic analysis, representation theory, and Lie theory to ROBERT J. STANTON, Department of Mathematics, The Ohio State University, 231 West 18th Avenue, Columbus, OH 43210-1174; email: `stanton@math.ohio-state.edu`

*Logic to STEFFEN LEMPP, Department of Mathematics, University of Wisconsin, 480 Lincoln Drive, Madison, Wisconsin 53706-1388; email: `lempp@math.wisc.edu`

*Ordinary differential equations, and applied mathematics to PETER W. BATES, Department of Mathematics, Michigan State University, East Lansing, MI 48824-1027; email: `bates@math.msu.edu`

*Partial differential equations to GUSTAVO PONCE, Department of Mathematics, South Hall, Room 6607, University of California, Santa Barbara, CA 93106; email: `ponce@math.ucsb.edu`

*Probability and statistics to KRZYSZTOF BURDZY, Department of Mathematics, University of Washington, Box 354350, Seattle, Washington 98195-4350; email: `burdzy@math.washington.edu`

*Real analysis and partial differential equations to DANIEL TATARU, Department of Mathematics, University of California, Berkeley, Berkeley, CA 94720; email: `tataru@math.berkeley.edu`

All other communications to the editors should be addressed to the Managing Editor, ROBERT GURALNICK, Department of Mathematics, University of Southern California, Los Angeles, CA 90089-1113; email: `guralnic@math.usc.edu`.

Titles in This Series

851 **Jie Wu,** On maps from loop suspensions to loop spaces and the shuffle relations on the Cohen groups, 2006

850 **Siegfried Echterhoff, S. Kaliszewski, John Quigg, and Iain Raeburn,** A categorical approach to imprimitivity theorems for C^*-dynamical systems, 2006

849 **Katsuhiko Kuribayashi, Mamoru Mimura, and Tetsu Nishimoto,** Twisted tensor products related to the cohomology of the classifying spaces of loop groups, 2006

848 **Bob Oliver,** Equivalences of classifying spaces completed at the prime two, 2006

847 **Eric T. Sawyer and Richard L. Wheeden,** Hölder continuity of weak solutions to subelliptic equations with rough coefficients, 2006

846 **Victor Beresnevich, Detta Dickinson, and Sanju Velani,** Measure theoretic laws for lim–sup sets, 2006

845 **Ehud Friedgut, Vojtech Rödl, Andrzej Ruciński, and Prasad V. Tetali,** A Sharp threshold for random graphs with a monochromatic triangle in every edge coloring, 2006

844 **Amadeu Delshams, Rafael de la Llave, and Tere M. Seara,** A geometric mechanism for diffusion in Hamiltonian systems overcoming the large gap problem: Heuristics and rigorous verification on a model, 2006

843 **Denis V. Osin,** Relatively hyperbolic groups: Intrinsic geometry, algebraic properties, and algorithmic problems, 2006

842 **David P. Blecher and Vrej Zarikian,** The calculus of one-sided M-ideals and multipliers in operator spaces, 2006

841 **Enrique Artal Bartolo, Pierrette Cassou-Noguès, Ignacio Luengo, and Alejandro Melle Hernández,** Quasi-ordinary power series and their zeta functions, 2005

840 **Sławomir Kołodziej,** The complex Monge-Ampère equation and pluripotential theory, 2005

839 **Mihai Ciucu,** A random tiling model for two dimensional electrostatics, 2005

838 **V. Jurdjevic,** Integrable Hamiltonian systems on complex Lie groups, 2005

837 **Joseph A. Ball and Victor Vinnikov,** Lax-Phillips scattering and conservative linear systems: A Cuntz-algebra multidimensional setting, 2005

836 **H. G. Dales and A. T.-M. Lau,** The second duals of Beurling algebras, 2005

835 **Kiyoshi Igusa,** Higher complex torsion and the framing principle, 2005

834 **Keníchi Ohshika,** Kleinian groups which are limits of geometrically finite groups, 2005

833 **Greg Hjorth and Alexander S. Kechris,** Rigidity theorems for actions of product groups and countable Borel equivalence relations, 2005

832 **Lee Klingler and Lawrence S. Levy,** Representation type of commutative Noetherian rings III: Global wildness and tameness, 2005

831 **K. R. Goodearl and F. Wehrung,** The complete dimension theory of partially ordered systems with equivalence and orthogonality, 2005

830 **Jason Fulman, Peter M. Neumann, and Cheryl E. Praeger,** A generating function approach to the enumeration of matrices in classical groups over finite fields, 2005

829 **S. G. Bobkov and B. Zegarlinski,** Entropy bounds and isoperimetry, 2005

828 **Joel Berman and Paweł M. Idziak,** Generative complexity in algebra, 2005

827 **Trevor A. Welsh,** Fermionic expressions for minimal model Virasoro characters, 2005

826 **Guy Métivier and Kevin Zumbrun,** Large viscous boundary layers for noncharacteristic nonlinear hyperbolic problems, 2005

825 **Yaozhong Hu,** Integral transformations and anticipative calculus for fractional Brownian motions, 2005

824 **Luen-Chau Li and Serge Parmentier,** On dynamical Poisson groupoids I, 2005

TITLES IN THIS SERIES

- 823 **Claus Mokler,** An analogue of a reductive algebraic monoid whose unit group is a Kac-Moody group, 2005
- 822 **Stefano Pigola, Marco Rigoli, and Alberto G. Setti,** Maximum principles on Riemannian manifolds and applications, 2005
- 821 **Nicole Bopp and Hubert Rubenthaler,** Local zeta functions attached to the minimal spherical series for a class of symmetric spaces, 2005
- 820 **Vadim A. Kaimanovich and Mikhail Lyubich,** Conformal and harmonic measures on laminations associated with rational maps, 2005
- 819 **F. Andreatta and E. Z. Goren,** Hilbert modular forms: Mod p and p-adic aspects, 2005
- 818 **Tom De Medts,** An algebraic structure for Moufang quadrangles, 2005
- 817 **Javier Fernández de Bobadilla,** Moduli spaces of polynomials in two variables, 2005
- 816 **Francis Clarke,** Necessary conditions in dynamic optimization, 2005
- 815 **Martin Bendersky and Donald M. Davis,** V_1-periodic homotopy groups of $SO(n)$, 2004
- 814 **Johannes Huebschmann,** Kähler spaces, nilpotent orbits, and singular reduction, 2004
- 813 **Jeff Groah and Blake Temple,** Shock-wave solutions of the Einstein equations with perfect fluid sources: Existence and consistency by a locally inertial Glimm scheme, 2004
- 812 **Richard D. Canary and Darryl McCullough,** Homotopy equivalences of 3-manifolds and deformation theory of Kleinian groups, 2004
- 811 **Ottmar Loos and Erhard Neher,** Locally finite root systems, 2004
- 810 **W. N. Everitt and L. Markus,** Infinite dimensional complex symplectic spaces, 2004
- 809 **J. T. Cox, D. A. Dawson, and A. Greven,** Mutually catalytic super branching random walks: Large finite systems and renormalization analysis, 2004
- 808 **Hagen Meltzer,** Exceptional vector bundles, tilting sheaves and tilting complexes for weighted projective lines, 2004
- 807 **Carlos A. Cabrelli, Christopher Heil, and Ursula M. Molter,** Self-similarity and multiwavelets in higher dimensions, 2004
- 806 **Spiros A. Argyros and Andreas Tolias,** Methods in the theory of hereditarily indecomposable Banach spaces, 2004
- 805 **Philip L. Bowers and Kenneth Stephenson,** Uniformizing dessins and Belyĭ maps via circle packing, 2004
- 804 **A. Yu Ol'shanskii and M. V. Sapir,** The conjugacy problem and Higman embeddings, 2004
- 803 **Michael Field and Matthew Nicol,** Ergodic theory of equivariant diffeomorphisms: Markov partitions and stable ergodicity, 2004
- 802 **Martin W. Liebeck and Gary M. Seitz,** The maximal subgroups of positive dimension in exceptional algebraic groups, 2004
- 801 **Fabio Ancona and Andrea Marson,** Well-posedness for general 2×2 systems of conservation law, 2004
- 800 **V. Poénaru and C. Tanas,** Equivariant, almost-arborescent representation of open simply-connected 3-manifolds; A finiteness result, 2004
- 799 **Barry Mazur and Karl Rubin,** Kolyvagin systems, 2004
- 798 **Benoît Mselati,** Classification and probabilistic representation of the positive solutions of a semilinear elliptic equation, 2004
- 797 **Ola Bratteli, Palle E. T. Jorgensen, and Vasyl' Ostrovs'kyĭ,** Representation theory and numerical AF-invariants, 2004

For a complete list of titles in this series, visit the
AMS Bookstore at **www.ams.org/bookstore/**.